# Clinical Applications of Immunomics

For other titles published in this series, go to
www.springer.com/series/7528

**Immunonomics Reviews**
**An Official Publication of the International Immunomics**
**Society**

This peer-reviewed book series offers insight on immunology for
21st century. The technological revolution has borne advances in
high-throughput instrumentation and information technology,
initiating a renaissance for biomathematics, and biostatistics.
Cross-fertilization between genomics and immunology has led to a
new field called immunomics, transforming the way in which
theoretical, clinical and applied immunology are practiced.
Immunomics Reviews will cover integrative approaches and
applications to the theory and practice of immunology and
explore synergistic effects resulting from a combination of
technological advances and the latest analytical tools with the
traditional fields of basic and clinical immunology.

Andras Falus

Editor

# Clinical Applications of Immunomics

 Springer

*Editor*
Andras Falus
Semmelweis University
Budapest, Hungary
faland@dgci.sote.hu

ISBN: 978-0-387-79207-1     e-ISBN: 978-0-387-79208-8
DOI: 10.1007/978-0-387-79208-8

Library of Congress Control Number: 2008934033

Printed on acid-free paper

springer.com

# Contents

# Contributors

Marcos J.C. Alcocer
Division of Nutritional Sciences, University of Nottingham, Loughborough, UK

Agnese Antognoli
Section of Cancer Research, Department of Experimental Pathology, University of Bologna, Bologna, Italy

Heimo Breiteneder
Department of Pathophysiology, Center of Physiology and Pathophysiology, Medical University of Vienna, Vienna, Austria

Vladimir Brusic
Cancer Vaccine Center, Dana-Farber Cancer Institute, Boston, MA 02115, USA, vladimir_brusic@dfci.harward.edu

Stefania Croci
Section of Cancer Research, Department of Experimental Pathology, University of Bologna, Bologna, Italy

Steven K. Dower
Cardiovascular Research Unit, University of Sheffield, Sheffield, UK

Franco H. Falcone
Division of Molecular and Cellular Science, School of Pharmacy, University of Nottingham, Nottingham, NG7 2RD, UK, Franco.Falcone@nottingham.ac.uk

Andras Falus
Department of Genetics, Cell and Immunobiology, Semmelweis University, Budapest, Hungary, Nagyvárad tér 4, 1089 Budapest, faland@dgci.sote.hu

Efrosini Fostieri
Hellenic Pasteur Institute, Department of Biochemistry, Athens, Greece, fostieri@pasteur.gr

Bernhard F. Gibbs
Medway School of Pharmacy, Universities of Kent and Greenwich, Kent ME4 4TB, UK

Anne S. De Groot
Director, Immunology and Informatics Institute, Center for Vaccine Research
and Design, Brown University, CEO, EpiVax, Inc., Providence, RI, USA,
AnnieD@brown.edu

Hongtao Guan
Cardiovascular Research Unit, University of Sheffield, Sheffield, UK

Heinrich Haas
Medigene AG, Martinsried, Germany, H.Haas@medigene.com

Helmut Haas
Division of Cellular Allergology, Research Centre Borstel, Leibniz Centre
for Medicine and Biosciences, D-23845 Borstel, Germany

Endre Kiss-Toth
Cardiovascular Research Unit, University of Sheffield, Royal Hallamshire
Hospital, Sheffield, S10 2JF, UK, e.kiss-toth@sheffield.ac.uk

Kalliopi Kostelidou
Hellenic Pasteur Institute, Department of Biochemistry, Athens, Greece,
k.kostelidou@pasteur.gr

Konstantinos Lazaridis
Hellenic Pasteur Institute, Department of Biochemistry, Athens, Greece,
klazaridis@pasteur.gr

Jing Lin
Division of Nutritional Sciences, University of Nottingham, Loughborough,
LE12 5RD, UK

Grazia Maria Liuzzi
University of Bari, Department of Biochemistry and Molecular Biology, 70126
Bari, Italy, m.g.liuzzi@biologia.uniba.it

Pier-Luigi Lollini
Section of Cancer Research, Department of Experimental Pathology, University of Bologna, I-40126 Bologna, Italy, pierluigi.lollini@unibo.it

Francesco M. Marincola
Infectious Disease and Immunogenetics Section, Department of Transfusion
Medicine, Clinical Center, National Institutes of Health, Bethesda, MD, USA,
FMarincola@cc.nih.gov

William Martin
EpiVax, Inc., Providence, RI, USA

Julie A. McMurry
EpiVax, Inc., Providence, RI, USA

Leonard Moise
EpiVax, Inc., Institute for Immunology and Informatics, University of Rhode
Island, Providence, RI, USA

Santo Motta
Department of Mathematics and Computer Science, University of Catania,
Italy

Annalisa Murgo
Section of Cancer Research, Department of Experimental Pathology, University of Bologna, Bologna, Italy

Sakunthala Muthugounder
Childrens Hospital, Los Angeles, CA 90027, USA, SMuthugounder@chla.usc.edu

Giordano Nicoletti
Laboratory of Oncologic Research, Rizzoli Orthopaedic Institutes, Bologna,
Italy

Arianna Palladini
Section of Cancer Research, Department of Experimental Pathology, University
of Bologna, Bologna, Italy

Francesco Pappalardo
Department of Mathematics and Computer Science, University of Catania,
Catania, Italy

Christian Radauer
Department of Pathophysiology, Center of Physiology and Pathophysiology, Medical University of Vienna, Vienna, Austria, Christian.Radauer@meduniwien.ac.at

Timothy Ravasi
Department of Bioengineering, Jacobs School of Engineering, University
of California-San Diego, La Jolla, CA 92093,USA, travasi@bioeng.ucsd.edu

Mepur H. Ravindranath
Pacific Clinical Research, Santa Monica, CA 90404, USA, mepurravi@yahoo.com

Neil Renault
Division of Nutritional Sciences, University of Nottingham, Loughborough
LE12 5RD, UK

Paolo Riccio
University of Basilicata, Department of Biology D.B.A.F., 85100 Potenza,
Italy, paolo.riccio@unibas.it

Rocco Rossano
University of Basilicata, Department of Biology D.B.A.F., 85100 Potenza,
Italy, rocco.rossano@unibas.it

Marianna Sabatino
Infectious Disease and Immunogenetics Section, Department of Transfusion
Medicine, Clinical Center, National Institutes of Health, Bethesda, MD, USA

Gabi Schramm
Division of Cellular Allergology, Research Centre Borstel, Leibniz Centre
for Medicine and Biosciences, D-23845 Borstel, Germany

Senthamil R. Selvan
Hoag Cancer Center, Newport Beach, CA 92663, USA, senthamil.selvan@
hoaghospital.org

Anastasia Sideri
Hellenic Pasteur Institute, Department of Biochemistry, Athens, Greece,
asideri@pasteur.gr

Socrates J. Tzartos
Hellenic Pasteur Institute, Department of Biochemistry; University of Patras,
Department of Pharmacy, Athens, Greece, tzartos@pasteur.gr

Peter Valent
Department of Internal Medicine I, Division of Hematology and Hemostaseology,
Medical University of Vienna, A-1090Vienna, Austria, peter.valent@meduniwien.
ac.at

Mauno Vihinen
Institute of Medical Technology, University of Tampere; Tampere University
Hospital, FI-33520 Tampere, Finland, mauno.vihinen@uta.fi

Ena Wang
Infectious Disease and Immunogenetics Section, Department of Transfusion
Medicine, Clinical Center, National Institutes of Health, Bethesda, MD, USA,
FMarincola@cc.nih.gov

# Introduction: Clinical Immunomics; A New Paradigm for Translational Research

V. Brusic and A. Falus

Rapid improvement in accessibility to molecular databases as well as availability of high-throughput genomic, proteomic, and other '*omics*' methodologies are forcing a considerable shift in research and development strategies for biomedicine. The recent change in research paradigm focusing on biology as system science is still difficult to grasp. *Systems biology*, a systematic study of complex interactions in biological systems, is currently closely related to the development and application of bioinformatics and biostatistics tools to genomic and proteomic data. Clinical immunology translates achievements of immunological research to medically relevant applications, that is, diagnosis, prevention, and therapy. It covers a broad area including complex disorders of the immune system (immunodeficiencies, autoimmune diseases, allergy, and lymphoproliferative disorders), immune responses (to pathogens, cancers, transfusion, and transplantation), and immunotherapies (Béné et al. 2000). The complexity of the human immune system has several sources: combinatorial variability, plasticity, degeneracy and adaptivity of the immune system; interactions of the immune system and other self cells, tissues, and organs; the variability of pathogens, self, and environmental antigens; and the presence of multiple regulatory pathways. This complexity ensures that huge amounts of data must be produced and analyzed for deciphering the workings of the immune system. To effectively use these huge databanks, we need increasingly sophisticated tools of biostatics, bioinformatics, and mathematical modeling. The integrative approaches combining multiple tools are particularly important in clinical research comprising large cohorts of usually non-homogenous groups where disease phenotype, clinical progression, molecular profiles, and patient characteristics show huge variation.

Clinical immunology is not clearly differentiated as a clinical specialty because it involves a number of medical disciplines. Nevertheless,

V. Brusic
Cancer Vaccine Center, Dana-Farber Cancer Institute, 77 Avenue Louis Pasteur,
HIM 401, Boston, MA 02115, USA,
e-mail: vladimir_brusic@dfci.harvard.edu

immunological approaches are increasingly being used as means of medical intervention (Béné et al. 2000).

Clinical immunology has a long history. Early writings date back to ancient Greece as far as 2400 years ago when Thucydides described the concept of acquired immunity to an infectious disease: "... *for the same man was never attacked twice – never at least fatally*" and the use of animal models for medical research: "*But of course the effects which I have mentioned could best be studied in a domestic animal like the dog*" (Thucydides 1982). The first text recognized (by the WHO) as scientific treatise on infectious disease is a 1100 years old essay by Rhazes, a Persian physician, describing discoveries about smallpox and measles (The Islamic Medical Manuscript Collection 2003). The later stages of the European scientific revolution of late 17th through 19th century brought both the advanced technologies (such as microscope) and the transformation of scientific ideas in chemistry and biology to provide the foundation of modern western medicine. The 19th century saw the development of successful vaccines, while 20th century produced large quantities of knowledge of fine details describing the cellular, molecular, and genetic basis of immunity.

A major driver for the advancement of immunology is the expansion of knowledge of structural and functional elements of the immune system at the cellular, organ, organism, and population level. This knowledge is accumulated, thanks to scientific and technological progress, which continues unabated; the volume of scientific information is estimated to double every 15 years (Lukasiewicz 1994). The key enabling technologies of genomics (Falus 2005), proteomics (Purcell and Gorman 2004; Brusic et al. 2007), and bioinformatics (Schönbach et al. 2007), and systems biology (including such genomic pathway analysis) (Tegnér et al. 2006) provide large quantities of data describing molecular profiles of various physiological and pathological states. Advanced methods for quantification of immune responses provide means for detailed study of human immune pathology and complex host–pathogen interactions. Flow cytometry enables measurement and characterization of individual cells and molecules representing various experimental states, for example, measurement of antigen-specific immune responses (Li Pira et al. 2007). Improvement of assays for immune monitoring (e.g., multiparametric flow cytometry, nanotechnology for quantitation of cytokine production, ELISPOT, intra-cytoplasmic cytokine staining, and mRNA as well as micro-RNA based assays) continuously expands our ability to measure profiles of cytokines and other molecules that direct and modulate immune responses (Sachdeva and Asthana 2007). Latest developments in laser scanning cytometry allow the measurement and analysis of effector function of individual cells in situ thus representing molecular and cellular events in physiological and pathological states (Harnett 2007).

This volume brings together examples of various topics in clinical immunology, and various tools of immunomics. This collection of articles is not a complete collection of works in this field. Rather, it is a starting point where examples of various immunomics approaches are studied for advancement of knowledge and translation of these results into new products, methods, and therapies. Translating basic immunology advances into medical applications increasingly requires multidisciplinary approach and teams comprising clinicians at the bedside, basic immunologists at the laboratory bench, engineers who develop advanced instrumentation, along with biostatisticians and bioinformaticians who perform data analyses and interpretations. Immunomics, therefore, is a powerful new technology that combines basic and clinical immunology with high-throughput instrumentation and bioinformatics for the analysis and interpretation of the data. Immunomics is similar to genomics and proteomics in that a major challenge is the understanding and manipulating genes and proteins involved in the functioning of the immune system. In addition, immunomics must address factors arising from the complex microenvironment affecting the immune function, as well as external challenges arising from pathogen diversity. Immunomics screening of markers of the immunologic status will in near future be used to determine who should be enrolled in a particular clinical trial and follow-up of these markers throughout the course of therapy (Tremoulet and Albani 2005). The likely major advances of immunomics will be seen first in the fields of vaccines and high-throughput diagnostics. In addition, immunomic approaches herald the development of the new generation of vaccines and immunotherapies to be tailored precisely to both the genetic make-up of the human population and of the disease profile, be it cancer, allergy, or infection (Brusic and August 2004).

This volume has 11 chapters covering a range of immunomic topics. In Chapter "Integrative Systems Approaches to study Innate Immunity", Tim Ravasi describes systems approach to the study of cellular aspects of innate immunity, focusing on macrophages. He offers an insight into applications of systems biology, in which all main components and their interactions within a biological system are measured and then assembled into modules for further study This approach makes no assumptions about underlying mechanisms – the measurement is direct; the disadvantage is that it the scale of information that can be obtained is enormous. Systems biology uncovers relations between entities in a biological system and their regulation. Macrophages are in our primary line of defense against pathogens, and they also mediate the pathology of infectious, inflammatory, and malignant disease, and therefore understanding the control of their function is expected to translate into rational development of therapies.

In Chapter "*Immunomics: At the Forefront of Innate Immunity Research*", Guan and Kiss-Toth describe the immunomics of innate immunity as a novel viewpoint in immunology research, integrating the approaches of cellular immunology, bioinformatics, genomics, proteomics, immuno-informatics, and other related scientific fields, with the aim to derive integrated models of immune modulatory processes. This chapter focuses on the system-based approaches to characterizing in detail the molecular mechanisms of regulatory processes in innate immune responses. The authors have demonstrated the power of integrated approach to characterization of master cytokines, their receptors, signaling pathways, and identification of novel components of innate immunity.

In Chapter "*Epitope-Based Immunome-Derived Vaccines: A Strategy for Improved Design and Safety*", De Groot et al. have explored the immunomics applications in vaccine science. They discuss the combination of bioinformatics prediction tools and an array of experimental models (biochemical assays and animal models). Two case studies, including tularemia and human papilloma virus are presented to demonstrate the utility of immunomics for epitope-based subunit vaccine development.

In Chapter "*Immunodeficiencies and Immunome: Diseases and Information Services*", Mauno Vihinen introduces the concept of Essential Human Immunome, and informatics resources for storage and computational analysis of genes and proteins of the immunome. This chapter focuses on primary immunodeficiencies and the analysis of related immunome entries from some 5000 patients. These resources assist health professionals to select suitable genetic and clinical tests for immunodeficiencies.

In Chapter "*Immunomics of Immune Rejection*", Wang et al. discuss the combination of high-throughput screening of samples representing autologous tumor rejection, clearance of pathogen, acute allograft rejection, and flares of autoimmunity. The use of systems biology helps identify common elements between these pathologies and precise identification of factors that balance host–target interactions. They investigate the interactions of innate and adaptive arms of human immune system and their effects to different disease scenarios.

In Chapter "*Spectrum, Function, and Value of Targets Expressed in Neoplastic Mast Cells*", Valent explores a number of attempts made to identify novel targets and to develop targeted drugs for mast cell leukemia. In the current paper, emerging new molecular targets expressed in neoplastic mast cells are discussed in light of novel therapeutic concepts, availability of drugs, and forthcoming clinical trials.

In Chapter "*Structure, Allergenicity and Cross-Reactivity of Plant Allergens*", Radauer and Breiteneder provided a detailed review of plant allergens, their structure, allergenicity, and allergic cross-reactivity. Grouping of allergens into structural families enables the identification of shared molecular properties of allergens and the basis for the prediction of allergenicity and analysis of cross-reactivity. The authors have made a case for allergy immunomics that combines allergology, structural biology, and bioinformatics.

In Chapter "*The Live Basophil Allergen Array (LBAA): A Pilot Study*", Falcone et al. have described a study where protein microarrays were used to profile activation of basophil cells. The activation of basophils was measured by detecting basophil activation surface marker CD63, as an indirect measurement of basophil degranulation. Combining protein arrays with functional cell-based assays provides a novel method for detection of allergic sensitization. The authors also discussed the limitations and potential pitfalls of the usage of live basophil allergen array in basophil immunobiology studies.

In Chapter "*Emerging Therapies for the Treatment of Autoimmune Myasthenia Gravis*", Kostelidou et al. have written about emerging therapies for the treatment of myasthenia gravis, an autoimmune disease. They have described various treatment modalities. The range of therapeutic approaches is stunning but because of lack of knowledge of causative agents. This field offers an ideal ground for the employment of various immunomics approaches.

In Chapter "*New Diagnostic and Therapeutic Options for the Treatment of Multiple Sclerosis*", Riccio et al. have reviewed new diagnostic and therapeutic options for multiple sclerosis, a multifactorial inflammatory autoimmune disease. The immunomics approach combines multiple approaches: molecular biology for blocking tissue damage through blocking the activity of matrix metalloproteinases, nanotechnology for formulation and delivery of therapeutic agents, combined with healthy and functional foods. This paper illustrates the holistic nature of immunomics where latest technological advances in combination with lifestyle (diet) modification can exert a profound effect on the course of an autoimmune disease.

In Chapter "*Glycoimmunomics of Human Cancer: Relevance to Monitoring Biomarkers of Early Detection and Therapeutic Response*", Ravindranath, Muthugounder and Selvan describe glycomics tools applied to clinical cancer immunology. They have reviewed the developments in early detection and therapeutic responses using gangliosides, a class of glycoantigens that are over-expressed on the tumor cells, relative to their expression on normal cells. Gangliosides are released from tumor cells into both tumor microenvironment and the bloodstream. They act as immunomodulatory agents in the tumor microenvironment, while in the blood they can be used as biomarkers for diagnostics and prognostics in oncology. Because they are recognized as immunogens, gangliosides represent suitable targets for both diagnosis and therapy.

In Chapter "*Translational Immunomics of Cancer Immunoprevention*", Nicoletti et al. introduced the concept of translational immunomics and applications in immunoprevention of cancer. They have combined genomics and mouse models of mammary tumor for efficient identification of oncoantigens, tumor profiling, and immune monitoring. The combination of mathematical modeling and in vivo immunization with the triplex (tre-component) vaccine of mice that spontaneously develop mammary tumors enabled optimization of vaccine scheduling and minimization of the number of vaccine injections.

Clinical immunomics covers a broad range of diseases and this volume is an attempt to bring together work representing a spectrum of immunomic applications to showcase key technologies that will shape clinical immunology in the near future. The articles in this issue show examples of immunomics approaches to cancer, autoimmunity, allergy, and primary immunodeficiencies, along with those that address more basic aspects of large-scale studies of immunology. A common theme in this volume and key concepts deal with various combinations of bioinformatics and biostatistics, instrumentation, molecular biology, animal models, and clinical samples for advancement of clinical immunology. The combination of these approaches enables the identification of complex association networks between immunologic and clinical outcomes. Identification of clinical outcome features and their precise measurement are essential for decision making on the selection of optimal therapy and modification of therapy by assessment of early responses (Tremoulet and Albani 2005). Translation of technological and scientific advances into actual clinical solutions remains a huge task and immunomics offers means for systematic analysis and speeding-up the selection, discovery, design, and validation of new diagnostic and therapeutic products for 21st century.

# References

Béné MC, Stockinger H, Capel PJ, Chapel H, Knapp W. Clinical immunology: a unified vision for Europe. *Immunol Today* 2000; 21(5):210–211.

Brusic V, August JT. The changing field of vaccine development in the genomics era. *Pharmacogenomics.* 2004; 5(6):597–600.

Brusic V, Marina O, Wu CJ, Reinherz EL. Proteome informatics for cancer research: from molecules to clinic. *Proteomics.* 2007; 7(6):976–991.

Falus A. (ed). Immunogenomics and Human Disease. Wiley, 2005.

Harnett MM. Laser scanning cytometry: understanding the immune system in situ. *Nat Rev Immunol.* 2007; 7(11):897–904.

Li Pira G, Kern F, Gratama J, Roederer M, Manca F. Measurement of antigen specific immune responses: 2006 update. *Cytometry B Clin Cytom.* 2007; 72(2):77–85.

Lukasiewicz J. *The Ignorance Explosion*: Understanding Industrial Civilization. McGill-Queen's University Press,1994.

Purcell AW, Gorman JJ. Immunoproteomics: Mass spectrometry-based methods to study the targets of the immune response. *Mol Cell Proteomics.* 2004; 3(3):193–208.

Sachdeva N, Asthana D. Cytokine quantitation: technologies and applications. *Front Biosci.* 2007; 12:4682–4695.

Schönbach C, Ranganathan S., Brusic V. (eds). *Immunoinformatics.* Springer 2007.

Tegnér J, Nilsson R, Bajic VB, Björkegren J, Ravasi T. Systems biology of innate immunity. *Cell Immunol.* 2006;244(2):105–109.

The Islamic Medical Manuscript Collection. *NLM Newsline* 2003; 58(1):10–14. www.nlm.nih. gov/pubs/nlmnews/janmar03/58n1newsline.pdf

Thucydides. The History of the Peloponnesian War. *The Internet Classics Archive* 1928; The Second Book, ChapVI, pp.12–16, at http://classics.mit.edu/Thucydides/pelopwar.2.second. html

Tremoulet AH, Albani S. Immunomics in clinical development: bridging the gap. *Expert Rev. Clin. Immunol.* 2005; 1(1):3–6.

# Integrative Systems Approaches to Study Innate Immunity

Timothy Ravasi

**Abstract** Integrative Systems Biology has emerged as an exciting research approach in molecular biology and functional genomics that involves a systematic use of genomic, proteomic, and metabolomic technologies for the construction of network based models of biological processes. These endeavors, collectively referred to as systems biology establish a paradigm by which to systematically interrogate, model, and iteratively refine our knowledge of the regulatory events within a cell. Here we discuss the latest experimental and computational advances in Integrative Systems Biology, with particular emphasis in approaches specifically designed to identify transcriptional networks governing the macrophage immune response to Lipopolysaccaride (LPS). We show examples where using this approach we are not only able to infer a global macrophage transcriptional network, but also time-specific sub-networks that are dynamically active across the LPS response. Another exiting aspect of this new approach to molecular sciences is that can be applied virtually to any mammalian systems.

**Keywords** Systems Biology · Integrative biology · Transcription regulatory networks · complex systems · Genomics · Proteomics · Protein-protein interaction networks · Gene regulatory circuits · Innate Immunity · Macrophages

The most important and crucial aspect of Systems Biology approaches is, without any doubt, the systems in consideration.

Innate immunity can be seen as a complex multicellular system where macrophages play a major role in orchestrating it. Macrophages are part of the mononuclear phagocyte system (MPS), a family of cells comprising bone marrow progenitors, blood monocytes and tissue macrophages. Macrophages are a

T. Ravasi
Department of Bioengineering, Jacobs School of Engineering, University of California, San Diego, 9500 Gilman Drive, La Jolla, CA 92093, USA,
e-mail: travasi@bioeng.ucsd.edu

A. Falus (ed.), *Clinical Applications of Immunomics*,
DOI: 10.1007/978-0-387-79208-8_1, © Springer Science+Business Media, LLC 2009

1

major cell population in most of the tissues in the body, and their numbers increase further in inflammation, wounding and malignancy (Hume 2006; Hume et al. 2002). Through their endocytic and cytotoxic activities, they provide a first line of defence against pathogens in the innate immune system. They contribute to antigen processing and presentation, and in turn their effector functions are activated in response to products of the acquired immune system, notably antibodies (via Fc receptors) and T cell products (e.g. interferon-gamma, interleukin 4, etc.). Apart from their roles in immunity, macrophages also contribute to many aspects of homeostasis, vascularisation, normal development, tumour progression and wound healing in part through their adaptation to recognise and remove cells undergoing apoptosis (Lichanska and Hume 2000) and through their secretion of growth factors and proteolytic enzymes.

Mononuclear phagocytes, as they appear in tissues, share a number of features.

- Stellate morphology and ultrastructural evidence of endocytic activity observed by light and electron microscopy.
- Expression of histochemical markers such as non-specific esterase, lysosomal hydrolases and ecto-enzymes.
- Evidence of endocytic activity.
- The presence of cell surface proteins (such as F4/80, CD14 and CSF-1R) defined by monoclonal antibodies.

The precise anatomical location of these cells supports a physiological role for tissue macrophages in the development, structure and homeostasis of organs. Not all macrophages express the F4/80 marker. To find an alternative, we have produced the MacGreen transgenic mouse lines in which macrophage-restricted expression of an EGFP transgene is directed by the promoter of the *CSF-1R* (*c-fms*) gene. The proliferation, differentiation and survival of macrophages are controlled by macrophage colony-stimulating factor (CSF-1), and the receptor is expressed in a myeloid-restricted manner. These mice provide a resource for many aspects of macrophage biology, including the isolation of macrophage populations from specific organs(MacDonald et al. 2005; Sasmono et al. 2003).

Our ability to monitor this population of cells in real time and in vivo raises the interesting question of exactly how their numbers are monitored, controlled and replenished, the cellular "systems biology" of the mononuclear phagocyte system. At least in the normal steady state, the availability of CSF-1 appears to be regulated via macrophage-dependent endocytosis and destruction, which provides a simple link between macrophage numbers and the production of cells by the marrow. Accordingly, the level of CSF-1 rises substantially in mice that lack the CSF-1 receptor (Dai et al. 2002). The control is undoubtedly more complex since CSF-1 expression is regulated in many different tissues (Sweet and Hume 2003), and at least some macrophage populations may be controlled by local proliferation (Hume 2006). Furthermore, several other growth factors can regulate macrophage production from bone marrow (Hume 2006).

Mammalian cells are complex systems. Their functions require the appropriate interactions of millions of individual components in appropriate order in time and space. Although biological scientists commonly study cells as if they were static entities, individual cells never exist in a steady state. Like an entire multicellular organism, each individual cell is born by cell division, adapts to its environment to carry out a particular function that is determined by its genetically programmed response to that environment and then dies. If all the components of a biological system interact with each other, the complexity would be impenetrable. However, as in most complex systems, individual components within a cell interact and organise in defined modules that we recognise as structures, pathways and regulatory frameworks. The traditional approach to understanding cell biology has been to identify those individual modules, dissect their components and then attempt to determine how they can be reassembled. Systems biology can be considered an alternative approach, in which one actually measures all the components and their interactions within a biological system and then assembles them into modules (Aderem 2005). The advantage of this approach is that it makes no assumptions; the disadvantage is that it is constrained by the scale of information that can be obtained.

To understand the function of a cell in perfect detail, we would need to know the identity of every macromolecule and metabolite, its location, abundance and chemical status across the lifetime of each cell. Since cells in a population influence each other's behaviour in many ways, we would also need to know the same information about its neighbours, every other cell in the body. This is clearly not possible. However, the problem can simplified if one takes the view that the entire control framework of a mammalian cell is encoded in its genome and exerted through the selective production of RNA, the process of transcription. If we know the identity of every transcript produced by the genome and the way that transcription changes with time, we can infer the fundamental control pathways. Certainly, different tissues express distinct transcriptional profiles – transcriptional lineage markers – that can be used to identify subpopulations of cells in complex tissues, for example, the stem cell populations profiled in the Stem Cell Gene Anatomy Project (see www.scgap.org), (Challen et al. 2005, 2004). The process of lineage differentiation is arguably the process of restricting the transcriptional network of a cell to those genes that define its function. Ultimately, the life of that cell can be viewed by a transcriptional network.

Transcription is itself controlled by RNA; the precise way this feedback control operates may be very complex. RNA may act directly on the induction, processing or stability of another transcript. Non-coding RNAs are a major, regulated, output of the mammalian genome (Katayama et al. 2005; Ravasi et al. 2006). Alternatively, the RNA may encode a protein. That protein may participate directly in transcription control in the nucleus, or it may be a secreted protein that directly or indirectly initiates a signalling cascade that ultimately regulates transcription. Hence, the control system of biological systems is ultimately a network of linked transcriptional switches. The completion of multiple eukaryote genome sequences and the rapid progress towards definition of the complete transcriptional output (the transcriptome) have given us access to all of the components of the network. The capacity to analyse the complete transcriptional profile of cells and the advent of technologies that allow us to capture protein/RNA and DNA interactions within this transcriptional context offers us the tools to identify the connectivities within the network and, ultimately, the control system.

There are many reasons why macrophages are an ideal cell type for applying a systems approach to transcription control in a mammalian context.

(1) They can be obtained as reasonably homogeneous primary cell cultures, from peripheral blood monocytes in humans or by cultivation of bone marrow cells or peritoneal lavage in laboratory animals.
(2) There are cell lines available in both mouse and human that replicates many of the characteristics of primary macrophages and which can be transiently or stably transfected.
(3) The range of mammalian biology/pathways that can be studied using macrophages is substantial; they can be regarded in significant measure as the archetypal cells.
(4) They alter their function in distinct ways in response to many different extracellular signals acting through multiple distinct signalling pathways, and most importantly, these changes in the macrophage regulatory networks can be measured over time (Rehli et al. 2005; Schroder et al. 2004; Sester et al. 2005; Wells et al. 2003a).
(5) There are substantial natural, or introduced, genetic influences on the function of macrophages (Beutler 2005; Beutler et al. 2005; Fortier et al. 2005; Wells et al. 2003a). These provide perturbations of the network, which can be analysed to establish the connections between nodes.
(6) Since macrophages are both our primary line of defence against pathogens and mediators of much of the pathology of infectious, inflammatory and malignant disease, a fundamental understanding of the way their function is controlled is likely to translate into rational development of human therapies.
(7) Their differentiation from progenitor cells, which involves stable epigenetic changes, can also be studied in detail *in vitro* (Tagoh et al. 2002).

## 1 Transcriptome Data Sets and the Macrophage Transcriptional Network

Macrophages as a cell population provide one of the most complex sources of transcripts in any cell type. In fact, we have performed comparative transcriptional profiling of LPS-stimulated mouse macrophages compared to 17-day embryos and embryonic stem cells. The three cell populations in combination cover the large majority of the protein-coding transcripts on a complex microarray platform. The complexity of macrophages as a source of transcripts was also exploited by the FANTOM consortium in its efforts to identify the full diversity of transcripts encoded by the mouse genome (Wells et al. 2003b). Arising from that effort, we have recently analysed the impact of alternative initiation, splicing and polyadenylation on the set on a selected part on the macrophage proteome, focussing specifically on signalling molecules that contribute to the LPS response (Wells et al. 2006). This analysis built upon the original finding that LPS acts upon an internal promoter in the IRAK-2 signalling molecule to generate inducible expression of a natural feedback repressor, a component of control network mentioned above. Using splicing arrays, we confirmed the expression of many additional examples of variant forms of signalling molecules that could form part of the control architecture.

Our own analysis has focussed on a particular cellular system. Activation of macrophages by LPS is across a time course from 1 to 48 hours.

Aside from extensive array profiling on a number of different platforms (Calvano et al. 2005; Nilsson et al. 2006; Tegner and Bjorkegren 2007; Tegner et al. 2006), symatlas.gnf.org, we have also used this system in the context of analysis of the promoters of the human and mouse genomes.

The transcriptional output of any cell is influenced by the interaction between nuclear proteins and DNA sequences in the vicinity of the gene (promoters, enhancer and repressors). The identification of promoters is dependent upon accurate determination of the transcription start site. Recently, this has become possible on a genome-wide scale using CAGE, essentially high throughput 5' RACE/SAGE, technology (Carninci et al. 2006). This technology, combined with sequences of full-length cDNAs and other polling technologies (diTags), has enabled the first genome-wide view of transcription initiation. To expedite access to this information, a Web portal is available at the RIKEN Genome Science Institute (http://fantom.gsc.riken.go.jp). For the analysis of macrophages, CAGE sequencing has been performed in multiple independent libraries across the time course of activation by LPS, providing an index of relative promoter use. Such data highlights the fact that a large proportion of the protein-coding genes in the genome have multiple promoters, expressed in different cell types and tissues.

The CAGE data provide us with a survey of the set of transcription start sites (TSS), utilised specifically in macrophages. Importantly, CAGE tags also provide a measure of frequency of promoter usage – information lacking from normalised EST or FL-cDNA datasets. The sequences flanking these start sites can be extracted and analysed as a class for the presence of over-represented sequence motifs for broad classes of transcription factors. Not surprisingly, given the known architecture of TATA-less myeloid promoters (Rehli et al. 1999; Ross et al. 1998), such an analysis reveals over-representation of purine-rich motifs recognised by the Ets transcription factor family amongst constitutive macrophage-specific genes, as well as sites for the LPS-inducible transcription factor NFkappaB amongst inducible genes (Carninci et al. 2006). A more penetrating analysis can be using either experimentally validated binding sites for known transcription factors (e.g. Transfac, Jaspar) or a priori motif detection (Calvano et al. 2005; Nilsson et al. 2006; Tegner and Bjorkegren 2007; Tegner et al. 2006). The final piece in the puzzle is the identity of the transcription factors that are actually present and likely to bind the candidate motifs in the promoters.

The FANTOM project has also provided a comprehensive set of candidate transcriptional regulators encoded by the mouse genome. Transcription factors were amongst the most highly alternatively spliced class in the F3 dataset – many new variants of known TF have been discovered (Kummerfeld and Teichmann 2006; Ravasi et al. 2003). Aside from known transcription factors, additional candidates were identified based upon the presence of domains associated with DNA binding or other nuclear functions. The set of known functional transcription factors in macrophages has been reviewed elsewhere (Himes and Hume 2003), but was anticipated to be far from comprehensive qRT-PCR analysis as carried out to identify the complete set of transcripts encoding nuclear proteins that is expressed by macrophages across the time course of response to LPS (Calvano et al. 2005; Nilsson et al. 2006; Tegner and Bjorkegren 2007; Tegner et al. 2006). The increased sensitivity of qRT-PCR is important, because many transcription factors are expressed at relatively low levels around the detection thresholds for microarrays. An interesting finding in this global analysis is that around 60% (850/1439) of all the transcripts encoding nuclear proteins were detectable in macrophages (Calvano et al. 2005; Nilsson et al. 2006; Tegner and Bjorkegren 2007; Tegner et al. 2006). The notable exclusions were members of the gene families known to be involved in developmental processes wherein they are expressed transiently in much defined embryonic locations. They include most members of the Hox, Sox, Gata, Neurog, Lhx and Tbx families. This observation alone suggests that other nuclear proteins that are not expressed in macrophages will probably also have functions in development that require absolutely stringent control.

Taking all of the pieces of information together, it is possible to construct a first-pass transcriptional control network for the response of macrophages to LPS. The analysis links transcription factor expression from qRT-PCr, the promoter architectures (i.e. presence/absence of transcription factor binding sites) of all the transcripts expressed in macrophages (the macrophage transcriptome) and the semantic approaches to infer regulatory circuits (Calvano et al. 2005; Nilsson et al. 2006; Tegner and Bjorkegren 2007; Tegner et al. 2006). The predicted network has the characteristics of a scale-free network, with a relatively small number of nodes that are highly connected. Amongst these predicted nodes are most of the known players in LPS responses, the Ets family, NF-kappaB and interferon-response element binding proteins. (Fig. 1 shows a representation of an integrative approach used to infer transcription regulatory network in mammals.)

Gene expression profiles describe the steady-state mRNA levels in the cell – the outcome of the regulatory network. Although network structure could in principle be inferred only from expression data (di Bernardo et al. 2005; Gardner et al. 2003), it is a very challenging task due to the small sample sizes

**Fig. 1** Integrative approach to infer transcription regulatory networks. (**A**) The main players are defined based on their expression in the systems (nodes of the network). (**B**) Promoter mapping associated with TFBS predictions are used to infer regulatory interaction and combined with datasets for physical interactions such as protein–protein interactions (edges of the network). (**C**) Nodes and edges are combined and represented as network topology. (**D**) The network is annotated and validated based on previous knowledge or experimental validation (*See* Color Insert)

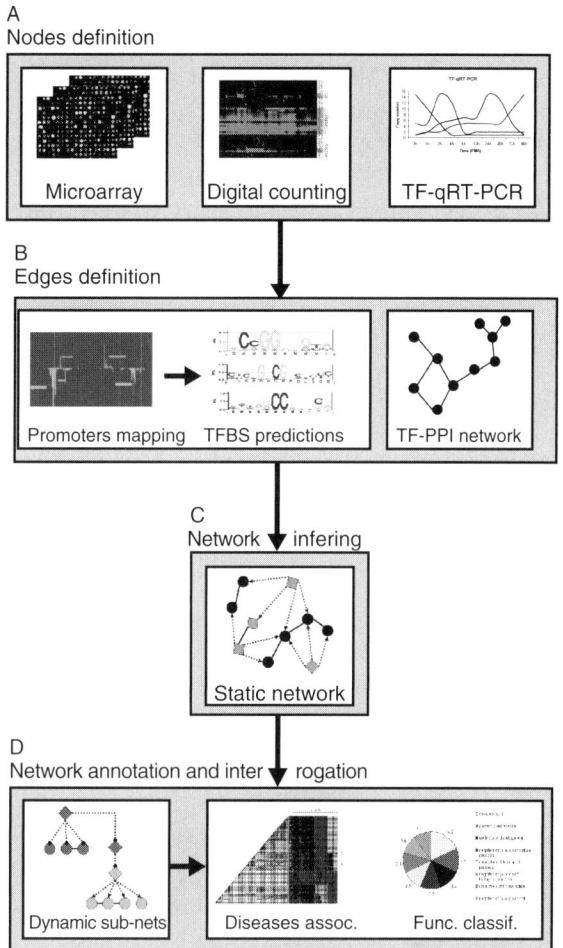

(number of genes greatly exceeds the number of measurements per gene) and large amount of noise in expression profiles. Nowadays, with the availability of large-scale physical interactions datasets, integration of these data with expression data provides a more direct route for reconstructing gene regulatory networks.

However, all different data sources have their own limitations. Currently, both gene expressions GWLA and PPI are noisy (Birney et al. 2007; Hart et al. 2006; Rual et al. 2005; Suzuki et al. 2003; Yu et al. 2004). Differentially expressed genes from replicate micro-array experiments typically overlap by 70–75% whereas the overlap between replicate ChIP-chip experiments is even lower, usually less than 50% (Harbison et al. 2004; Lee et al. 2002; Ren et al. 2000). Because gene expression and TR location data provide complementary information, integration of the two data sources can emphasize the functional part of the network and thus makes the inferred network more biologically relevant. This observation provides the rationale for several recently developed integrative approaches. Bar-Joseph et al. developed an iterative method in which coherent gene expression profile is used to include low-confidence but true ChIP-chip targets in regulatory modules (Bar-Joseph et al. 2003). Gao et al. used multiple linear regressions to model gene expression ratio as a result of TF binding to gene promoters as measured with ChIP-chip (Gao et al. 2004). Similarly, Liao et al. used two sets of linear equations to simultaneously model the regulatory strength of TFs on their targets as well as the activity levels of the TF themselves (Liao et al. 2003). In comparison, the linear regression approach of Gao and colleagues can only model TF activity levels (Gao et al. 2004).

Most cellular functions are carried out by protein complexes, such as ribosome, splicesosome and proteasome. So far, little is known about how protein complexes are regulated. Protein interaction data could also be integrated with gene expression profiles to discover regulated protein complexes and to study the regulatory dynamics of protein complexes (de Lichtenberg et al. 2005; Ho et al. 2002; Hsing et al. 2004; Mann et al. 2001; Simonis et al. 2004; Tan et al. 2007).

A regulatory network, defined from several different data sources as outlined above, is useful in different ways. Clearly, the network structure suggests novel mechanistic hypothesis, which eventually has to be experimentally tested as a final validation step. However, before this step is taken, it is mandatory to consider that networks are condition- and state-dependent, that is, different parts of the network will be active under different conditions. For example, a cell that is exposed to a particular compound or a physiological condition such as stress will produce two different activity patterns. Therefore, a static network defined from different data sources has to be evaluated and projected onto the specific condition of interest. Such a network projection can be performed in space (over different organs/ tissues) and/or in time (in response to a stimulation), as shown in Fig. 2 for a macrophage-specific gene regulatory circuit activated by LPS stimulation.

The tissue-specific mRNA expression patterns of a gene can offer important clues to its physiological function. Previously, Su et al. have generated a large compendium of gene expression profiles of 79 human and 61 mouse tissues (Su et al. 2004). Apart from normal tissues, a large number of disease-specific expression profiles (cancer, immune disorders, neurological disease, etc.) have also been generated over the past decades. These expression datasets, which in essence constitute an activity-defined "fingerprint" of a disease state, provide unprecedented opportunities to study the transcriptional regulatory networks in mammals. The key to success is to adopt a system to integrate these expression profiles with the static (interactome) network and phenotypic data. Recently, as a proof-of-concept, several groups have adopted this strategy to study the regulatory networks in human diseases, including cancer (Segal et al. 2005; Tomlins et al. 2007), innate immunity and inflammation. For example, Segal and coworkers used a large compendium of gene expression profiles to define shared and cancer subtype-specific modules (Segal et al. 2005). Similarly, Paulsson and coworkers have defined metabolic interaction networks and

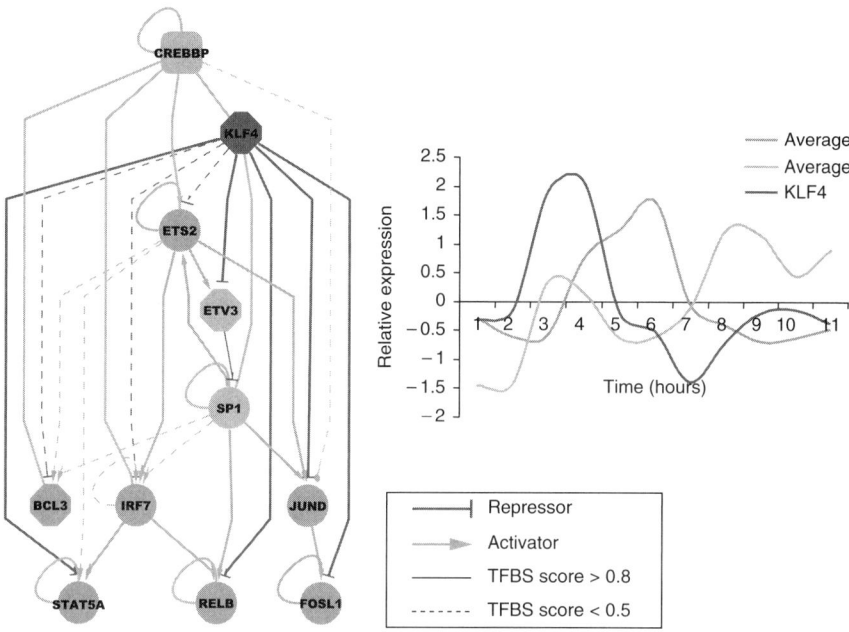

**Fig. 2** Gene regulatory circuits activities in macrophage after stimulation with LPS. By integrating several datasets (expression, PPI, previous knowledge and promoter analysis), it is possible to infer dynamics and hierarchy of gene regulation; this particular case shows a beautiful feedback loop where the main players are the transcription factors KLF4, SP1 and ETV3. Green arrowed edges = gene activation, red arrowed edges = gene repression, grey edge = protein–protein interactions. The trickiness of the edges is proportional to the TFBS prediction confidence score. In the right panel is shown the expression profile of TFs and regulated genes composing the network (*See* Color Insert)

used these static networks, together with expression data sampled from different conditions, to define the disease states of relevance for diabetes and cardiovascular disease (Duarte et al. 2007).

The pre-genomic era was characterised in some measure by a race to acquire information. The post-genomic era saw a secondary race to identify the transcriptional outputs of mammalian genomes, and the impact of proteomic, and ultimately a metabolomic, revolution is probably still to come.

A major emerging challenge of network biology is to systematically compare and contrast biological networks over different species, conditions, cell types, disease states or points in time. For this purpose, methods are being developed to compare/contrast protein interaction networks to predict regulatory interactions and to identify conserved interaction complexes and pathways. Although most of the previous researches have focused on protein interaction networks, many methods can be extended to compare transcriptional regulatory networks. An intriguing future application of these integrative networks may be to probe the functional role of SNPs, which are now rapidly being assembled for a range of diseases.

Finally, it is clear that data-integration methods need to be formulated in a proper statistical framework. For example, for each data type it is essential to develop statistical measures that indicate the reliability of the data, thus setting the stage for a weighted integrative method where each data source contributes in proportion to its internal quality and relevance to the question at hand.

**Acknowledgments** Timothy Ravasi was supported by the National Institute of Mental Health (NIMH), grant 2P30MH062261-07.

# References

Aderem, A.: Systems biology: its practice and challenges. *Cell 121:* 511–3, 2005

Bar-Joseph, Z., Gerber, G. K., Lee, T. I., Rinaldi, N. J., Yoo, J. Y., Robert, F., Gordon, D. B., Fraenkel, E., Jaakkola, T. S., Young, R. A., and Gifford, D. K.: Computational discovery of gene modules and regulatory networks. *Nat Biotechnol 21:* 1337–42, 2003

Beutler, B.: The Toll-like receptors: analysis by forward genetic methods. *Immunogenetics 57:* 385–92, 2005

Beutler, B., Georgel, P., Rutschmann, S., Jiang, Z., Croker, B., and Crozat, K.: Genetic analysis of innate resistance to mouse cytomegalovirus (MCMV). *Brief Funct Genomic Proteomic 4:* 203–13, 2005

Birney, E. and Stamatoyannopoulos, J. A. and Dutta, A. and Guigo, R. and Gingeras, T. R. and Margulies, E. H. and Weng, Z. and Snyder, M. and Dermitzakis, E. T. and Thurman, R. E. and Kuehn, M. S. and Taylor, C. M. and Neph, S. and Koch, C. M. and Asthana, S. and Malhotra, A. and Adzhubei, I. and Greenbaum, J. A. and Andrews, R. M. and Flicek, P. and Boyle, P. J. and Cao, H. and Carter, N. P. and Clelland, G. K. and Davis, S. and Day, N. and Dhami, P. and Dillon, S. C. and Dorschner, M. O. and Fiegler, H. and Giresi, P. G. and Goldy, J. and Hawrylycz, M. and Haydock, A. and Humbert, R. and James, K. D. and Johnson, B. E. and Johnson, E. M. and Frum, T. T. and Rosenzweig, E. R. and Karnani, N. and Lee, K. and Lefebvre, G. C. and Navas, P. A. and Neri, F. and Parker, S. C. and Sabo, P. J. and Sandstrom, R. and Shafer, A. and Vetrie, D. and Weaver, M. and Wilcox, S. and Yu, M. and Collins, F. S. and Dekker, J. and Lieb, J. D. and Tullius, T. D. and Crawford, G. E. and Sunyaev, S. and Noble, W. S. and Dunham, I. and Denoeud, F. and Reymond, A. and Kapranov, P. and Rozowsky, J. and Zheng, D. and Castelo, R. and Frankish, A. and Harrow, J. and Ghosh, S. and Sandelin, A. and Hofacker, I. L. and Baertsch, R. and Keefe, D. and Dike, S. and Cheng, J. and Hirsch, H. A. and Sekinger, E. A. and Lagarde, J. and Abril, J. F. and Shahab, A. and Flamm, C. and Fried, C. and Hackermuller, J. and Hertel, J. and Lindemeyer, M. and Missal, K. and Tanzer, A. and Washietl, S. and Korbel, J. and Emanuelsson, O. and Pedersen, J. S. and Holroyd, N. and Taylor, R. and Swarbreck, D. and Matthews, N. and Dickson, M. C. and Thomas, D. J. and Weirauch, M. T. and Gilbert, J., et al.: Identification and analysis of functional elements in 1% of the human genome by the ENCODE pilot project. *Nature 447:* 799–816, 2007

Calvano, S. E., Xiao, W., Richards, D. R., Felciano, R. M., Baker, H. V., Cho, R. J., Chen, R. O., Brownstein, B. H., Cobb, J. P., Tschoeke, S. K., Miller-Graziano, C., Moldawer, L. L., Mindrinos, M. N., Davis, R. W., Tompkins, R. G., and Lowry, S. F.: A network-based analysis of systemic inflammation in humans. *Nature 437:* 1032–7, 2005

Carninci, P., Sandelin, T.A., Lenhard, B., Katayamal, S., Shimokawa, K., Ponjavic, P., Semple, C.A., Taylor, M.S., Engstrom, P., Frith, M., Forrest, A.R.R., Alkema, W.B., Tan, S.L., Plessy, C., Kodzius, K., Ravasi, T. et al. Genome-wide analysis of mammalian promoter architecture and evolution. *Nat Genet.* 38(6):626–35, 2006

Challen, G., Gardiner, B., Caruana, G., Kostoulias, X., Martinez, G., Crowe, M., Taylor, D. F., Bertram, J., Little, M., and Grimmond, S. M.: Temporal and spatial transcriptional programs in murine kidney development. *Physiol Genomics 23:* 159–71, 2005

Challen, G. A., Martinez, G., Davis, M. J., Taylor, D. F., Crowe, M., Teasdale, R. D., Grimmond, S. M., and Little, M. H.: Identifying the molecular phenotype of renal progenitor cells. *J Am Soc Nephrol 15:* 2344–57, 2004

Dai, X. M., Ryan, G. R., Hapel, A. J., Dominguez, M. G., Russell, R. G., Kapp, S., Sylvestre, V., and Stanley, E. R.: Targeted disruption of the mouse colony-stimulating factor 1 receptor gene results in osteopetrosis, mononuclear phagocyte deficiency, increased primitive progenitor cell frequencies, and reproductive defects. *Blood 99:* 111–20, 2002

de Lichtenberg, U., Jensen, L. J., Brunak, S., and Bork, P.: Dynamic complex formation during the yeast cell cycle. *Science 307:* 724–7, 2005

di Bernardo, D., Thompson, M. J., Gardner, T. S., Chobot, S. E., Eastwood, E. L., Wojtovich, A. P., Elliott, S. J., Schaus, S. E., and Collins, J. J.: Chemogenomic profiling on a genome-wide scale using reverse-engineered gene networks. *Nat Biotechnol 23:* 377–83, 2005

Duarte, N. C., Becker, S. A., Jamshidi, N., Thiele, I., Mo, M. L., Vo, T. D., Srivas, R., and Palsson, B. O.: Global reconstruction of the human metabolic network based on genomic and bibliomic data. *Proc Natl Acad Sci USA 104:* 1777–82, 2007

Fortier, A., Min-Oo, G., Forbes, J., Lam-Yuk-Tseung, S., and Gros, P.: Single gene effects in mouse models of host: pathogen interactions. *J Leukoc Biol 77:* 868–77, 2005

Gao, F., Foat, B. C., and Bussemaker, H. J.: Defining transcriptional networks through integrative modeling of mRNA expression and transcription factor binding data. *BMC Bioinformatics 5:* 31, 2004

Gardner, T. S., di Bernardo, D., Lorenz, D., and Collins, J. J.: Inferring genetic networks and identifying compound mode of action via expression profiling. *Science 301:* 102–5, 2003

Harbison, C. T., Gordon, D. B., Lee, T. I., Rinaldi, N. J., Macisaac, K. D., Danford, T. W., Hannett, N. M., Tagne, J. B., Reynolds, D. B., Yoo, J., Jennings, E. G., Zeitlinger, J., Pokholok, D. K., Kellis, M., Rolfe, P. A., Takusagawa, K. T., Lander, E. S., Gifford, D. K., Fraenkel, E., and Young, R. A.: Transcriptional regulatory code of a eukaryotic genome. *Nature 431:* 99–104, 2004

Hart, G. T., Ramani, A. K., and Marcotte, E. M.: How complete are current yeast and human protein-interaction networks? *Genome Biol 7:* 120, 2006

Himes, S. R. and Hume, D. A.: Transcription factors that regulate macrophage development. *Handb Exp Pharmacol 158:* 11–40, 2003

Ho, Y., Gruhler, A., Heilbut, A., Bader, G. D., Moore, L., Adams, S. L., Millar, A., Taylor, P., Bennett, K., Boutilier, K., Yang, L., Wolting, C., Donaldson, I., Schandorff, S., Shewnarane, J., Vo, M., Taggart, J., Goudreault, M., Muskat, B., Alfarano, C., Dewar, D., Lin, Z., Michalickova, K., Willems, A. R., Sassi, H., Nielsen, P. A., Rasmussen, K. J., Andersen, J. R., Johansen, L. E., Hansen, L. H., Jespersen, H., Podtelejnikov, A., Nielsen, E., Crawford, J., Poulsen, V., Sorensen, B. D., Matthiesen, J., Hendrickson, R. C., Gleeson, F., Pawson, T., Moran, M. F., Durocher, D., Mann, M., Hogue, C. W., Figeys, D., and Tyers, M.: Systematic identification of protein complexes in Saccharomyces cerevisiae by mass spectrometry. *Nature 415:* 180–3, 2002

Hsing, M., Bellenson, J. L., Shankey, C., and Cherkasov, A.: Modeling of cell signaling pathways in macrophages by semantic networks. *BMC Bioinformatics 5:* 156, 2004

Hume, D. A.: The mononuclear phagocyte system. *Curr Opin Immunol 18:* 49–53, 2006

Hume, D. A., Ross, I. L., Himes, S. R., Sasmono, R. T., Wells, C. A., and Ravasi, T.: The mononuclear phagocyte system revisited. *J Leukoc Biol 72:* 621–7, 2002

Katayama, S., Tomaru, Y., Kasukawa, T., Waki, K., Nakanishi, M., Nakamura, M., Nishida, H., Yap, C. C., Suzuki, M., Kawai, J., Suzuki, H., Carninci, P., Hayashizaki, Y., Wells, C., Frith, M., Ravasi, T., Pang, K. C., Hallinan, J., Mattick, J., Hume, D. A., Lipovich, L., Batalov, S., Engstrom, P. G., Mizuno, Y., Faghihi, M. A., Sandelin, A., Chalk, A. M., Mottagui-Tabar, S., Liang, Z., Lenhard, B., and Wahlestedt, C.: Antisense transcription in the mammalian transcriptome. *Science 309:* 1564–6, 2005

Kummerfeld, S.K. and Teichmann, S.A. DBD: a transcription factor prediction database. *Nucleic Acids Res. 34*, 2006

Lee, T. I., Rinaldi, N. J., Robert, F., Odom, D. T., Bar-Joseph, Z., Gerber, G. K., Hannett, N. M., Harbison, C. T., Thompson, C. M., Simon, I., Zeitlinger, J., Jennings, E. G., Murray, H. L., Gordon, D. B., Ren, B., Wyrick, J. J., Tagne, J. B., Volkert, T. L., Fraenkel, E., Gifford, D. K., and Young, R. A.: Transcriptional regulatory networks in Saccharomyces cerevisiae. *Science 298:* 799–804, 2002

Liao, J. C., Boscolo, R., Yang, Y. L., Tran, L. M., Sabatti, C., and Roychowdhury, V. P.: Network component analysis: reconstruction of regulatory signals in biological systems. *Proc Natl Acad Sci U S A 100:* 15522–7, 2003

Lichanska, A. M. and Hume, D. A.: Origins and functions of phagocytes in the embryo. *Exp Hematol 28:* 601–11, 2000

MacDonald, K. P., Rowe, V., Bofinger, H. M., Thomas, R., Sasmono, T., Hume, D. A., and Hill, G. R.: The colony-stimulating factor 1 receptor is expressed on dendritic cells during differentiation and regulates their expansion. *J Immunol 175:* 1399–405, 2005

Mann, M., Hendrickson, R. C., and Pandey, A.: Analysis of proteins and proteomes by mass spectrometry. *Annu Rev Biochem 70:* 437–73, 2001

Nilsson, R., Bajic, V.B., Katayama, K., Suzuki, H., Reid, J.F., Sweet, M.J., Gariboldi, M., Carninci, P., Hayashizaki, Y., Hume, D.A., Tegner, J., and Ravasi, T.: Transcriptional network dynamics governing macrophage activation. *Genomics* 88(2): 133–42, 2006

Ravasi, T., Huber, T., Zavolan, M., Forrest, A., Gaasterland, T., Grimmond, S., and Hume, D. A.: Systematic characterization of the zinc-finger-containing proteins in the mouse transcriptome. *Genome Res 13:* 1430–42, 2003

Ravasi, T., Suzuki, H., Pang, K. C., Katayama, S., Furuno, M., Okunishi, R., Fukuda, S., Ru, K., Frith, M. C., Gongora, M. M., Grimmond, S. M., Hume, D. A., Hayashizaki, Y., and Mattick, J. S.: Experimental validation of the regulated expression of large numbers of non-coding RNAs from the mouse genome. *Genome Res 16:* 11–9, 2006

Rehli, M., Lichanska, A., Cassady, A. I., Ostrowski, M. C., and Hume, D. A.: TFEC is a macrophage-restricted member of the microphthalmia-TFE subfamily of basic helix-loop-helix leucine zipper transcription factors. *J Immunol 162:* 1559–65, 1999

Rehli, M., Sulzbacher, S., Pape, S., Ravasi, T., Wells, C. A., Heinz, S., Sollner, L., El Chartouni, C., Krause, S. W., Steingrimsson, E., Hume, D. A., and Andreesen, R.: Transcription factor Tfec contributes to the IL-4-inducible expression of a small group of genes in mouse macrophages including the granulocyte colony-stimulating factor receptor. *J Immunol 174:* 7111–22, 2005

Ren, B., Robert, F., Wyrick, J. J., Aparicio, O., Jennings, E. G., Simon, I., Zeitlinger, J., Schreiber, J., Hannett, N., Kanin, E., Volkert, T. L., Wilson, C. J., Bell, S. P., and Young, R. A.: Genome-wide location and function of DNA binding proteins. *Science 290:* 2306–9, 2000

Ross, I. L., Yue, X., Ostrowski, M. C., and Hume, D. A.: Interaction between PU.1 and another Ets family transcription factor promotes macrophage-specific Basal transcription initiation. *J Biol Chem 273:* 6662–9, 1998

Rual, J. F., Venkatesan, K., Hao, T., Hirozane-Kishikawa, T., Dricot, A., Li, N., Berriz, G. F., Gibbons, F. D., Dreze, M., Ayivi-Guedehoussou, N., Klitgord, N., Simon, C., Boxem, M., Milstein, S., Rosenberg, J., Goldberg, D. S., Zhang, L. V., Wong, S. L., Franklin, G., Li, S., Albala, J. S., Lim, J., Fraughton, C., Llamosas, E., Cevik, S., Bex, C., Lamesch, P., Sikorski, R. S., Vandenhaute, J., Zoghbi, H. Y., Smolyar, A., Bosak, S., Sequerra, R., Doucette-Stamm, L., Cusick, M. E., Hill, D. E., Roth, F. P., and Vidal, M.: Towards a proteome-scale map of the human protein-protein interaction network. *Nature 437:* 1173–8, 2005

Sasmono, R. T., Oceandy, D., Pollard, J. W., Tong, W., Pavli, P., Wainwright, B. J., Ostrowski, M. C., Himes, S. R., and Hume, D. A.: A macrophage colony-stimulating factor receptor-green fluorescent protein transgene is expressed throughout the mononuclear phagocyte system of the mouse. *Blood 101:* 1155–63, 2003

Schroder, K., Hertzog, P. J., Ravasi, T., and Hume, D. A.: Interferon-gamma: an overview of signals, mechanisms and functions. *J Leukoc Biol 75:* 163–89, 2004

Segal, E., Fricdman, N., Kaminski, N., Regev, A., and Koller, D.: From signatures to models: understanding cancer using microarrays. *Nat Genet 37 Suppl:* S38–45, 2005

Sester, D. P., Trieu, A., Brion, K., Schroder, K., Ravasi, T., Robinson, J. A., McDonald, R. C., Ripoll, V., Wells, C. A., Suzuki, H., Hayashizaki, Y., Stacey, K. J., Hume, D. A., and Sweet, M. J.: LPS regulates a set of genes in primary murine macrophages by antagonising CSF-1 action. *Immunobiology 210:* 97–107, 2005

Simonis, N., van Helden, J., Cohen, G. N., and Wodak, S. J.: Transcriptional regulation of protein complexes in yeast. *Genome Biol 5:* R33, 2004

Su, A. I., Wiltshire, T., Batalov, S., Lapp, H., Ching, K. A., Block, D., Zhang, J., Soden, R., Hayakawa, M., Kreiman, G., Cooke, M. P., Walker, J. R., and Hogenesch, J. B.: A gene atlas of the mouse and human protein-encoding transcriptomes. *Proc Natl Acad Sci U S A 101:* 6062–7, 2004

Suzuki, H., Saito, R., Kanamori, M., Kai, C., Schonbach, C., Nagashima, T., Hosaka, J., and Hayashizaki, Y.: The mammalian protein-protein interaction database and its viewing system that is linked to the main FANTOM2 viewer. *Genome Res 13:* 1534–41, 2003

Sweet, M. J. and Hume, D. A.: CSF-1 as a regulator of macrophage activation and immune responses. *Arch Immunol Ther Exp (Warsz) 51:* 169–77, 2003

Tagoh, H., Himes, R., Clarke, D., Leenen, P. J., Riggs, A. D., Hume, D., and Bonifer, C.: Transcription factor complex formation and chromatin fine structure alterations at the murine c-fms (CSF-1 receptor) locus during maturation of myeloid precursor cells. *Genes Dev 16:* 1721–37, 2002

Tan, K., Shlomi, T., Feizi, H., Ideker, T., and Sharan, R.: Transcriptional regulation of protein complexes within and across species. *Proc Natl Acad Sci USA 104:* 1283–8, 2007

Tegner, J. and Bjorkegren, J.: Perturbations to uncover gene networks. *Trends Genet 23:* 34–41, 2007

Tegner, J., Nilsson, R., Bajic, V. B., Bjorkegren, J., and Ravasi, T.: Systems biology of innate immunity. *Cell Immunol 244:* 105–9, 2006

Tomlins, S. A., Mehra, R., Rhodes, D. R., Cao, X., Wang, L., Dhanasekaran, S. M., Kalyana-Sundaram, S., Wei, J. T., Rubin, M. A., Pienta, K. J., Shah, R. B., and Chin-naiyan, A. M.: Integrative molecular concept modeling of prostate cancer progression. *Nat Genet 39:* 41–51, 2007

Wells, C. A., Chalk, A., Forrest, A., Taylor, D., Waddell, N., Schroder, K., Himes, S. R., Faulkner, G., Lo, S., Kasukawa, T., Kawaji, H., Kai, C., Kawai, J., Katayama, S., Carninci, P., Hayashizaki, Y., Hume, D. A., and Grimmond, S. M.: Alternative transcription of the toll-like receptor signalling cascade. *Genome Biol 7:* R10, 2006

Wells, C. A., Ravasi, T., Faulkner, G. J., Carninci, P., Okazaki, Y., Hayashizaki, Y., Sweet, M., Wainwright, B. J., and Hume, D. A.: Genetic control of the innate immune response. *BMC Immunol 4:* 5, 2003a

Wells, C. A., Ravasi, T., Sultana, R., Yagi, K., Carninci, P., Bono, H., Faulkner, G., Okazaki, Y., Quackenbush, J., Hume, D. A., and Lyons, P. A.: Continued discovery of transcriptional units expressed in cells of the mouse mononuclear phagocyte lineage. *Genome Res 13:* 1360–5, 2003b

Yu, H., Luscombe, N. M., Lu, H. X., Zhu, X., Xia, Y., Han, J. D., Bertin, N., Chung, S., Vidal, M., and Gerstein, M.: Annotation transfer between genomes: protein-protein interologs and protein-DNA regulogs. *Genome Res 14:* 1107–18, 2004

# Immunomics: At the Forefront of Innate Immunity Research

Hongtao Guan, Steven K Dower, and Endre Kiss-Toth

**Abstract** The function and mechanisms of action of the immune system have traditionally been studied at and above the cellular level. However, due to the rapid development of molecular biology technology and widening research tool-set of genetics, as well as the success of high-throughput sequencing projects, understanding of the immune system is increasingly aided by studies performed at the molecular level. To this end, "Immunomics" is used to describe a novel field of multidisciplinary science in immunology research with the aim to derive integrated models of immune-modulatory processes.

Innate immunity is regarded as the first line of defence in fighting infections and restoring the integrity of the organism through wound healing. This chapter will focus on the recent advances in "system based" attempts to understand the molecular mechanisms of regulatory processes in innate immune responses. It will also highlight the main technologies, which have successfully been employed to identify the key components of inflammatory signal processing. We will start with introducing some of the most important "master cytokines" and cytokine receptors. We will then focus on the TIR-coupled signalling pathways induced by members of the IL-1 cytokine family, as well as by Toll-like receptors (TLR), with a brief summary of our current understanding on some crucial components in this pathway. Finally, we will discuss advanced methods for identifying novel components of important signalling pathways, highlighting the exploitation of cDNA library screen methods in innate immunity research.

**Keywords** Toll/IL-1R · Inflammation · Immunomics · Innate immunity

E. Kiss-Toth
Cardiovascular Research Unit, University of Sheffield, Royal Hallamshire Hospital, Glossop road, Sheffield, S10 2JF, UK
e-mail: e.kiss-toth@sheffield.ac.uk

A. Falus (ed.), *Clinical Applications of Immunomics*,
DOI: 10.1007/978-0-387-79208-8_2, © Springer Science+Business Media, LLC 2009

# 1 Introduction

Innate immunity is one the most evolutionarily conserved systems, designed to protect the organism from viruses and bacterial infections, stress and many other types of attacks from the outside world. During the past decade, the capacity of molecular biology and information technology to produce and analyse data have grown exponentially, rapidly reforming many aspects of immunology research in the post-genomics era. As a result, scientific understanding of signalling networks governing the innate immunity response in human tissues and other organisms has evolved beyond recognition, compared to even just a decade ago. Many strategies have been designed over the years to identify novel proteins, which have a crucial role in innate immunity responses by regulating particular signalling pathways. These projects had many advantages, including the definition of novel drug targets, as exemplified by the recent success of anti-TNF therapy, as well as leading to a better, system-wide understanding of the molecular control of innate immunity. In the past few years, a new concept, Immunomics, has been adopted to define an emerging, multidisciplinary field of research (Schonbach, 2003).

Although rapid progress has been made to identify the proteins playing pivotal roles in the innate immunity–related signalling pathways (for example, TIR signalling pathways), the catalogue of proteins with a key regulatory function identified and studied is far from completed. Novel proteins need to be characterised to gain a more comprehensive picture of how signalling networks are regulated. Recently, many strategies have been designed to identify proteins that have crucial roles in innate immune response by regulating particular signalling pathways. These include methods studying protein–protein interactions based on function, such as yeast two-hybrid system, in vitro pull-down assays and cDNA library screening.

In this chapter, we will review innate immune signalling systems in a non-conventional way by using various components of these systems to illustrate the use of a range of gene discovery approaches, which contributed to our current models of innate immune signal transduction.

# 2 Cytokines

## 2.1 Interferons

Cytokines were found to have a vast diversity of biological functions. Half a century ago, the effects of heat-inactivated influenza virus on the growth of live virus in chicken egg membranes were noticed and the interfering substance was identified and was coined "interferon" (Isaacs and Lindenmann, 1957; Isaacs et al., 1957). The specific interfering substance identified in these early studies is today known as 'Type I interferon'. Since then, a substantial body of further research on interferons (IFNs) has thrived and an immense knowledge of the biology of IFNs and the signalling pathways they regulate have accumulated. The members, receptors and regulatory proteins in the signalling pathways induced by IFNs have been thoroughly reviewed in depth (Biron, 2001; Stark et al., 1998; Theofilopoulos et al., 2005; van Boxel-Dezaire et al., 2006). Synthesis and secretion of IFNs are induced in response to stimuli sensing by the innate immune system mediated by one or more Toll-like receptors (TLRs) (Beutler et al., 2003; Toshchakov et al., 2002). Here we will focus on some of the most fruitful strategies used for identifying novel proteins in the IFN-mediated signalling pathways.

First, we need to introduce the concept of forward and reverse genetics. Broadly, the former strategy was used to search for novel genes/proteins and also used in large cDNA library screening projects (function-to-gene approach), whereas the latter strategy was mainly used to study the function of known genes (gene-to-function approach). For example, the discovery and cloning of the IFNα receptor are using the forward genetics strategy. Here, the clones from a human cDNA library were transferred into mouse cells, which are non-responsive to the human cytokine. Selecting for cells sensitive to human IFNα led to the cloning of the human receptor chain (Uze et al., 1990).

The isolation, cloning and characterisation of the IFNα receptor initiated an array of further studies on the IFN signalling pathway. The next step was the identification of co-factors binding to the intracellular region of IFNα receptor and the cloning of intermediate factors that transmit the signals initiated by the receptor–ligand interaction. Further, many regulators, which activate or repress the signal, and transcription factors, which activate or suppress the transcription of the expression of IFN target genes, were also identified. Amongst different forward genetics approaches, high-frequency mutagenesis in the human cell line 2fTGH was used to dissect the IFNα signalling pathway (John et al., 1991; Lutfalla et al., 1995; McKendry et al., 1991). An *Escherichia coli* selectable marker gene *gpt*, which is tightly regulated by a human IFNα inducible gene 6–16, was expressed in 2fTGH cells. Therefore, selection for or against IFN-mediated expression of *gpt* in these cells could be undertaken following mutagenesis in the cells. Complementation of mutants by Stark and colleagues (year) led to the identification of critical genes and subunits of transcription factors, which regulate the response to IFNs, including the functional $E_\alpha$ and $E_\gamma$ subunits of the transcription factor E. By using the same mutagenesis and complementation strategy, they also cloned and characterised the IFNαR2 (Lutfalla et al., 1995).

## 2.2 Interleukins (ILs)

Interleukins (ILs) are a group of cytokines, secreted by a wide variety of cells and play important roles in the both the innate and the adaptive immune systems.

### 2.2.1 Traditional IL-1 Cytokines

IL-1, a collective name defined in 1979 at the Second International Lymphokine Workshop (Aarden et al, 1979), is one of the most pleiotropic cytokines and mediates responses affecting the expression of well over 90 genes, including genes for other cytokines, cytokine receptors, acute phase reactants, growth factors, tissue-remodelling enzymes, extracellular matrix components and adhesion molecules (O'Neill, 1995). Further, these affected genes all directly function in the immune and inflammatory defence system.

The IL-1 family consists of three well-characterised proteins, IL-1α, IL-1β and IL-1 receptor antagonist (IL-1RA). When the first two IL-1 cDNAs were cloned in 1984, it was still unclear how many isoforms of IL-1 s exist. A cDNA encoding for murine IL-1 was cloned from the murine macrophage line P388D (a line that was induced to synthesise high levels of IL-1 mRNA) (Lomedico et al., 1984). The first human IL-1 cDNA was isolated from human blood monocytes (Auron et al., 1984). After the molecular characterisation of these clones, it became apparent that there are two agonistic isoforms (α and β) of IL-1 (March et al., 1985). The pro-IL-1α is a 271-aa protein and the mature IL-1α is a 159-aa protein. The pro-IL-1β is a 269-aa protein and the mature form is a 153-aa protein. By contrast, there are two isoforms of secreted IL-1Ra: one is a 22-kD glycosylated protein, and the other a 17-kD non-glycosylated protein. The cDNA of IL-1ra was first cloned in 1990 from a human monocyte library (Eisenberg et al., 1990). IL-1α and IL-1β are agonists and, after binding to the IL-1 receptor, induce a variety of cellular signals. IL-1RA does not induce cellular signal after binding to the IL-1 receptor and hence acts as an inhibitor of IL-1 activity (review in Roux-Lombard, 1998). Human IL-1α and IL-1β amino acid sequences are 26% homologous, and the nucleotide sequences of their genes are 45% homologous (Gubler et al., 1986; March et al., 1985).

The regulation mechanisms of the expression of IL-1α and IL-1β were intensively studied. For example, transcription binding sites with functional importance, responding to cAMP, LPS and NF-κB, have been characterised on the promoter of IL-1β gene (Shirakawa et al., 1993; Tsukada et al., 1994). Expression of both IL-1 agonists can be induced by a variety of stimuli, such as LPS, GM-CSF, IFNγ and TNFα (review in Roux-Lombard, 1998). In addition, expression of the IL-1 s can also be suppressed by a variety of other cytokines (such as TGF-β), including other members of the IL family, such as IL-4, IL-6, IL-10 and IL-13 (Chantry et al., 1989; de Waal Malefyt et al., 1991; Schindler et al., 1990; Vannier et al., 1992). Similarly, IL-1Ra can also be regulated by cytokines such as IL4, IL6, IL-10, IL-13 and GM-CSF, as well as by LPS (review in Roux-Lombard, 1998). Based on the sequence homology, and tertiary structure, IL-18 is also considered by many as a member of the IL-1 family of ligands. IL-18 was shown to signal through binding to IL-1R-related protein (IL-1Rrp1 or IL-18R) (Born et al., 1998).

A systematic nomenclature was later proposed, and hence the traditional IL-1 cytokine family members IL-1α, IL-1β, IL-1Ra and IL-18 became IL-1F1, IL-1F2, IL-1F3 and IL-1F4, respectively (Sims et al., 2001).

### 2.2.2 Novel IL-1 Cytokines

Recently, seven new members of the IL-1 family were identified by exploiting DNA database mining for homologues to IL-1, and they were termed as IL-1F5-11 (review in Barksby et al., 2007), with IL-33 (also called IL-1F11) identified at the latest (Schmitz et al., 2005). IL-1F6, 8, 9 and 11 were shown to be predominantly expressed in the skin. Unlike IL-1α and IL-1β, haematopoietic cells are not the main source of the novel IL-1 family members, IL-1F5-11. Functionally, IL-1F6, 8 and 9 all act via the IL-1/TLR family receptor IL-1Rrp2, but not IL-1RI. They can all activate NF-κB and mitogen-activated protein kinases (MAPK) pathways, leading to upregulation of IL-6 and IL-8 in responsive cells (Towne et al., 2004). Therefore, IL-1F6, 8 and 9 are all activators. In contrast, IL-1F5 was considered to be a negative regulator for NF-κB and MAPK pathways. The mechanisms of the regulation of these novel cytokines, IL-1F5-11, are still not fully understood.

## 3 Cytokine Receptors

### 3.1 IL-1 Receptors

The major signalling receptor for IL-1 is the type I IL-1 receptor (IL-1RI). IL-1RI was cloned by a direct expression strategy from the screening expression libraries generated from mouse and human T-cell RNA (Sims et al., 1989, 1988). A key feature of cDNA libraries used in these experiments was that they were constructed in an expression vector containing the SV40 origin of replication (to facilitate continuous replication of the plasmid DNA in specific mammalian cell lines), followed by transfection into COS cells. The plasmids containing full-length IL-1RI cDNAs can be picked by their ability to exert high-level binding of $^{125}$I labelled IL-1 on transfected COS cells detected on X-ray film (Sims et al., 1988). The human IL-1RI was identified and sub-cloned using mouse IL-1R1 cDNA as a probe to screen an oligo-(dT)-primed cDNA library made from clone 22 RNA (a CD4$^+$ CD8$^-$ human T-cell line) (Acres et al., 1987). IL-1RI is a transmembrane monomeric molecule, whose cytoplasmic domain does not have any intrinsic kinase activity allowing a signal to be transmitted (Sims et al., 1988). In addition, it was found that the human IL-1RI is very closely related to the mouse IL-1RI in sequence (69% identical at the amino acid level and 75% nucleotide level) and overall structure.

Three years later, an improved expression cloning method was used to clone the type II IL-1R (IL-1RII) (McMahan et al., 1991). Instead of using the previous SV40 based system, a plasmid (pDC406, containing a fragment of the origin of replication from EBV genome) -host (a host cell line, CV1/EBNA, constitutively expressing the EBV replication protein EBNA-1) replication system derived from Epstein-Barr virus was used. This modification, compared to the original method for cloning the IL-1R1, resulted in a two- to threefold increase in the number of detectable transfected cells. Therefore, the revised method had a higher sensitivity in identifying positive cells (McMahan et al., 1991).

It was found that the extracellular portion of IL-1RII was highly homologous to that of IL-1RI. However, the IL-1RII has a short cytoplasmic tail (29 amino acids), in contrast to the longer cytoplasmic tail of IL-1RI (∼215 amino acid), and was found not to signal upon IL-1 binding. As the function of the cytokine and the initiation/transmission of IL-1RI induced signals were still bewildering scientists, an observation was reported that the cytoplasmic domain of IL-1RI showed a highly significant homology with the cytoplasmic region of the *Drosophila melanogaster* protein Toll (Gay and Keith, 1991).

Soon after the isolation of the transmembrane IL-1RI and IL-1RII, soluble receptors for IL-1 were purified and characterized. Both the IL-1-RI and IL-1RII have soluble forms (IL-1sRI and IL-1sRII), due to the proteolytic cleavage of the extracellular portion of the membrane receptors IL-1RI and IL-1RII. These soluble IL-1Rs can be found in the serum and the urine of healthy subjects, as well as in various inflammatory biological fluids (Sims et al., 1994; Symons et al., 1991). These soluble receptors bind to different members of the IL-1 family with different affinities and therefore act as inhibitors, but by a different mechanism to that of IL-1Ra. IL-1sRI binds preferentially to IL-1α and IL-1Ra; by contrast, IL-1sRII binds preferentially to IL-1β.

Since the cloning of IL-1RI and IL-1RII, a number of proteins were reported to have homology to IL-1RI and/or Toll in the cytoplasmic domains critical for signalling. This cluster of proteins was subsequently defined as the Toll/IL-1 receptor family (TIR). One subgroup contains members that are homologous to IL-1RI, including IL-1R accessory protein (IL-1AcP) (Greenfeder et al., 1995), and the putative interferon-γ-inducing factor (also know as IL-1-γ or IL-18) receptor, IL-1 receptor-related protein (IL-1Rrp) (Parnet et al., 1996). The extracellular regions of all the proteins in this subgroup have immunoglobulin-like structure. In contrast, the other subgroup contains members homologous to Toll and all contain leucine-rich repeats extracellularly, including another *Drosophila* protein 18-wheeler (Williams et al., 1997), mammalian TLRs and a number of plant proteins involved in immunity (Baker et al., 1997; Bent, 1996).

## 3.2  Toll-Like Receptors

Through studies of interferon secretion, it was found that synthesis and secretion of IFNs are induced in response to stimuli sensing by the innate immune system, which is mediated by one or more toll-like receptors (TLRs) (Beutler et al., 2003; Toshchakov et al., 2002). TLRs are a class of 13-membrane proteins, which promote innate immune responses. Again, a forward genetic approach was proved to be a powerful tool in identifying the critical genes here.

Systematic screens for mutations that change the spatial pattern of the cuticle of the *Drosophila* larva were carried out to identify a body of gene products. These genes have specific functions in embryonic pattern formation. The *Toll* gene was one of those identified in *Drosophila melanogaster* and was initially defined as a functionally important gene in embryogenesis to establish the dorsal–ventral axis (Gerttula et al., 1988). Later on, *Toll* gene was found to have a role in the immunity of *D. melanogaster* to resist fungal infections (Lemaitre et al., 1996). Through bioinformatics, a homologous sequence in the EST database derived from human fetal liver/spleen library was identified by searching a sequence profile of the *Drosophila Toll* signalling domain in 1997. This fragment of a human cDNA was characterised, cloned and later defined as one of the human TLRs (Medzhitov et al., 1997). Since then, another 12 mammalian TLRs were gradually identified and cloned. Amongst other bioinformatics exercises, the use of Blast search for homologues and CLUSTAL W alignment algorithms led to the identification and cloning of TLRs1-5 (Rock et al., 1998).

TLRs form a family of evolutionarily conserved molecules, their origins backdating the separation of plants and animals. The most conserved protein motif among the TLRs is the TIR domain, which is shared by TLRs, the interleukin-1 receptors (IL-1R) and plant disease resistance genes (Beutler and Rehli, 2002). Once again, bioinformatics/computational methods proved to be of immense value in identifying these similarities. The origins of the TLR family date back more than 100 million years; some TLRs can be tracked back to the origins of vertebrate life (Beutler and Rehli, 2002). Sequence analysis of TLR and IL-1-R1 showed a less conserved, 200-AA cytoplamsic domain between the two families, which mediates signalling following receptor–ligand binding. Multiple alignment of the toll homology (TH) module (an extended version of the TIR domain), using CLUSTAL W program, reveals an interesting evolutionary trait showing that the TH sequences diversify from insects, plants and vertebrates, which include the human TLRs. Four main branches separate the plant proteins: the MyD88 adaptor proteins, IL-1Rs, and Toll-like proteins containing both the *Drosophila* and human TLRs (Rock et al., 1998).

At present, a collection of vertebrate and invertebrate homologues of the fly Toll receptor have been identified. Vertebrate TLRs can be divided into six subfamilies, based on their primary sequence (review in Chen et al., 2007). In mammals, the 13 TLRs have been shown to recognise specific ligands (or distinct types of pathogens): dsRNA is the ligand for TLR3, ssRNA for TLR7 (mouse) or TLR8 (human), lipopolysaccharide (LPS) for TLR4, and prokaryotic unmethylated CpG-DNA for TLR9.

### 3.2.1 Identification of Lps Locus

The identification of mouse *Lps* locus spanned a few decades and can be traced back to over half a century ago when diversity in inflammatory responses of different mouse strains to empirical challenge with purified LPS was demonstrated for the first time (review in Morrison and Ryan, 1979). Decreased sensitivity to LPS lethal effects was observed in C3H/HeJ, C57BL/10ScCr and its progenitor C57BL/10ScN (also known as C57BL/10ScNCr) mouse strains (Coutinho et al., 1977; Sultzer, 1968). In addition, further studies showed that the progenitors of the C3H/HeJ mouse strain, C3H/He and C3H/HeN (a closely related substrain), as well as several other C3H sublines, were normally responsive to LPS, indicating that the hyporesponsive trait arose by a spontaneous mutation (Glode and Rosenstreich, 1976).

Following the observation of the LPS hyporesponsiveness, the genetic basis responsible for this phenotype was studied. By using a series of backcross experiments, evidence that LPS hyporesponsiveness in C3H/HeJ mice was due to mutation of a single gene was obtained (Watson and Riblet, 1974). Since then, the symbol $Lps$ was defined to represent this locus, with the C3H/HeJ mutant allele coined $Lps^d$ (defective LPS response) and the normal allele $Lps^n$ (normal LPS response). In addition, after the observation of intermediate to high responsiveness to LPS, it was concluded that $Lps^n$ is inherited either dominantly or co-dominantly with $Lps^d$ (Scibienski, 1981).

To further characterise the $Lps$ locus, genetic mapping by a three-point backcross using C57BL/6 J mice carrying the chromosome markers revealed homology between mouse chromosome 4 ($Mmu4$) and human chromosome 9 ($Hsa9$), which harbour the $Lps$ locus (Peiffer-Schneider et al., 1997; Qureshi et al., 1996; Watson et al., 1978). A substantial body of data were accumulated over time leading to the hypothesis that the $Lps$ gene encoded a membrane protein that functioned as a receptor for LPS to initiate the signal for cellular activation (Bright et al., 1990; Chen et al., 1990; Lei and Morrison, 1988a,b).

Advances in the development of comprehensive sets of informative microsatellite markers and large-insert DNA libraries, e.g. BAC and YAC clones, made it possible for several groups to undertake a positional cloning approach to identify the gene encoded by the $Lps$ locus. This strategy enables precise genetic localisation and DNA cloning of the candidate gene, thus identifying the transcript of $Lps$. The gene encoding Toll-like receptor 4 ($Tlr4$) was identified as a candidate for $Lps$ (Poltorak et al., 1998a,b). The mouse $Tlr4$ gene contains one open reading frame of 2505 nucleotides, and encodes a protein of 835 amino acids. In addition, the protein contains an extracellular domain formed by a tandem arrangement of 22 leucine-rich repeat motifs, connected by a single transmembrane span to an intracellular signalling domain that shares homology with the IL-1R. The $Tlr4$ gene in LPS-resistant C3H/HeJ mice revealed a single missense mutation, a C to A mutation at nucleotide 2135, predicting a replacement of proline for histidine at position 712 within the intracellular signalling domain. After further experiments (in one such experiment, LPS responsiveness was examined in mice that had been rendered deficient for $Tlr4$), this mutation of $Tlr4$ gene was concluded to be the genetic basis of $Lps^d$ locus (Hoshino et al., 1999; Poltorak et al., 1998a; Qureshi et al., 1999).

# 4  TIR Signalling Pathways (Downstream Components)

The IL-1 signalling pathway in mammalian cells is similar to that in *Drosophila*. The target for the Toll pathway during development in *Drosophila* is Dorsal, a transcription factor of the Rel/NF-κB family. Cactus, homologous to mammalian I-κB, binds to Dorsal in *Drosophila* to retain it in the cytoplasm (Anderson et al., 1985a,b). In addition, it was reported that in *Drosophila* Dorsal activation is mediated by the protein kinase Pelle (Galindo et al., 1995; Grosshans et al., 1994). The mammalian homologue of Pelle is IRAK (IL-1R-associated kinase) (Cao et al., 1996), highlighting that this signalling pathway is highly conserved between fly and human.

Over the years, a large body of knowledge has accumulated on TIR-induced signalling pathways. Most TIRs induce an overlapping/closely related set of pathways (Fig. 1). This is not a surprise, since most TLRs and IL1Rs share a similar intracellular domain, the TIR domain. Upon ligand binding, the TIR domain recruits the adaptor protein, myeloid differentiation primary response gene 88 (MyD88). Activated MyD88 recruits IRAK-4, and activates it. Activated IRAK-4, in turn, recruits and activates another kinase, IRAK-1. The TNF receptor–associated factor (TRAF)-6 then associates to form an IRAK-1/ TRAF-6 complex, which dissociates from IRAK-4. The IRAK-1/TRAF-6 complex then recruits another protein complex containing TGF-β activated kinase 1 (TAK1), TAK1-biding protein 1 (TAB1) and TAB2. Both TAK and TAB were originally identified by yeast two-hybrid system (Shibuya et al., 1996; Yamaguchi et al., 1995). Following this event, IRAK1 is degraded, and the remaining complex traffics to the cytoplasm and recruits ubiquitin-conjugating enzymes Ubc13 and Uev1A. These activate TAK1, ultimately leading to the activation of nuclear factor kappa light polypeptide gene enhancer in B cells (NF-κB), which translocates into the nucleus to initiate changes in the transcription of a host of genes.

## 4.1  MyD88

MyD88 is a universal adaptor protein as it is used by all TLRs (except TLR3) and the IL-1 receptor to activate NF-κB. This protein was identified based on primary sequence homology to the TIR domain of the IL-1R (Lord et al., 1990). The cloned cDNA was then used to characterise the biological function of MyD88. As sequencing data in public databases accumulated, several MyD88 orthologues were identified, including Mal (Horng et al., 2001), TRIF (Yamamoto et al., 2002) and TRAM (Yamamoto et al., 2002). Most of these adaptors are used only by a subset of TIR receptors. For instance, Mal is necessary to recruit MyD88 to TLR2 and TLR4. TLRs can use both MyD88-dependent and independent pathways. MyD88 has an N-terminal death domain (DD) and a C-terminal Toll domain. Yeast two-hybrid interaction analysis suggests that

**Fig. 1** Mediators of
inflammatory signal
transduction (*See* Color
Insert)

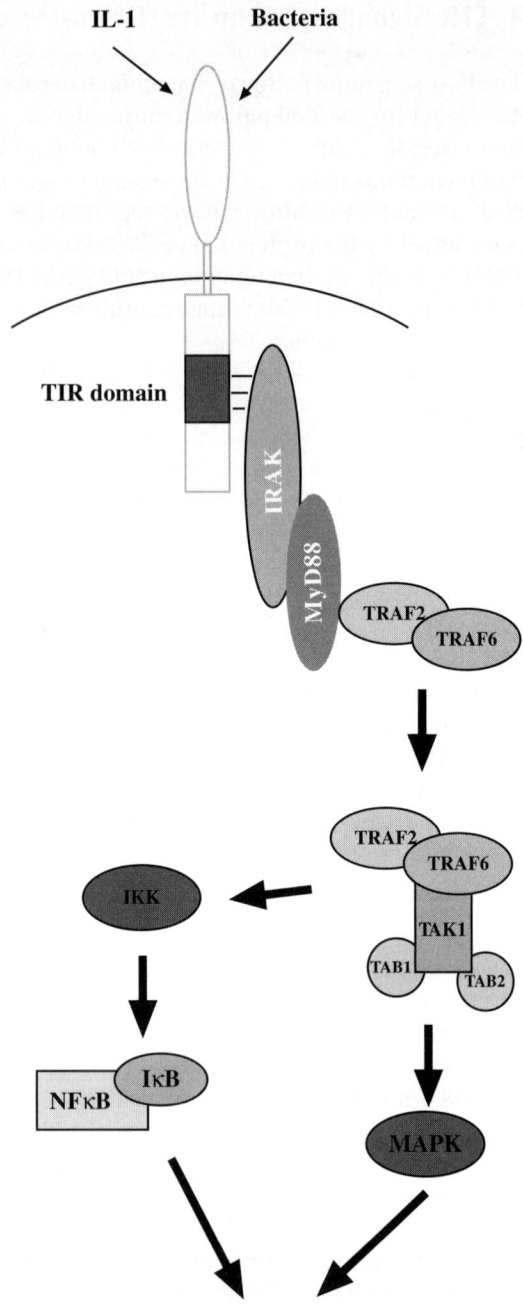

MyD88 forms homodimers through DD–DD and Toll–Toll interactions to function as an adaptor connecting Toll- and death domain–containing proteins in the TIR signalling pathway. In addition, overexpression of MyD88 was shown to activate both NF-κB and JNK, which mimics many aspects of the IL1-induced cellular response (Burns et al., 1998). Functional studies showed that MyD88 activates the NF-$k$B activity, in the same pathway as TRAF6 and IRAKs. The N-terminal of the MyD88 (DD) was proved to be crucial to activate the NF-κB, since the truncated version of the protein (aa 152–296) cannot induce any NF-κB activity, as measured by luciferase reporter assay. It has also been reported that MyD88 binds to IRAK through the DD regions of both proteins (Muzio et al., 1997), which leads to IRAK activation through an yet ill-defined mechanism.

## 4.2 IRAKs

To date, four IRAK family proteins were characterised and studied, namely IRAK1, IRAK2, IRAK-M and IRAK-4. The IRAK family was firstly identified biochemically as a serine/threonine kinase activity, associated with the IL-1R (Cao et al., 1996), and later on IRAK1 was found to be critical in mediating signals upon IL-1 and LPS treatment by studying the macrophages derived from IRAK-1 knockout mice (Swantek et al., 2000).It is clear now that the four members of the IRAK family have distinct functions and each has specific downstream targets, and therefore each of the IRAK family members activates distinct cellular responses upon challenge (review in Huang et al., 2005).

The original isolation of IRAK1 by Cao et al. was carried out by biochemical purification (e.g. co-immunoprecipitation) of the IL-1RI interacting proteins. IRAK1 was shown to bind to the IL-1R on the TIR domain. Upon IL-1 or LPS challenge, IRAK1 (unmodified, ~85kD) was shown to be phosphorylated and ubiquitinated (modified, ~100kD), and subsequently its cellular distribution changed, with the majority of the modified IRAKs accumulated in the nucleus (Huang et al., 2004; Li et al., 2001). IRAK1 was shown to form complexes with MyD88 and TRAF6 (Medzhitov et al., 1998), subsequently mediating IL-1/LPS-induced NF-κB activation. Besides acting as a bridge to transmit the cellular response from IL-1/LPS-bound TIR to the activation of NF-κB, IRAK1 was also shown to stimulate the IL-10 expression by binding to the IL-10 promoter element and thus stimulating the IL-10 promoter (Huang et al., 2004).

Since the cloning of IRAK1, the other three members, IRAK2, IRAK-M and IRAK4, were all originally identified by searching EST databases for homologues of IRAK1 (Li et al., 2002; Muzio et al., 1997; Wesche et al., 1999). IRAK2, apart from binding to MyD88 and TRAF6, was also shown to interact with TLR intracellular adaptor molecule Mal/TIRAP, whereas IRAK1 was unable to form similar complexes (Fitzgerald et al., 2001; Fitzgerald et al., 2003). IRAK-M was shown to be a negative regulator of the NF-κB activity upon IL-1/LPS stimulation, demonstrated by an experiment where IRAK-M$^{-/-}$ mice showed increased inflammatory responses to *Salmonella typhimurium* infection (Kobayashi et al., 2002). By contrast, IRAK4 knockout mice showed severe impairment in NF-κB activation upon stimulation (Suzuki et al., 2002).

## 4.3 TRAFs

The TRAF family is a group of adaptor proteins with six members, mediating a variety of cellular signals from cell surface receptors, including TNF and TIR receptor families, which lead to the activation of NF-κB and MAPKs. The individual TRAF members share a conserved C-terminal domain, which is called TRAF domain. This TRAF domain is further subdivided into a more divergent N-proximal (TRAF-N) and a highly conserved C-proximal (TRAF-C) sub-domain. Homo- or heterodimers of TRAFs bind to each other through this TRAF domain. TRAF4 and TRAF6 were originally identified by cDNA (and EST expression) library screening. TRAF1 and TRAF2 were isolated and molecularly cloned using biochemical purification and the yeast two-hybrid system (Rothe et al., 1994). TRAF2 is the first protein binding to MyD88. This complex then recruits TRAF6 and subsequently activates NF-κB and MAPK pathways (review in Lee and Lee, 2002).

## 4.4 NF-κB/I-κBs

The activation of the family of NF-κB transcription factors is central in innate immune responses and, therefore it is an intensively studied signalling pathway. In mammalian cells, there are five NF-κB family members, RelA (p65), RelB, c-Rel, p50 (NF-κB1) and p52 (NF-κB2). All NF-κB family members contain an N-terminal domain of approximately 300 amino acids called the Rel-homology domain, which mediates DNA binding and dimerisation. These proteins can form homo- and heterodimers, with the complex p65/p50 the most common form.

In resting cells, the p50 and p65 NF-κB proteins are retained in the cytoplasm by a family of inhibitory proteins, inhibitors of NF-κB (I-κBs). Interacting with I-κB masks the nuclear localisation signal located in the Rel homology domain of RelA (p65). In mammalian cells, there are three main I-κBs: I-κBα, I-κBβ and I-κBε. Upon cellular stimulation such as IL-1 or LPS challenge, the I-κB proteins get phosphorylated and subsequently degraded. The NF-κB complexes can then translocate into the nucleus to initiate transcription. The active NF-κB transcription factor binds to the promoter region of the target genes, thus activating their transcription and ultimately leading to the inflammatory cellular responses (review, (Perkins, 2007)). Research has found that for I-κBα the masking effect is limited and the NF-κB/ I-κB complexes can shuttle into the nucleus even in the absence of cellular stimulation. In addition, I-κBα contains a nuclear export sequence, which causes the rapid export of the NF-κB/I-κB complexes back to the cytoplasm (Hayden and Ghosh, 2004).

A variety of signalling pathways can activate NF-κB. These were classified into two sets: classical (or called canonical) and non-canonical pathways (Perkins, 2007). In classical pathways, the NF-κB can be activated in response to a number of inflammatory stimuli, including stress, TNFα, IL-1, engagement of the T-cell receptor (TCR) and bacterial products such as LPS. The common feature of activating NF-κB through signals initiated by all the above stimuli is that the I-κBα is rapidly phosphorylated at S32 and S36 through activated IKK complex containing all the three components IKKα, IKKβ and IKKγ, and subsequently degraded. In addition, RelA(p65)/p50 is the dominant complex (Hayden and Ghosh, 2004). There are also alternative (non-canonical) pathways that lead to the activation of NF-κB. These include the CD40 and lymphotoxin-β receptors, B-cell-activating factor of the TNF family (BAFF), and LPS and latent membrane protein-1 (LMP1) of Epstein-Barr virus (Bonizzi and Karin, 2004). In these cases, NF-κB-induced kinase (NIK) activates IKKα, leading to the formation of p52 from its precursor p100. Thus, p52/RelB heterodimers become the dominant form and regulate a different set of the target genes from the p65/p50 complex in the classical pathway (Bonizzi and Karin, 2004).

NF-κB plays a pivotal role in immune and inflammatory responses (as well as many other physiological processes such as carcinogenesis). It controls the expression of a large number of target genes, which regulate inflammation, stress, innate immunity and adaptive immunity for the host to defence, as well as cell proliferation and apoptosis. Among many NF-κB-responsive genes, a number of proteins, chemokines (e.g. MIP-1α), pro-inflammatory cytokines (IL-1, IL-6 and TNFα), adhesion molecules (e.g. E-selectin and VCAM-1) can be expressed, following the activation of the NF-κB (reviewed in Yates and Gorecki, 2006).

## 4.5 IKKs

After the mechanisms of phosphorylation and degradation of I-κBs were studied, the upstream complexes responsible for the phosphorylation and ubiquitination of I-κBs were the targets to be identified. Using chromatography, Chen et al. (1996) identified a complex of approximately 700 kDa, leading to the phosphorylation of I-κB proteins. Subsequently, the complex was fractionated and the components of the complex were identified, sub-cloned and named I-κB kinase (IKK) -α, -β and -γ. IKKα and IKKβ were shown to form heterodimers, and both directly phosphorylate the critical residues of I-κBα at S32 and S36 (Lee et al., 1998; Zandi et al., 1997). Initially, the IKK complex was found in unstimulated cells, and later was identified from TNFα-treated cells. Both IKKα and IKKβ were identified in stimulated and unstimulated cells, indicating their constitutive expression. IKKγ, also called NEMO, was found to bind with complex IKKα and IKKβ to mediate signals from upstream proteins on the various signalling pathways, such as TIR and TNF pathways. The IKKα, IKKβ and IKKγ can be activated by extracellular stimuli, which activate the classical (canonical) pathway through IL-1R/TLR, TNFR and TCR (T cell receptor).

Apart from the above classical IKK proteins, a new family of IKK-like kinases have recently been discovered. This group of kinases includes IKKε (also named IKKi) and TBK1 (also called NAK) (Peters and Maniatis, 2001). They were found not to directly phosphorylate I-κB; instead, they may function upstream of IKKs. IKKε and TBK1 were shown to be activated in PMA-treated cells (Tojima et al., 2000).

# 5 Identification of Key Components of TIR Signalling

## 5.1 Methods for Identifying Novel Components of Signalling Pathway

Although our knowledge of proteins involved in innate immunity has been rapidly expanding, there are still unknown mechanisms and proteins playing crucial roles in the immune signalling pathways. Hence, identification and isolation of novel proteins is an important topic in Immunomics.

A variety of methods were used for characterisation of the proteins with novel function in both innate immunity and adaptive immunity. Yeast two-hybrid system aided the identification of novel proteins interacting with ITAMs-containing IgE protein, (Osborne et al., 1995). Mass spectrometry was used to characterise and clone the IKK-1 and IKK-2 (Mercurio et al., 1997). The development of "protein-fragment complementation assay" enabled the isolation of novel proteins of the PKB signalling pathway (Remy and Michnick, 2004). Finally, protein micro-array was key in the development of antibody arrays, which were successfully used to identify both known and unknown proteins in cancer cells exposed to ionizing radiation (Sreekumar et al., 2001)

## 5.2 cDNA Library Screening as a Powerful Tool to Explore Signalling Networks

As discussed above, screening of mammalian expression libraries led to the identification of key components of innate immune signal transduction systems. As an example, IL-1R was first cloned (Sims et al., 1988) using a modified technique, a single-cell autoradiography after radioligand binding, based on the methods established by Aruffo and Seed (1987). However, these screens do not directly provide functional information for the genes identified, resulting in a need to undertake substantial studies to characterize gene function. Whilst this is feasible at a limited scale, approaches to rapidly annotate function of a large number of gene products are required to complement the "information boom" provided by genome and EST sequencing programs. Thus, we have recently developed a novel screening approach, termed "transcription expression cloning", to enable rapid functional assignment of the proteins expressed from cDNA libraries. This strategy exploits the observation that overexpression of most signalling intermediates mimics the effect of extracellular agonists (Deng and Karin, 1994; Muzio et al., 1997; Tojima et al., 2000), and thus the downstream response can be detected by a specific and sensitive reporter system. Thus, components of signalling networks can be identified by co-transfecting cells with cDNA expression libraries and an agonist-responsive, inducible reporter (Fig. 2). The method has the capacity to screen entire cDNA expression libraries and has the potential of detecting all/most gene products, which activate the reporter following overexpression. To maximise the performance of screens based on this strategy, the use of an additional, constitutive reporter is advisable. This produces a signal that is dependent on transfection efficiency, the precision of liquid handling, etc. As a readout, relevant to studies in innate immune signal transduction, an inducible promoter construct of the IL-8 chemokine gene was coupled to transcriptional reporters, such as destabilised EGFP or luciferase. The former was used to design, optimise and validate the method by using confocal microscopy to detect EGFP expression at a single-cell level (Kiss-Toth et al., 2000). Testing the bioactivity of several known components of the IL-1/TNF signalling network revealed that most elements give detectable reporter activation upon overexpression. Furthermore, the GFP system allows the detection of additional aspects of the bioactivity of the transfected cDNA in addition to its impact on the reporter activity. Some of the tested cDNAs had profound effects on cell morphology (Kiss-Toth et al., 2000). This observed phenotypic difference suggests that preliminary information related to effects on viability can be obtained directly in this system, in addition to that obtained on reporter activities.

Using luciferase as a reporter, such a platform is relatively straightforward to automate. A high-throughput colony picker, a liquid-handling robot and a high-capacity luminometer allow the screening of entire expression libraries ($\sim 10^6$ clones) in less than a year. Compared to other genomics/proteomics strategies, the reporter-based screen can be run by a relatively small group. All these features can make this latest version of expression cloning an attractive addition to genome-wide gene-function annotation techniques.

Several applications of this strategy at a global scale were published (Guan et al., 2006; Guan et al., 2007 in press, Kiss-Toth et al., 2004, 2006). Iourgenko et al. (2003) also used the pIL-8 reporter/HeLa system we designed to screen a non-redundant set of 20,000 full-length human cDNAs for the activators of this promoter (Iourgenko et al., 2003). They reported that 64 cDNAs induced the IL-8 reporter by more than fivefold. The verified active cDNAs included

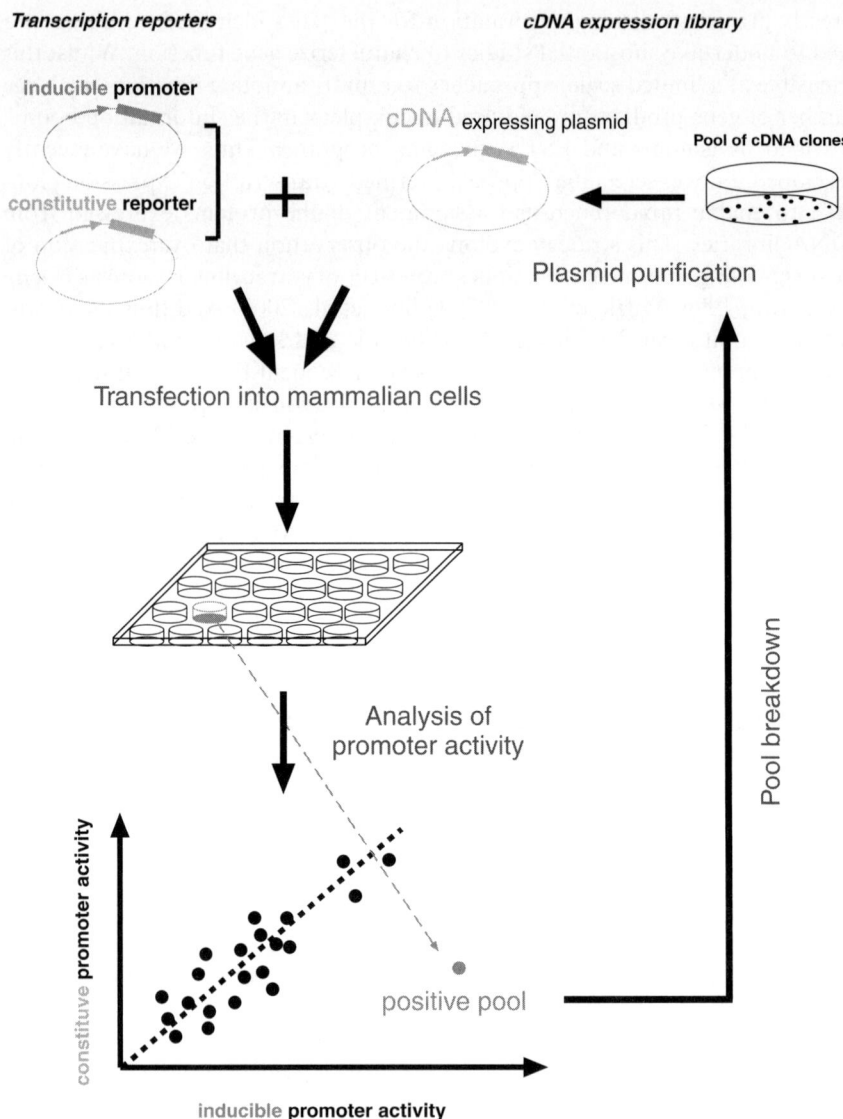

**Fig. 2** Transcription expression cloning (*See* Color Insert)

one to three copies of 28 unique genes. A further screen on the non-redundant gene set from the same group (41) detected 129 cDNAs with AP-1 inducer activity. An intriguing aspect of the data is that about 25% of the hits represent novel, previously uncharacterised genes, and a further 50% of genes had known activities other than those detected by the screens. The specificity of the screen by Chanda et al. (2003) was confirmed by testing a number of potential oncogenes in a proliferation assay. As a logical step forward, they mapped the interactions between several components of AP-1 activation network by co-transfecting an siRNA construct against an mRNA encoding "A" protein and an expression construct for "B" gene. If the reporter is activated, "B" functions downstream of "A". The success of this mapping exercise makes the transcription reporter-based assay system a viable approach to generate functional interaction maps for entire signalling networks. Our group has also undertaken a transcriptome-wide functional screening exercise, using the IL-8 reporter system (Kiss-Toth et al., 2006), which led to the identification of several genes of previously unknown function, including the MAPK regulator tribbles and the miRNA, BIC/mir155, which have recently been shown to be essential in immune function (Rodriguez et al., 2007).

However, the screening platform described above is bias towards genes, which has the greatest induction capacity of the reporter used. Often, such activity profile does not reflect the ranking of biological importance. An inherent feature of physiological signal transduction systems is the presence of feedback loops, which enable precise spatio/temporal modulation of signal processing and consequent cellular activation. We believe that incorporating of such feedback mechanisms into cDNA screening approaches may lead to the identification of previously uncharacterised, regulatory components of signal transduction. Therefore, we have recently undertaken proof of concept studies and shown that a transcriptional reporter system with a built-in feedback regulation is capable of detecting transcriptional activators of inflammatory signal transduction systems. We have used an IRES vector to construct the potential regulator (Y) to one of the expression cassettes and the reporter (EGFP/Luciferase) to the other, so that the potential regulator is transcribed in the same transcript as the reporter. This transcript is under the control of an inflammatory signal–responsive promoter of the IL-8 gene. We have tested this system by using well-characterised proteins such as RelA, TRAF6, MyD88 and IκBα, and proved that it can be used to select specific proteins in the TIR signalling pathway (Guan et al., 2006, et al., 2007). As expected, the expression of the regulator in the cells is self-regulated, nearer to the physiological concentration, and therefore this method has the potential to identify weak to intermediate activators/repressors in the TIR signalling pathway. Large-scale screens exploiting this method are currently under way.

# 6 Concluding Remarks

Understanding of the rules governing the regulation of intracellular signal processing systems requires the concerted effort of a number of disciplines, ranging from high-throughput detection of cellular expression profiles at the mRNA and protein levels, functional annotation of transcriptomes to building *in silico* tools and models of signalling to analyse the data generated by the experimental approaches. As a result of the complexity of signal processing, as our knowledge and experimental tools advance, we continuously need to re-evaluate our models of these processes. For instance, the recent recognition of miRNAs in the post-transcriptional regulation of transcriptomes is currently leading to a complete re-evaluation of our understanding of gene expression. Consequently, the ways we approach these fundamental problems is evolving as well, leading to the amalgamation of knowledge and approaches from a variety of disciplines and the emergence of novel terms, describing these new fields. A fine example of such novel research fields is "Immunomics", which brings together many traditional aspects of immunology research and combines it with molecular biology and bioinformatics, to name only a few, giving us tools to develop advanced, comprehensive models of the immune system. In addition to the scientific impact these "-omics" approaches have, they also prove invaluable for the identification of novel targets for future drug research.

# References

Aarden L A, Brunner T K, Cerrottini J C, Dayer J M, De Weck A L, Dinarello C A, Di Sabato G, Farrar J J, Gery I, Gillis S, Handschumacher R E, Henney C S, Hoffmann M K, Koopman W J, Krane S M, Lachman L B, Lefkowits I, Mishell R I, Mizel S B, Oppenheim J J, Paetkau V, Plate J, Rollinghoff M, Rosenstreich D, Rosenthal A S, Rosenwasser L J, Schimpl A, Shim H S, Simon P L, Smith K A, Wagner H, Watson J D, Wecker E, Wood D D. 1979. Revised nomenclature for antigen-non specific T cell proliferation helper factors (letter). *J. Immunol. 123*, 2928.

Acres, R. B., Larsen, A., and Conlon, P. J. (1987). IL 1 expression in a clone of human T cells. J Immunol *138*, 2132–2136.

Anderson, K. V., Bokla, L., and Nusslein-Volhard, C. (1985a). Establishment of dorsal-ventral polarity in the Drosophila embryo: The induction of polarity by the Toll gene product. Cell *42*, 791–798.

Anderson, K. V., Jurgens, G., and Nusslein-Volhard, C. (1985b). Establishment of dorsal-ventral polarity in the Drosophila embryo: Genetic studies on the role of the Toll gene product. Cell *42*, 779–789.

Aruffo, A., and Seed, B. (1987). Molecular cloning of a CD28 cDNA by a high-efficiency COS cell expression system. Proc Natl Acad Sci U S A *84*, 8573–8577.

Auron, P. E., Webb, A. C., Rosenwasser, L. J., Mucci, S. F., Rich, A., Wolff, S. M., and Dinarello, C. A. (1984). Nucleotide sequence of human monocyte interleukin 1 precursor cDNA. Proc Natl Acad Sci U S A *81*, 7907– 7911.

Baker, B., Zambryski, P., Staskawicz, B., and Dinesh-Kumar, S. P. (1997). Signaling in plant-microbe interactions. Science *276*, 726–733.

Barksby, H. E., Lea, S. R., Preshaw, P. M., and Taylor, J. J. (2007). The expanding family of interleukin-1 cytokines and their role in destructive inflammatory disorders. Clin Exp Immunol *149*, 217–225.

Bent, A. F. (1996). Plant Disease Resistance Genes: Function Meets Structure. Plant Cell *8*, 1757–1771.

Beutler, B., Du, X., and Hoebe, K. (2003). From phenomenon to phenotype and from phenotype to gene: Forward genetics and the problem of sepsis. J Infect Dis*187 Suppl 2*, S321–326.

Beutler, B., and Rehli, M. (2002). Evolution of the TIR, tolls and TLRs: Functional inferences from computational biology. Curr Top Microbiol Immunol *270*, 1–21.

Biron, C. A. (2001). Interferons alpha and beta as immune regulators – a new look. Immunity *14*, 661–664.

Bonizzi, G., and Karin, M. (2004). The two NF-kappaB activation pathways and their role in innate and adaptive immunity. Trends Immunol *25*, 280–288.

Born, T. L., Thomassen, E., Bird, T. A., and Sims, J. E. (1998). Cloning of a novel receptor subunit, AcPL, required for interleukin-18 signaling. J Biol Chem *273*, 29445–29450.

Bright, S. W., Chen, T. Y., Flebbe, L. M., Lei, M. G., and Morrison, D. C. (1990). Generation and characterization of hamster-mouse hybridomas secreting monoclonal antibodies with specificity for lipopolysaccharide receptor. J Immunol *145*, 1–7.

Burns, K., Martinon, F., Esslinger, C., Pahl, H., Schneider, P., Bodmer, J. L., Di Marco, F., French, L., and Tschopp, J. (1998). MyD88, an adapter protein involved in interleukin-1 signaling. J Biol Chem *273*, 12203–12209.

Cao, Z., Henzel, W. J., and Gao, X. (1996). IRAK: A kinase associated with the interleukin-1 receptor. Science *271*, 1128–1131.

Chanda, S. K., White, S., Orth, A. P., Reisdorph, R., Miraglia, L., Thomas, R. S., DeJesus, P., Mason, D. E., Huang, Q., Vega, R., et al (2003). Genome-scale functional profiling of the mammalian AP-1 signaling pathway. Proc Natl Acad Sci U S A *100*, 12153–12158.

Chantry, D., Turner, M., Abney, E., and Feldmann, M. (1989). Modulation of cytokine production by transforming growth factor-beta. J Immunol *142*, 4295–4300.

Chen, R., Alvero, A. B., Silasi, D. A., and Mor, G. (2007). Inflammation, cancer and chemoresistance: Taking advantage of the toll-like receptor signaling pathway. Am J Reprod Immunol *57*, 93–107.

Chen, T. Y., Bright, S. W., Pace, J. L., Russell, S. W., and Morrison, D. C. (1990). Induction of macrophage-mediated tumor cytotoxicity by a hamster monoclonal antibody with specificity for lipopolysaccharide receptor. J Immunol *145*, 8–12.

Chen, Z. J., Parent, L., and Maniatis, T. (1996). Site-specific phosphorylation of IkappaBalpha by a novel ubiquitination-dependent protein kinase activity. Cell *84*, 853–862.

Coutinho, A., Forni, L., Melchers, F., and Watanabe, T. (1977). Genetic defect in responsiveness to the B cell mitogen lipopolysaccharide. Eur J Immunol *7*, 325–328.

de Waal Malefyt, R., Abrams, J., Bennett, B., Figdor, C. G., and de Vries, J. E. (1991). Interleukin 10(IL-10) inhibits cytokine synthesis by human monocytes: An autoregulatory role of IL-10 produced by monocytes. J Exp Med *174*, 1209–1220.

Deng, T., and Karin, M. (1994). c-Fos transcriptional activity stimulated by H-Ras-activated protein kinase distinct from JNK and ERK. Nature *371*, 171–175.

Eisenberg, S. P., Evans, R. J., Arend, W. P., Verderber, E., Brewer, M. T., Hannum, C. H., and Thompson, R. C. (1990). Primary structure and functional expression from complementary DNA of a human interleukin-1 receptor antagonist. Nature *343*, 341–346.

Fitzgerald, K. A., Rowe, D. C., Barnes, B. J., Caffrey, D. R., Visintin, A., Latz, E., Monks, B., Pitha, P. M., and Golenbock, D. T. (2003). LPS-TLR4 signaling to IRF-3/7 and NF-kappaB involves the toll adapters TRAM and TRIF. J Exp Med *198*, 1043–1055.

Fitzgerald, K. A., Palsson-McDermott, E. M., Bowie, A. G., Jefferies, C. A., Mansell, A. S., Brady, G., Brint, E., Dunne, A., Gray, P., Harte, M. T., et al (2001). Mal (MyD88-adapter-like) is required for Toll-like receptor-4 signal transduction. Nature *413*, 78–83.

Galindo, R. L., Edwards, D. N., Gillespie, S. K., and Wasserman, S. A. (1995). Interaction of the pelle kinase with the membrane-associated protein tube is required for transduction of the dorsoventral signal in Drosophila embryos. Development *121*, 2209–2218.

Gay, N. J., and Keith, F. J. (1991). Drosophila Toll and IL-1 receptor. Nature *351*, 355–356.

Gerttula, S., Jin, Y. S., and Anderson, K. V. (1988). Zygotic expression and activity of the Drosophila Toll gene, a gene required maternally for embryonic dorsal-ventral pattern formation. Genetics *119*, 123–133.

Glode, L. M., and Rosenstreich, D. L. (1976). Genetic control of B cell activation by bacterial lipopolysaccharide is mediated by multiple distinct genes or alleles. J Immunol *117*, 2061–2066.

Greenfeder, S. A., Nunes, P., Kwee, L., Labow, M., Chizzonite, R. A., and Ju, G. (1995). Molecular cloning and characterization of a second subunit of the interleukin 1 receptor complex. J Biol Chem *270*, 13757–13765.

Grosshans, J., Bergmann, A., Haffter, P., and Nusslein-Volhard, C. (1994). Activation of the kinase Pelle by Tube in the dorsoventral signal transduction pathway of Drosophila embryo. Nature *372*, 563–566.

Guan, H., Holland, K., Qwarnstrom, E., Dower, S. K., and Kiss-Toth, E. (2006). Feedback loops in intracellular signal processing and their potential for identifying novel signalling proteins. Cell Immunol *244*, 158–161.

Guan, H., Kiss-toth, E., and Dower, SK. (2007). Analysis of innate immune signal transduction with autocatalytic expression vectors. Journal of Immunological Methods. *330*(1–2), 96–108. Doi: 10.1016/j.jim.2007.11.002

Gubler, U., Chua, A. O., Stern, A. S., Hellmann, C. P., Vitek, M. P., DeChiara, T. M., Benjamin, W. R., Collier, K. J., Dukovich, M., Familletti, P. C., and et al. (1986). Recombinant human interleukin 1 alpha: Purification and biological characterization. J Immunol *136*, 2492–2497.

Hayden, M. S., and Ghosh, S. (2004). Signaling to NF-kappaB. Genes Dev *18*, 2195–2224.

Horng, T., Barton, G. M., and Medzhitov, R. (2001). TIRAP: An adapter molecule in the Toll signaling pathway. Nat Immunol *2*, 835–841.

Hoshino, K., Takeuchi, O., Kawai, T., Sanjo, H., Ogawa, T., Takeda, Y., Takeda, K., and Akira, S. (1999). Cutting edge: Toll-like receptor 4 (TLR4)-deficient mice are hyporesponsive to lipopolysaccharide: Evidence for TLR4 as the Lps gene product. J Immunol *162*, 3749–3752.

Huang, Y., Li, T., Sane, D. C., and Li, L. (2004). IRAK1 serves as a novel regulator essential for lipopolysaccharide-induced interleukin-10 gene expression. J Biol Chem *279*, 51697–51703.

Huang, Y. S., Misior, A., and Li, L. W. (2005). Novel role and regulation of the interleukin-1 receptor associated kinase (IRAK) family proteins. Cell Mol Immunol *2*, 36–39.

Iourgenko, V., Zhang, W., Mickanin, C., Daly, I., Jiang, C., Hexham, J. M., Orth, A. P., Miraglia, L., Meltzer, J., Garza, D., Chirn, G. W., McWhinnie, E., Cohen, D., Skelton, J., Terry, R., Yu, Y., Bodian, D., Buxton, F. P., Zhu, J., Song, C., and Labow, M. A. (2003). Identification of a family of cAMP response element-binding protein coactivators by genome-scale functional analysis in mammalian cells. Proc Natl Acad Sci U S A *100*, 12147–12152.

Isaacs, A., and Lindenmann, J. (1957). Virus interference. I. The interferon. Proc R Soc Lond B Biol Sci *147*, 258–267.

Isaacs, A., Lindenmann, J., and Valentine, R. C. (1957). Virus interference. II. Some properties of interferon. Proc R Soc Lond B Biol Sci *147*, 268–273.

John, J., McKendry, R., Pellegrini, S., Flavell, D., Kerr, I. M., and Stark, G. R. (1991). Isolation and characterization of a new mutant human cell line unresponsive to alpha and beta interferons. Mol Cell Biol *11*, 4189–4195.

Kiss-Toth, E., Guesdon, F. M., Wyllie, D. H., Qwarnstrom, E. E., and Dower, S. K. (2000). A novel mammalian expression screen exploiting green fluorescent protein-based transcription detection in single cells. J Immunol Methods *239*, 125–135.

Kiss-Toth, E., Qwarnstrom, E. E., and Dower, S. K. (2004). Hunting for genes by functional screens. Cytokine Growth Factor Rev *15*, 97–102.

Kiss-Toth, E., Wyllie, D. H., Holland, K., Marsden, L., Jozsa, V., Oxley, K. M., Polgar, T., Qwarnstrom, E. E., and Dower, S. K. (2006). Functional mapping and identification of novel regulators for the Toll/Interleukin-1 signalling network by transcription expression cloning. Cell Signal *18*, 202–214.

Kobayashi, K., Hernandez, L. D., Galan, J. E., Janeway, C. A., Jr., Medzhitov, R., and Flavell, R. A. (2002). IRAK-M is a negative regulator of Toll-like receptor signaling. Cell *110*, 191–202.

Lee, F. S., Peters, R. T., Dang, L. C., and Maniatis, T. (1998). MEKK1 activates both IkappaB kinase alpha and IkappaB kinase beta. Proc Natl Acad Sci U S A 95, 9319–9324.

Lee, N. K., and Lee, S. Y. (2002). Modulation of life and death by the tumor necrosis factor receptor-associated factors (TRAFs). J Biochem Mol Biol 35, 61–66.

Lei, M. G., and Morrison, D. C. (1988a). Specific endotoxic lipopolysaccharide-binding proteins on murine splenocytes. I. Detection of lipopolysaccharide-binding sites on splenocytes and splenocyte subpopulations. J Immunol 141, 996–1005.

Lei, M. G., and Morrison, D. C. (1988b). Specific endotoxic lipopolysaccharide-binding proteins on murine splenocytes. II. Membrane localization and binding characteristics. J Immunol 141, 1006–1011.

Lemaitre, B., Nicolas, E., Michaut, L., Reichhart, J. M., and Hoffmann, J. A. (1996). The dorsoventral regulatory gene cassette spatzle/Toll/cactus controls the potent antifungal response in Drosophila adults. Cell 86, 973–983.

Li, S., Strelow, A., Fontana, E. J., and Wesche, H. (2002). IRAK-4: A novel member of the IRAK family with the properties of an IRAK-kinase. Proc Natl Acad Sci U S A 99, 5567–5572.

Li, X., Commane, M., Jiang, Z., and Stark, G. R. (2001). IL-1-induced NFkappa B and c-Jun N-terminal kinase (JNK) activation diverge at IL-1 receptor-associated kinase (IRAK). Proc Natl Acad Sci U S A 98, 4461–4465.

Lomedico, P. T., Gubler, U., Hellmann, C. P., Dukovich, M., Giri, J. G., Pan, Y. C., Collier, K., Semionow, R., Chua, A. O., and Mizel, S. B. (1984). Cloning and expression of murine interleukin-1 cDNA in Escherichia coli. Nature 312, 458–462.

Lord, K. A., Hoffman-Liebermann, B., and Liebermann, D. A. (1990). Nucleotide sequence and expression of a cDNA encoding MyD88, a novel myeloid differentiation primary response gene induced by IL6. Oncogene 5, 1095–1097.

Lutfalla, G., Holland, S. J., Cinato, E., Monneron, D., Reboul, J., Rogers, N. C., Smith, J. M., Stark, G. R., Gardiner, K., Mogensen, K. E., and et al. (1995). Mutant U5A cells are complemented by an interferon-alpha beta receptor subunit generated by alternative processing of a new member of a cytokine receptor gene cluster. Embo J 14, 5100–5108.

March, C. J., Mosley, B., Larsen, A., Cerretti, D. P., Braedt, G., Price, V., Gillis, S., Henney, C. S., Kronheim, S. R., Grabstein, K., and et al. (1985). Cloning, sequence and expression of two distinct human interleukin-1 complementary DNAs. Nature 315, 641–647.

McKendry, R., John, J., Flavell, D., Muller, M., Kerr, I. M., and Stark, G. R. (1991). High-frequency mutagenesis of human cells and characterization of a mutant unresponsive to both alpha and gamma interferons. Proc Natl Acad Sci U S A 88, 11455–11459.

McMahan, C. J., Slack, J. L., Mosley, B., Cosman, D., Lupton, S. D., Brunton, L. L., Grubin, C. E., Wignall, J. M., Jenkins, N. A., Brannan, C. I., and et al. (1991). A novel IL-1 receptor, cloned from B cells by mammalian expression, is expressed in many cell types. Embo J 10, 2821–2832.

Medzhitov, R., Preston-Hurlburt, P., and Janeway, C. A., Jr. (1997). A human homologue of the Drosophila Toll protein signals activation of adaptive immunity. Nature 388, 394–397.

Medzhitov, R., Preston-Hurlburt, P., Kopp, E., Stadlen, A., Chen, C., Ghosh, S., and Janeway, C. A., Jr. (1998). MyD88 is an adaptor protein in the hToll/IL-1 receptor family signaling pathways. Mol Cell 2, 253–258.

Mercurio, F., Zhu, H., Murray, B. W., Shevchenko, A., Bennett, B. L., Li, J., Young, D. B., Barbosa, M., Mann, M., Manning, A., and Rao, A. (1997). IKK-1 and IKK-2: Cytokine-activated IkappaB kinases essential for NF-kappaB activation. Science 278, 860–866.

Morrison, D. C., and Ryan, J. L. (1979). Bacterial endotoxins and host immune responses. Adv Immunol 28, 293–450.

Muzio, M., Ni, J., Feng, P., and Dixit, V. M. (1997). IRAK (Pelle) family member IRAK-2 and MyD88 as proximal mediators of IL-1 signaling. Science 278, 1612–1615.

O'Neill, L. A. (1995). Towards an understanding of the signal transduction pathways for interleukin 1. Biochim Biophys Acta *1266*, 31–44.

Osborne, M. A., Dalton, S., and Kochan, J. P. (1995). The yeast tribrid system – genetic detection of trans-phosphorylated ITAM-SH2-interactions. Biotechnology (N Y) *13*, 1474–1478.

Parnet, P., Garka, K. E., Bonnert, T. P., Dower, S. K., and Sims, J. E. (1996). IL-1Rrp is a novel receptor-like molecule similar to the type I interleukin-1 receptor and its homologues T1/ST2 and IL-1R AcP. J Biol Chem *271*, 3967–3970.

Peiffer-Schneider, S., Schutte, B. C., Murray, J. C., Frees, K. L., Williamson, K., Leysens, N. J., and Schwartz, D. A. (1997). Exclusion of Ifa and Ifb as the Lps gene and mapping of three markers near the Lps locus. Mamm Genome *8*, 785–786.

Perkins, N. D. (2007). Integrating cell-signalling pathways with NF-kappaB and IKK function. Nat Rev Mol Cell Biol *8*, 49–62.

Peters, R. T., and Maniatis, T. (2001). A new family of IKK-related kinases may function as I kappa B kinase kinases. Biochim Biophys Acta *1471*, M57–62.

Poltorak, A., He, X., Smirnova, I., Liu, M. Y., Van Huffel, C., Du, X., Birdwell, D., Alejos, E., Silva, M., Galanos, C., et al. (1998a). Defective LPS signaling in C3H/HeJ and C57BL/10ScCr mice: Mutations in Tlr4 gene. Science *282*, 2085–2088.

Poltorak, A., Smirnova, I., He, X., Liu, M. Y., Van Huffel, C., McNally, O., Birdwell, D., Alejos, E., Silva, M., Du, X., et al. (1998b). Genetic and physical mapping of the Lps locus: Identification of the toll-4 receptor as a candidate gene in the critical region. Blood Cells Mol Dis *24*, 340–355.

Qureshi, S. T., Lariviere, L., Leveque, G., Clermont, S., Moore, K. J., Gros, P., and Malo, D. (1999). Endotoxin-tolerant mice have mutations in Toll-like receptor 4 (Tlr4). J Exp Med *189*, 615–625.

Qureshi, S. T., Lariviere, L., Sebastiani, G., Clermont, S., Skamene, E., Gros, P., and Malo, D. (1996). A high-resolution map in the chromosomal region surrounding the Lps locus. Genomics *31*, 283–294.

Remy, I., and Michnick, S. W. (2004). A cDNA library functional screening strategy based on fluorescent protein complementation assays to identify novel components of signaling pathways. Methods *32*, 381–388.

Rock, F. L., Hardiman, G., Timans, J. C., Kastelein, R. A., and Bazan, J. F. (1998). A family of human receptors structurally related to Drosophila Toll. Proc Natl Acad Sci U S A *95*, 588–593.

Rodriguez, A., Vigorito, E., Clare, S., Warren, M. V., Couttet, P., Soond, D. R., van Dongen, S., Grocock, R. J., Das, P. P., Miska, E. A., et al. (2007). Requirement of bic/microRNA-155 for normal immune function. Science *316*, 608–611.

Rothe, M., Wong, S. C., Henzel, W. J., and Goeddel, D. V. (1994). A novel family of putative signal transducers associated with the cytoplasmic domain of the 75 kDa tumor necrosis factor receptor. Cell *78*, 681–692.

Roux-Lombard, P. (1998). The interleukin-1 family. Eur Cytokine Netw *9*, 565–576.

Schindler, R., Mancilla, J., Endres, S., Ghorbani, R., Clark, S. C., and Dinarello, C. A. (1990). Correlations and interactions in the production of interleukin-6 (IL-6), IL-1, and tumor necrosis factor (TNF) in human blood mononuclear cells: IL-6 suppresses IL-1 and TNF. Blood *75*, 40–47.

Schmitz, J., Owyang, A., Oldham, E., Song, Y., Murphy, E., McClanahan, T. K., Zurawski, G., Moshrefi, M., Qin, J., Li, X., et al. (2005). IL-33, an interleukin-1-like cytokine that signals via the IL-1 receptor-related protein ST2 and induces T helper type 2-associated cytokines. Immunity *23*, 479–490.

Schonbach, C. (2003). From immunogenetics to immunomics: Functional prospecting of genes and transcripts. Novartis Found Symp *254*, 177–188; discussion 189–192, 216–122, 250–172.

Scibienski, R. J. (1981). Immunologic properties of protein-lipopolysaccharide complexes. IV. Circumventing suppression in immunologically tolerant animals. Cell Immunol *58*, 293–301.

Shibuya, H., Yamaguchi, K., Shirakabe, K., Tonegawa, A., Gotoh, Y., Ueno, N., Irie, K., Nishida, E., and Matsumoto, K. (1996). TAB1: An activator of the TAK1 MAPKKK in TGF-beta signal transduction. Science *272*, 1179–1182.

Shirakawa, F., Saito, K., Bonagura, C. A., Galson, D. L., Fenton, M. J., Webb, A. C., and Auron, P. E. (1993). The human prointerleukin 1 beta gene requires DNA sequences both proximal and distal to the transcription start site for tissue-specific induction. Mol Cell Biol *13*, 1332–1344.

Sims, J. E., Acres, R. B., Grubin, C. E., McMahan, C. J., Wignall, J. M., March, C. J., and Dower, S. K. (1989). Cloning the interleukin 1 receptor from human T cells. Proc Natl Acad Sci U S A *86*, 8946–8950.

Sims, J. E., Giri, J. G., and Dower, S. K. (1994). The two interleukin-1 receptors play different roles in IL-1 actions. Clin Immunol Immunopathol *72*, 9–14.

Sims, J. E., March, C. J., Cosman, D., Widmer, M. B., MacDonald, H. R., McMahan, C. J., Grubin, C. E., Wignall, J. M., Jackson, J. L., Call, S. M., and et al. (1988). cDNA expression cloning of the IL-1 receptor, a member of the immunoglobulin superfamily. Science *241*, 585–589.

Sims, J. E., Nicklin, M. J., Bazan, J. F., Barton, J. L., Busfield, S. J., Ford, J. E., Kastelein, R. A., Kumar, S., Lin, H., Mulero, J. J., et al. (2001). A new nomenclature for IL-1-family genes. Trends Immunol *22*, 536–537.

Sreekumar, A., Nyati, M. K., Varambally, S., Barrette, T. R., Ghosh, D., Lawrence, T. S., and Chinnaiyan, A. M. (2001). Profiling of cancer cells using protein microarrays: Discovery of novel radiation-regulated proteins. Cancer Res *61*, 7585–7593.

Stark, G. R., Kerr, I. M., Williams, B. R., Silverman, R. H., and Schreiber, R. D. (1998). How cells respond to interferons. Annu Rev Biochem *67*, 227–264.

Sultzer, B. M. (1968). Genetic control of leucocyte responses to endotoxin. Nature *219*, 1253–1254.

Suzuki, N., Suzuki, S., Duncan, G. S., Millar, D. G., Wada, T., Mirtsos, C., Takada, H., Wakeham, A., Itie, A., Li, S., et al. (2002). Severe impairment of interleukin-1 and Toll-like receptor signalling in mice lacking IRAK-4. Nature *416*, 750–756.

Swantek, J. L., Tsen, M. F., Cobb, M. H., and Thomas, J. A. (2000). IL-1 receptor-associated kinase modulates host responsiveness to endotoxin. J Immunol *164*, 4301–4306.

Symons, J. A., Eastgate, J. A., and Duff, G. W. (1991). Purification and characterization of a novel soluble receptor for interleukin 1. J Exp Med *174*, 1251–1254.

Theofilopoulos, A. N., Baccala, R., Beutler, B., and Kono, D. H. (2005). Type I interferons (alpha/beta) in immunity and autoimmunity. Annu Rev Immunol *23*, 307–336.

Tojima, Y., Fujimoto, A., Delhase, M., Chen, Y., Hatakeyama, S., Nakayama, K., Kaneko, Y., Nimura, Y., Motoyama, N., Ikeda, K., et al. (2000). NAK is an IkappaB kinase-activating kinase. Nature *404*, 778–782.

Toshchakov, V., Jones, B. W., Perera, P. Y., Thomas, K., Cody, M. J., Zhang, S., Williams, B. R., Major, J., Hamilton, T. A., Fenton, M. J., and Vogel, S. N. (2002). TLR4, but not TLR2, mediates IFN-beta-induced STAT1alpha/beta-dependent gene expression in macrophages. Nat Immunol *3*, 392–398.

Towne, J. E., Garka, K. E., Renshaw, B. R., Virca, G. D., and Sims, J. E. (2004). Interleukin (IL)-1F6, IL-1F8, and IL-1F9 signal through IL-1Rrp2 and IL-1RAcP to activate the pathway leading to NF-kappaB and MAPKs. J Biol Chem *279*, 13677–13688.

Tsukada, J., Saito, K., Waterman, W. R., Webb, A. C., and Auron, P. E. (1994). Transcription factors NF-IL6 and CREB recognize a common essential site in the human prointerleukin 1 beta gene. Mol Cell Biol *14*, 7285–7297.

Uze, G., Lutfalla, G., and Gresser, I. (1990). Genetic transfer of a functional human interferon alpha receptor into mouse cells: Cloning and expression of its cDNA. Cell *60*, 225–234.

van Boxel-Dezaire, A. H., Rani, M. R., and Stark, G. R. (2006). Complex modulation of cell type-specific signaling in response to type I interferons. Immunity *25*, 361–372.

Vannier, E., Miller, L. C., and Dinarello, C. A. (1992). Coordinated antiinflammatory effects of interleukin 4: Interleukin 4 suppresses interleukin 1 production but up-regulates gene expression and synthesis of interleukin 1 receptor antagonist. Proc Natl Acad Sci U S A *89*, 4076–4080.

Watson, J., Kelly, K., Largen, M., and Taylor, B. A. (1978). The genetic mapping of a defective LPS response gene in C3H/HeJ mice. J Immunol *120*, 422–424.

Watson, J., and Riblet, R. (1974). Genetic control of responses to bacterial lipopolysaccharides in mice. I. Evidence for a single gene that influences mitogenic and immunogenic respones to lipopolysaccharides. J Exp Med *140*, 1147–1161.

Wesche, H., Gao, X., Li, X., Kirschning, C. J., Stark, G. R., and Cao, Z. (1999). IRAK-M is a novel member of the Pelle/interleukin-1 receptor-associated kinase (IRAK) family. J Biol Chem *274*, 19403–19410.

Williams, M. J., Rodriguez, A., Kimbrell, D. A., and Eldon, E. D. (1997). The 18-wheeler mutation reveals complex antibacterial gene regulation in Drosophila host defense. Embo J *16*, 6120–6130.

Yamaguchi, K., Shirakabe, K., Shibuya, H., Irie, K., Oishi, I., Ueno, N., Taniguchi, T., Nishida, E., and Matsumoto, K. (1995). Identification of a member of the MAPKKK family as a potential mediator of TGF-beta signal transduction. Science *270*, 2008–2011.

Yamamoto, M., Sato, S., Mori, K., Hoshino, K., Takeuchi, O., Takeda, K., and Akira, S. (2002). Cutting edge: A novel Toll/IL-1 receptor domain-containing adapter that preferentially activates the IFN-beta promoter in the Toll-like receptor signaling. J Immunol *169*, 6668–6672.

Yates, L. L., and Gorecki, D. C. (2006). The nuclear factor-kappaB (NF-kappaB): From a versatile transcription factor to a ubiquitous therapeutic target. Acta Biochim Pol *53*, 651–662.

Zandi, E., Rothwarf, D. M., Delhase, M., Hayakawa, M., and Karin, M. (1997). The IkappaB kinase complex (IKK) contains two kinase subunits, IKKalpha and IKKbeta, necessary for IkappaB phosphorylation and NF-kappaB activation. Cell *91*, 243–252.

# Epitope-Based Immunome-Derived Vaccines: A Strategy for Improved Design and Safety

Anne S. De Groot, Leonard Moise, Julie A. McMurry, and William Martin

**Abstract** Vaccine science has extended beyond genomics to proteomics and has come to also encompass 'immunomics,' the study of the universe of pathogen-derived or neoplasm-derived peptides that interface with B and T cells of the host immune system. It has been theorized that effective vaccines can be developed using the minimum essential subset of T cell and B cell epitopes that comprise the 'immunome.' Researchers are therefore using bioinformatics sequence analysis tools, epitope-mapping tools, microarrays, and high-throughput immunology assays to discover the minimal essential components of the immunome. When these minimal components, or epitopes, are packaged with adjuvants in an appropriate delivery vehicle, the complete package comprises an epitope-based immunome-derived vaccine. Such vaccines may have a significant advantage over conventional vaccines, as the careful selection of the components may diminish undesired side effects such as have been observed with whole pathogen and protein subunit vaccines. This chapter will review the pre-clinical and anticipated clinical development of computer-driven vaccine design and the validation of epitope-based immunome-derived vaccines in animal models; it will also include an overview of heterologous immunity and other emerging issues that will need to be addressed by vaccines of all types in the future.

**Keywords** Epitope · Immunome · Vaccines

## 1 Introduction

The availability of immunome-mining tools has fueled the design and development of vaccines by a process that has come to be termed 'reverse vaccinology,' 'vaccinomics,' 'immunome-derived vaccine' (IDV) design, or 'genome-derived

A.S. De Groot
Director, Immunology and Informatics Institute, Center for Vaccine Research and Design, University of Rhode Island; Brown University; CEO, EpiVax, Inc., Providence RI 02903, USA
e-mail: AnnieD@brown.edu

A. Falus (ed.), *Clinical Applications of Immunomics*,
DOI: 10.1007/978-0-387-79208-8_3, © Springer Science+Business Media, LLC 2009

vaccine' design (Rappuoli and Covacci 2003; Petrovsky and Brusic 2002; Pederson 1999; De Groot and Martin 2003; Doytchinova, Taylor and Flower 2003). This vaccine concept is based on the identification of a minimal set of antigens that induce a competent immune response to a pathogen or neoplasm. Recognition of antigens occurs through the presentation of B cell and T cell epitopes derived from the antigen, in the correct immunological milieu. In its minimal form, an IDV would contain only adjuvanated B cell and T cell epitopes in delivery vehicles such as liposomes. When these minimal components are packaged in an appropriate delivery vehicle, the complete package comprises an IDV.

Compared to traditional vaccines, IDVs have the potential to be safer and more effective since the vaccine focuses the protective immune response on the most essential antigenic elements of the pathogen/neoplasm. A number of IDVs have been tested in clinical trials (Elliott 2008; Gahery et al. 2006; Asjö et al. 2002; Kran et al. 2004). Because epitope-based IDVs are generally considered to be safe, when compared to other vectored or attenuated live vaccines, many have progressed rapidly from pre-clinical concept into clinical trials. In the cancer vaccine field, where epitope-based vaccines are well-established, many such vaccines are currently in Phase I/II clinical trials (Pietersz, Pouniotis and Apostolopoulos 2006).

This chapter will review the development and validation of IDVs; it will also provide a step-by-step guide to develop IDV using validated immunoinformatics tools. These tools have the potential for dramatically accelerating the development of new and improved vaccines for the emerging and existing infectious diseases. Whole-antigen-based IDVs will be covered briefly; however, the main focus will be on epitope-based IDVs and a description of the immunoinformatics tools that have been developed to accelerate the pre-clinical phase of vaccine discovery. So as to illustrate the process of pre-clinical vaccine development using these tools, two epitope-based IDVs case studies will be presented: (i) a genome-derived vaccine for Tularemia and (ii) an epitope-based HPV vaccine for adjunctive treatment of cancer.

## 2 Defining the Immunome

Application of molecular biology techniques led to the sequencing of genomes and improved definition of the proteome (expressed proteins). Even though the fundamental concept of the 'immunome' (the subset of fragments of expressed proteins that interface with the host immune system) is relatively well accepted, many important questions remain. Due to genetic variation and poorly understood determinants of antigen processing, it has become clear that immunomes can vary substantially from one host to the next, even when major histocompatibility complex (MHC) and antibody germline genes are shared. The extent of the overlap between immunomes (in different hosts) and the general size of the immunome-representing epitopes derived from a particular pathogen or cancer remain to be determined. Recently published studies are beginning to address thesc questions, partly due to the availability of algorithms that facilitate the identification of epitopes from whole genomes.

## 2.1 How Large Is the Immunome?

The size of the immunome has been puzzling vaccinologists for decades. Certainly, there are examples in the literature that 'a single epitope protects.' For example, Crowe et al. (2006) recently found that immunization with a single Th epitope provided a one-log reduction in influenza viral titers early in infection. A single epitope has also been shown to protect against viral disease in woodchucks (Menne et al. 1997), and multiple single epitopes have protected mice against an array of pathogens (An and Whitton 1997 and Olsen et al. 2000). A more diverse set of T cell epitopes appear to be critical to immune response to vaccinia (in mice): 49 Class I MHC epitopes were shown to contribute to the large majority of the CD8+ T cell immune response (Moutaftsi et al. 2006). Remarkably, it was found that 49 of these predicted epitopes (derived from more than 175,000 candidates) represented over 90% of the vaccinia-specific CD8 T cell repertoire. How this number might be extrapolated to humans is unknown, but the result is relevant because 49 epitopes is few enough to be easily packaged and delivered in a vaccine. An epitope-based IDV for genetically diverse populations of humans will almost certainly require more than that number of epitopes, particularly if the vaccine is intended to protect against complex bacteria or viruses, or against solid tumors presenting variable antigenic profiles. Harnessing the power of immunoinformatics accelerates the tasks of defining the immunome and of identifying and developing new vaccines for human diseases.

## 3 Steps in the Development of an Epitope-Based IDV

The task of developing an epitope-based IDV can be deconstructed into a series of achievable steps.

## 3.1 Select Protein Antigens of Interest

After selecting a target organism, the next step in the immunome-to-vaccine process is to identify, from within the target genome, a set of potentially antigenic genes/proteins. These sequences can then be screened using a variety of in silico, in vitro, and in vivo mechanisms or a combination of methods. In the case of small viral genomes, it may be possible to include the entire genome in the search universe. In the case of larger bacterial and viral pathogens, the traditional vaccine targets include surface proteins and secreted proteins (which can easily be found using *in silico* screening programs), toxins (which can be identified through homology matching), and virulence factors (which can be identified through the use of comparative genomics). However, many other types of proteins may also be worthy of consideration: in particular, proteins expressed in high amounts (such as viral capsid proteins), proteins overexpressed during growth or replication, or proteins overexpressed in response to stress conditions.

For cancer vaccines, comparisons between cancerous and normal tissue can uncover cancer-specific genes. One approach involves the use of mRNA derived from cancerous and normal tissue to probe DNA microarrays, followed by selection of genes that are upregulated in cancerous and not in normal tissue (Mathiassen et al. 2001; Sepkowitz 2001).

Using 'Reverse Vaccinology', a term recently coined by Rino Rappuoli, the immune memory of subjects who have successfully encountered and defeated a pathogen can be interrogated to identify the primary targets of a natural immune response. Rappuoli and colleagues identified novel vaccine targets using *in silico* techniques to screen the target genome for 'surface protein-like' sequences. Candidate proteins were then expressed in *Escherichia coli* and screened against human sera isolated from pathogen-exposed subjects. Reactive proteins were deemed relevant to immune response (Pizza et al. 2000).

A related approach for discovering candidate vaccine antigens involves analyzing the target pathogen's proteome in silico, using T cell epitope mapping tools. Putative T cell epitopes identified can then be screened against peripheral blood mononuclear cells (PBMC) isolated from human subjects who have been infected with the target pathogen (or who have the target cancer). T cell reaction to a particular peptide epitope, typically measured by ELISA or ELISpot assay, implies that the protein from which the peptide was derived, expressed, processed, and presented to the immune system in the course of a 'natural' immune response. Using this method, measuring immune response to an epitope reveals a protein antigen. Our group describes this approach as 'fishing for antigens using epitopes as bait.'

Other common in vitro techniques used to identify expressed or over-expressed genes/proteins include: 2D SDS-PAGE, (Kaufmann et al. 1992; Sonnenberg and Belisle 1997; Hernychova et al. 2001), mass spectrometry (Tomlinson, Jameson and Naylor 1996), and/or tandem mass spectrometry (Barnea et al. 2002).

In the context of vaccines against infectious diseases, it may be prudent to exclude proteins that are highly conserved across species; such proteins (including housekeeping genes) may cross-react with unrelated avirulent organisms or with self-proteins to which there may be pre-existing tolerance. However, conservation *across species* should not be confused with conservation *within species variants*. Sequence conservation within species variants is a highly desirable trait for vaccine components and one that the IDV approach is particularly well suited to harness. RNA viruses in particular (HCV, HIV, coronaviruses) are highly variable pathogens. In these cases, selecting epitopes that are conserved across variants or subtypes may allow for the development of a more broadly applicable vaccine. Alternatively, single proteins that are relatively well conserved, when compared to the balance of the target pathogen, can be selected as vaccine candidates. Recently, for example, one team has developed a hexon-epitope vaccine that may be effective against a range of adenoviruses, across serotypes (Leen et al. 2008).

Once critical antigens have been identified, the next steps in epitope-based IDV development are to select epitopes and confirm their immunogenicity.

## 3.2 Identifying B Cell Antigens

Once the adaptive immune system has been engaged, a humoral, or antibody-based response forms the first line of defense against most viral and bacterial pathogens. Antibodies recognize B cell epitopes composed of either linear peptide sequences or conformational determinants, which are present only in the three-dimensional form of the antigen. Several B cell epitope prediction tools, such as 3DEX and CEP, have been proposed and are in the process of being refined (Enshell-Seijffers et al. 2003; Schreiber et al. 2005; Kulkarni et al. 2005). Unfortunately, the computational resources and modeling complexity required to predict B cell epitopes are enormous. This complexity is due in part to the inherent flexibility in the complementarity determining regions (CDR) of the antibody and in part due to glycosylation, and other post-translational modifications can result in modification of B cell epitopes.

Although accurate B cell epitope mapping tools remain elusive, the selection of potent B cell antigens can be accelerated using T cell epitope mapping tools. When considering B cell antigens as potential subunit vaccines, it may be important to also consider their T cell epitope content since the quality and kinetics of the antibody response is dependent upon the presence of T help. B cell antigens, which contain a significant T help, may outperform B cell antigens lacking cognate help. And in some cases, an identified T cell epitope may contain a B cell epitope. Although different epitopes activate T and B cells, it has been widely reported that B cell epitopes have been shown to co-localize near, or overlap, Class II (Th, CD4+) epitopes (Graham et al. 1989; Rajnavolgyi et al. 1999).

## 3.3 Identifying T Cell Antigens

The adaptive immune system's second line of defense is the T lymphocyte. Class I-restricted cytotoxic T cells (CD8+ CTL) directly engage and attack infected host cells. Class II-restricted T helper cells mediate the growth and differentiation of both T effector cells and antibody-producing B lymphocytes. Both Class I and Class II T cells carry out their roles in response to T cell epitopes, small linear fragments derived from protein antigens, displayed on the surface of antigen-presenting cells (APC) by various alleles of MHC. While B cells and antibodies generally recognize epitopes on surface proteins only, T cells recognize epitopes derived from a variety of proteins.

Once taken up by APC, antigenic proteins are broken down by digestive enzymes. During this process very large numbers of peptide fragments are released. Any one of these fragments could be a T cell epitope, but only about 2% of all the fragments generated can implant themselves in the binding groove of the MHC molecule and be presented on the surface of the APC. One of the critical determinants of T cell epitope immunogenicity is the strength of epitope binding to MHC molecules (Lazarski et al. 2005). Peptides binding with higher affinity are more likely to be selected by MHC molecules and to be displayed on the cell surface where they can be recognized by T lymphocytes. Using a variety of methods including frequency analysis, support vector machines, hidden Markov models, and neural networks, researchers have developed highly accurate tools for modeling the MHC–peptide interface and for accurately predicting T cell epitopes. For a review of T cell epitope mapping tools, see De Groot and Berzofsky (2004) and the accompanying issue of Methods. What all these tools have in common is an ability to quickly screen large volumes of genomic sequences for putative epitopes; this preliminary screen reduces the search space dramatically, typically by at least 20-fold.

The ability to accurately predict T cell epitopes from raw genomic data is fundamental to the development of an IDV. However, even a highly accurate prediction is still only a prediction. Before including predicted epitopes in a candidate vaccine, it is important to validate their immunogenicity in vitro and in vivo.

### 3.3.1 In Vitro Assays: Peptide Binding Assays

Once identified, peptides representing the selected epitopes are then synthesized. HLA binding assays can be used to assess whether peptides derived from immunoinformatics analysis can bind to either MHC Class I or Class II by measuring the affinities of predicted epitope sequences for the HLA alleles in vitro. In vitro evaluation of MHC binding can be performed by quantifying the ability of exogenously added peptides to compete with a fluorescently labeled known MHC ligand (Steere et al. 2006) and can be adapted for high throughput (McMurry et al. 2007a). EpiVax routinely uses these high-throughput HLA binding assays to confirm epitope predictions in vitro. A concordance between HLA binding and immunogenicity is often observed (McMurry et al. 2005).

### 3.3.2 In Vitro Assays: Measuring T Cell Responses

Peptides are used to measure T cell responses in vitro; they can be of variable lengths (9–25). Peptides presented in the context of Class I MHC are generally limited to 9 or 10 amino acids in length, although some processing is believed to occur during the T cell assay and so 15-mers are also used for Class I assays.

In contrast with Class I epitopes, which are short and fit tightly in the bounded MHC molecule, Class II (T helper) epitopes lie within an open-ended groove in the MHC II. As such, a Class II epitope can shift within the groove, thereby accommodating MHC of various haplotypes. The only limit on the size of the peptide is its ability to remain in a linear conformation in the open-ended groove.

Both MHC Class I- and MHC Class II-restricted epitopes (targeting CD4+ and CD8+ T cells, respectively) are believed to be important for the development of effective vaccines. CD4+ T helper cells enhance and amplify cytotoxic T cell (CTL) immune responses and have been shown to be important in the development of CD8+ T cell memory to a range of pathogens (Ahlers et al. 2001). CTLs generally play a role in the containment of viral and bacterial infection (Plotnicky et al. 2003), and the prevalence of CTLs usually correlates with the rate of pathogen clearance.

Like peptides, whole antigens too can be used to measure T cell responses in vitro. The recognition of these antigens requires the presence of an APC that is capable of processing and presenting peptides derived from the antigen.

If blood from exposed individuals is available, the peptides validated as MHC ligands in binding assays can be tested for their reactivity with T cells, serum, or both. A positive immune response (as measured by ELISA, ELISpot, or intracellular cytokine staining) should be interpreted as a sign that the parent protein interfaces with host immune response in the course of natural infection or disease. Following confirmation, the peptides that stimulate a response can be considered vaccine candidates themselves or can be used to select the entire protein for use in a subunit vaccine. These candidates can then be incorporated into a vaccine delivery vehicle with an appropriate adjuvant.

ELISA and ELISpot are related methods for detecting T cell responses by the measurement of cytokines secreted by the T cells (gamma interferon, IL-2, and IL-4 are examples). The expansion (proliferation) of T cells in response to stimulation by peptide:MHC can be measured by (1) the dilution of a fluorescent dye in subsequent generations of cells (CFSE) and (2) the incorporation of a radioactive label in the proliferating cell's DNA (tritiated thymidine incorporation assay). Fluorescence activated cell sorting (FACS) and intracellular cytokine staining (ICS) are the most precise methodologies available for measuring and defining T cell response. For example, T cells that respond to a particular epitope can be directly labeled using tetramers (comprising MHC Class II: peptide complexes). Labeled cells can then be sorted and counted, and the phenotype of T cells that respond to the antigen can be determined using cell surface markers and ICS (Tobery et al. 2006).

## 3.4 Select Delivery Vehicle and Adjuvant

Factors extrinsic to processing, such as the cytokine milieu induced in response to a particular component of a vaccine (Krieg et al. 1998) or pathogen

(Ghosh et al. 1998), also play a role in the conditioning of the immune response. Thus, T cell epitopes may be necessary to drive immune response, but are not sufficient. Co-stimulatory molecules that provide a second signal, the right cytokine milieu and other factors directing the nature (Th1 vs. Th2) of the immune response, are also crucial (Shahinian et al. 1993; Kuchroo et al. 1995). Adjuvants provide this added 'boost' in the context of vaccines.

The same range of delivery vehicles that exist for conventional vaccines can be used for the development of IDVs and epitope-based IDVs. For example, IDVs and epitope-based IDVs can be formulated and delivered as pseudo-proteins or peptides in a carrier vehicle such as a liposome or viral-like protein (VLP); alternatively, the sequence of the IDV antigens or epitope string can be inserted into a viral or bacterial vector such as adenovirus or salmonella; alternatively, a DNA vaccine construct encoding the antigen(s) or epitopes can be developed. The choice of adjuvants for use in humans is relatively extensive and each adjuvant has advantages and disadvantages. The advantages and disadvantages of each type of vaccine delivery vehicle and adjuvant here is beyond the scope of this chapter; readers are referred to a review by Fraser et al. (2007).

## 3.5 Animal Model for Vaccine Efficacy

The next step in the development of epitope-driven IDV is to determine whether immunization provides competent immune response. The IDV or epitope-based IDV is administered and immune responses to the components are evaluated following immunization. Even though a range of animal models are used for the evaluation of vaccines, results from immunogenicity studies in these models should be interpreted with caution. Although their functions may be similar, the MHC of mice, rodents, and non-human primates differ from human MHC (known as HLA in the context of human immune response) at the amino acid level and these differences effect which epitopes can be presented. This helps to explain why different strains of mice (Balb/C, C57Bl/6) have different immune responses to pathogens as well as vaccines for those pathogens (Klitgaard et al. 2006). In particular, epitope-based vaccines that are developed using predicted human T cell epitope mapping tools can be tested only in murine models that are HLA transgenic.

Fortunately, a number of transgenic mouse strains that express the most common HLA A, HLA B and HLA DR, molecules have been developed. T cell responses in these mice correlate directly with T cell responses observed in infected/vaccinated humans (Man et al. 1995; Shirai et al. 1995). HLA transgenic mice are now routinely used to assay and optimize (human) epitope-driven vaccines in pre-clinical studies (Ishioka et al. 1999; Charo et al. 2001; Livingston et al. 2001). Despite the limited number of HLA Class II alleles for which Tg mice have been developed, comparisons of immunogenicity can be done to a high degree of accuracy in the mouse model for selected HLA Class II alleles (HLA DR 0101, 0301, 0401, 1501). Unfortunately, it appears these mice may have difficulty breeding due to poorly understood consequences of their transgenic heritage, limiting the use of this important model system.

## 3.6 Challenge Studies

The final step in the development of any vaccine is experimental validation of the immunogenicity and protective efficacy of computationally selected antigens. Currently, a series of experimental vaccines have shown efficacy in animal models and several IDVs are being tested in clinical studies. In the context of infectious disease, genome-derived vaccines that have progressed furthest along the vaccine development pipeline are generally based on whole proteins rather than epitopes (Rappuoli and Covacci 2003; Pizza et al. 2000). However, epitope-based IDVs are currently being developed for a range of infectious diseases by the authors' laboratory and by many others (Depla et al. 2008).

While it is common knowledge that subunit-based vaccines can protect against infection, similar success with epitope-based approaches is not as widely known. In addition to the studies previously cited, immunization of BALB/c mice with three doses of a peptide construct containing an H-2(d)-restricted cytotoxic T lymphocyte (CTL) epitope from a murine malaria parasite induced both T cell proliferation and a peptide-specific CTL response mediating nitric-oxide-dependent elimination of malaria-infected hepatocytes in vitro, as well as partial protection of BALB/c mice against sporozoite challenge (Franke et al. 2000). In a separate study, immunization of BALB/c and CBA mice with measles virus CTL epitopes resulted in the induction of epitope-specific CTL responses and conferred some protection against encephalitis following intra-cerebral challenge with a lethal dose of virus (Schadeck et al. 1999). These are just a few successful examples of many studies carried out in animal models; however, translation to prevention of disease in humans has been difficult to achieve.

## 3.7 Clinical Development

In contrast with whole-protein subunit vaccines, IDVs and epitope-based IDVs have taken longer to make the transition from animal model to the clinic, mainly due to the novelty of the concept and perhaps unfounded concerns that epitopes are not sufficient for the generation of effective immune response. Cancer therapy is an exception to this rule. As previously described, eptiope-based IDVs have been evaluated in the context of therapy against chronic infection or cancer (Ueda et al. 2004; Valmori et al. 2003).

In the cancer vaccine field, where the concept of epitope-driven vaccines is well established, many more peptide vaccines have successfully passed pre-clinical tests and are currently in Phase I/II clinical trials (Pietersz, Pouniotis and Apostolopoulos 2006). New approaches are emerging, which may improve the success rate and, indeed, the results from recent clinical trials prove the principle. One approach is to identify epitopes that are unique to the tumor (prostate, lung, colon) and to pre-screen the patient for response to the peptide. This approach, called personalized vaccination, takes into account the diversity of CTL epitope recognition among patients. Whereas the response rates to classical (non-personalized) peptide vaccines have been disappointing, responses to personalized vaccines (in a Phase I trial, conducted in Japan) have been as high as 11.1% in the advanced cancers and equal to or more than 20% in malignant glioma and cervical cancers, respectively (Itoh and Yamada 2006).

It is noteworthy that just a few epitope-driven vaccines against viral and microbial pathogens have reached the stage of Phase I or II efficacy trials in humans. For example, Bionor Immuno's HIV p24 gag peptide vaccine (Vacc-4X) was demonstrated to be safe and well tolerated in Phase I trials (Asjö et al. 2002) and dose-dependent and immunogenic in Phase II trials in Norway (Kran et al. 2004). Similarly, Nardin's epitope-based vaccine for malaria is moving along the clinical trial pathway (Nardin et al. 2000).

# 4 Epitope Mapping Tools for IDV

In this section, we describe the immunomics tools developed and used by the EpiVax vaccine development group in recent collaborations with Dr. Steve Gregory of Lifespan, Dr. Ousmane Koita of the University of Bamako, Mali and Dr. David Weiner of University of Pennsylvania.

## 4.1 EpiMatrix: T cell Epitope Mapping for IDV

T cell epitopes are linear peptides that bind to MHC molecules. Binding is mediated by the interactions between the R-groups of the amino acids in the peptide ligand and the pockets on the floor of the MHC binding groove. Because the MHC:peptide interaction is well characterized, pattern-matching algorithms can be used to screen protein sequences for peptides that will bind MHC. The authors of this report currently use the EpiMatrix system, a suite of epitope-mapping tools that has been validated by more than a decade of use in selecting putative epitopes for in vitro and in vivo studies (see references (De Groot et al. 1997; Bond et al. 2001; Dong et al. 2004; McMurry et al. 2005; Koita et al. 2006). The EpiMatrix algorithm is based on a set of Class I and Class II HLA matrices wherein individual frequencies of all 20 amino acids (aa) in each HLA pocket position are applied to the prediction of overlapping 9- and 10-mer peptides. In a typical analysis, protein antigens are parsed into overlapping 9-mer frames where each 9-mer overlaps the last by eight amino acids. Each 9-mer is then scored for predicted binding affinity to one or more Class I or Class II HLA alleles. In order to compare potential epitopes across multiple HLA alleles, EpiMatrix raw scores are converted to a normalized 'Z' scale. Peptides scoring above 1.64 on the EpiMatrix 'Z' scale (typically the top 5% of any given sample) are likely to be MHC ligands. Since Class II epitopes can be promiscuous, our approach to the prediction of Class II epitopes is to estimate the binding potential of each frame with respect to each of a panel of eight common Class II alleles (DRB1*0101, *0301, *0401, *0701, *0801, *1101, *1301, and *1501). Taken together, these alleles 'cover' the genetic backgrounds of most humans worldwide (Southwood et al. 1998) and they also represent the predominant types of 'pockets' for the most common MHC.

Recently, the EpiMatrix system has been utilized to measure the potential immunogenicity of whole proteins. In this context, EpiMatrix assesses the aggregate epitope density of a given protein with respect to the aggregate epitope density of a set of randomly generated pseudo-protein sequences of similar size (De Groot 2006). By correcting for the size and expected epitope density, the potential immunogenicity candidate vaccine antigens can be directly compared. Further immunogenicity of low scoring proteins may be enhanced by modifying the immunogenic region amino acid sequence so that it contains more T cell epitopes (see illustration of this approach in the HPV vaccine section, below).

## 4.2 ClustiMer: Finding Promiscuous T cell Epitopes

Peptides predicted to bind to multiple HLA alleles are known as promiscuous T cell epitopes. The ClustiMer algorithm is used to scan the output produced by the EpiMatrix and identifies the polypeptides predicted to bind to an unusually large number of HLA alleles. Briefly, the scores of each analyzed 9-mer are aggregated. High-scoring 9-mers are then extended at the N- and C-terminal flanks until the predicted epitope density of the promiscuous epitope falls below a given threshold value. This particular approach to mapping epitopes has also been useful for discovering 'EpiBars,' which may be a signature feature of highly immunogenic, promiscuous Class II epitopes. An example of a promiscuous T cell cluster containing an EpiBar (Tetanus Toxin 830–844) is shown in Fig. 1.

A single T cell epitope 'cluster' usually ranges from 9 to about 25 amino acids in length and can contain anywhere from 4 to 40 binding motifs. Using EpiMatrix as described above and ClustiMer, scores above 10 and, in particular, scores above 15 indicate significant immunogenic potential (De Groot 2006). Note the horizontal bar of high Z scores at position 308 in Fig. 3. Having observed this 'EpiBar' pattern to be characteristic of promiscuous epitopes, the authors have integrated the pattern into the prospective selection of clusters.

Tetanus Toxin Peptide AA 830-844

| Frame Start | AA Sequence | Frame Stop | DRB1*0101 Z score | DRB1*0301 Z score | DRB1*0401 Z score | DRB1*0701 Z score | DRB1*0801 Z score | DRB1*1101 Z score | DRB1*1301 Z score | DRB1*1501 Z score | HITS | Cluster Score |
|---|---|---|---|---|---|---|---|---|---|---|---|---|
| 830 | QYIKANSKF | 838 | | | | | | | | | | |
| 831 | YIKANSKFI | 839 | 2.68 | | 1.81 | 3.03 | | | | 1.64 | 4 | |
| 832 | IKANSKFIG | 840 | 1.38 | 2.83 | | | 2.79 | 2.12 | 3.13 | 2.18 | 5 | |
| 833 | KANSKFIGI | 841 | | | | | | | | | | |
| 834 | ANSKFIGIT | 842 | | | | | | | | | | |
| 835 | NSKFIGITE | 843 | | | | | | | | | | |
| 836 | SKFIGITEL | 844 | | | | 1.41 | | | | | | |

Deviation from expectation: 23.82

**Fig. 1** EpiBar-Typical EpiMatrix analysis. Z score ( Top 10% Top 5% Top 1% ) indicates the potential of a 9-mer frame to bind to a given HLA allele. All Z scores in the Top 5% (>1.64) are considered 'hits'. Though not hits, scores in the top 10% are considered elevated; scores below 10% are masked for simplicity. Frames containing four or more alleles scoring above 1.64 are colloquially referred to as 'EpiBars' (see frames 831:YIKANSKFI and 832: IKANSKFIG). This band-like pattern is characteristic of promiscuous epitopes. The tetanus toxin peptide scores are extremely high for all eight alleles in EpiMatrix; the deviation compared to expectation is + 23.82

Promiscuous epitopes also exist, to a certain degree, for Class I alleles. Some laboratories have demonstrated cross-presentation of peptides within HLA 'superfamilies' (such as the A3 superfamily: A11, A3, A31, A33, and A68) described by (Sette and Sidney 1998). The authors have confirmed cross-MHC binding and presentation to T cells in our HIV vaccine studies (De Groot et al. 2001).

# 5 Additional Vaccine Design Tools

## 5.1 Conservatrix: Finding Conserved T Cell Epitopes

One limitation of conventional vaccination, and to a lesser extent natural infection, is that the immune system focuses strongly on the most mutable immunogen of the virus – typically, the viral envelope. In the case of HIV and other viruses, vaccination with more conserved, subdominant epitopes has been shown to circumvent this hierarchy and potentiate cross-strain protection (Ostrowski et al. 2002; Nara and Lin 2005). In like manner, a conserved T helper–directed vaccine may provide a more 'democratic' way of stimulating immune response, increasing the number targets for T cell recognition, thereby providing T help to antibody response despite potential viral variability (Santra et al. 2002; Subbramanian et al. 2003; Scherle and Gerhard 1988; Scherle and Gerhard 1986; Russell and Liew 1979; Johansson et al 1987).

The genetic variability of some pathogens constitutes a significant challenge to the efforts to design a vaccine driven by cellular immune response (De Groot et al. 2001; De Groot et al. 2002). The authors have been involved in developing an HIV-1 vaccine that includes highly conserved (cross-clade) T cell epitopes. The Conservatrix algorithm, developed for this application, parses input sequences into component strings (the lengths of the strings may be determined by the operator) and then searches the input dataset for matching segments. Conservatrix may be used to compare strings derived from different strains of the same organism (hepatitis C, for example, or HIV) or to search a given sequence for a user-supplied target sequence. Target sequences may be input as specific sequences or as coded patterns. Thus, the operator can use 'wild cards', allowing for one or more of the amino acid residues in any given peptide sequence to be any amino acid [X], or a limited set of amino acids such as [L, V]. Results of each analysis are stored in a database and may be browsed or exported to another program for analysis.

By selecting highly conserved epitopes, regardless of their distance from the ancestral HIV-1 genome, we have identified sequences conserved for structural and functional reasons and are therefore less likely to be modified in the course of further evolution of HIV-1 (Peyerl et al. 2004; Koibuchi et al. 2005).

## 5.2  EpiAssembler: Immunogenic Consensus Sequence Epitopes

The problem of virus variability also significantly complicates the selection of epitopes that have a population-coverage advantage; such epitopes are termed 'clustered,' 'superfamily,' or 'promiscuous.' To address this problem, the authors developed EpiAssembler (De Groot et al. 2004) to identify sets of overlapping, conserved, and promiscuously immunogenic epitopes and assemble them into extended immunogenic consensus sequences (ICS) (see Fig. 2).

In theory, proper processing and presentation of these sequences would allow for the presentation of highly conserved peptides in the context of more than one MHC. The resulting peptide is not a 'pseudosequence' as such, since each constituent epitope occurs in its corresponding position in the native protein. Thus, while the full-length 'immunogenic consensus sequence' is not necessarily found in any one variant sequence, the peptide is more representative of the sequence universe.

In the case of HIV, for example, we used the ICS approach to design a peptide-based vaccine. While the full-composite ICS peptides happen to be exactly conserved in a few individual strains of HIV, each peptide represents a significant percentage of circulating strains, because every constituent overlapping epitope is conserved in a large number (range 893 to 2,254) of individual HIV-1 strains. As compared with immunogenic consensus sequences, randomly selected counterparts, on average, contain half as many binding motifs and cover a third fewer isolates. To develop vaccines of equivalent antigenic 'payload,' using conventional methods would be prohibitively expensive as it would require including multiple different variants of each antigen. This approach has been useful for identifying highly immunogenic epitopes for HIV vaccine design (De Groot et al. 2005). By focusing on conserved, MHC-promiscuous T helper epitopes, the ICS approach has the potential to efficiently overcome the genetic variability of both virus and host.

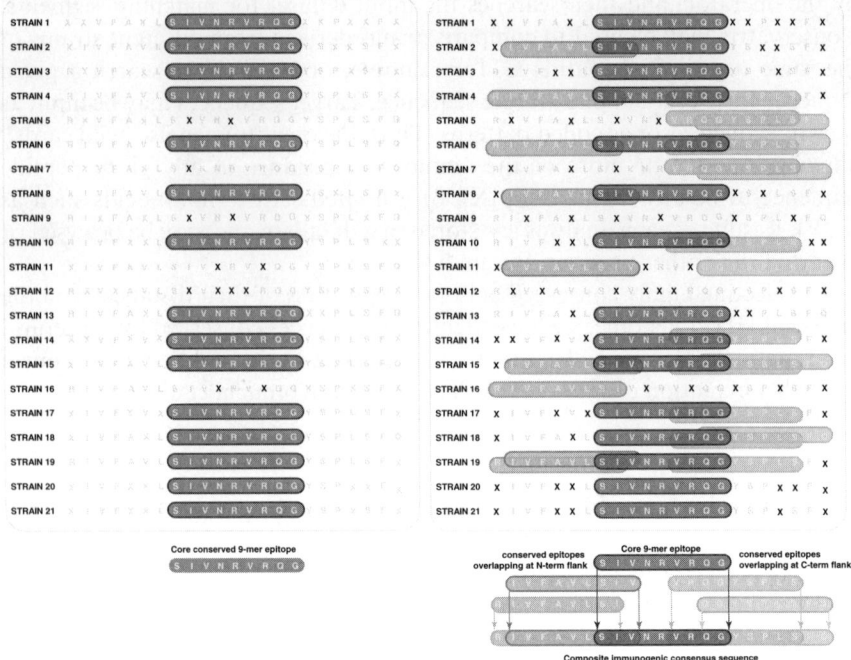

**Fig. 2** The ICS assembly operation was performed using EpiAssembler (Bill Martin, EpiVax, 2004). *Left panel*: each variant strain is first analyzed and a highly conserved, putatively promiscuous 9-mer is chosen as the core peptide. Mismatches with the selected epitope sequences are represented with the letter X. *Right panel*: Additional epitopes are then identified, which overlap with the natural N- and C-terminal flanking regions of the core 9-mer epitope. The overlap length requirement is decreased by one amino acid iteratively until reaching a minimum of three overlapping amino acids. If more than one suitable overlap is identified, the overlapping peptide with the higher overall EpiMatrix rank is selected. This process is repeated using the extended peptide as the new core sequence. The cycle can be repeated for the length of an entire protein or can be truncated when the peptide reaches a length that can be easily produced synthetically

## 5.3  Eliminating Cross-Reactivity (BlastiMer)

One of the advantages of IDV is that it is possible to omit deleteriously cross-reactive epitopes. Perhaps, the most famous example of an adverse effect due to cross-reactivity with self was observed following vaccination for Lyme disease with the Osp A protein. The vaccine has been recently re-engineered with the cross-reactive epitope removed (Willett et al. 2004). In the context of our own work, peptides selected for in vitro evaluation are evaluated for homology with human proteins by BLASTing the sequences against the human sequence database at GenBank (http://www.ncbi.nlm.nih.gov/).

BLASTiMer automates the process of submitting sequences to the websites featuring search engines such as the blast engine at NCBI (www.ncbi.nlm.nih.gov/blast). By default, BLASTiMer blasts sequences against all non-redundant

GenBank CDS translations, PDB, SwissProt, PIR, and PRF. BlastiMer assesses the homology between the submitted sequence and the sequence of proteins of other organisms. Patent BLAST, on the other hand, targets a database of sequences gleaned from patents. Users of either program may control all of the submission options available to interactive users at NCBI. In both cases, results are recorded in a database and can be browsed, exported, or summarized and rendered in a report format. According to the authors' standard practice, any peptide that shares greater than 80% identity with peptides contained in the human proteome is eliminated from consideration in a vaccine.

## 5.4  Vaccine CAD: Aligning Epitopes

A number of methods for enhancing epitope-based vaccines have been described and implemented (Thomson et al. 1998; Rodriguez and Whitton 2000). One approach is to align the epitopes in a protein or DNA vaccine construct as a 'string of beads' without any intervening sequences or 'spacers' in a DNA plasmid encoding the individual epitopes (An et al. 2000). However, the lack of 'natural flanking sequences' – has raised concern that their proteolytic processing may be compromised, and that junctional epitopes, peptides other that the specific peptides of interest, may be generated as a result of processing (Godkin et al. 2001). To address this concern, the authors developed Vaccine-CAD (see Fig. 3), an algorithm that incorporates the evaluation of

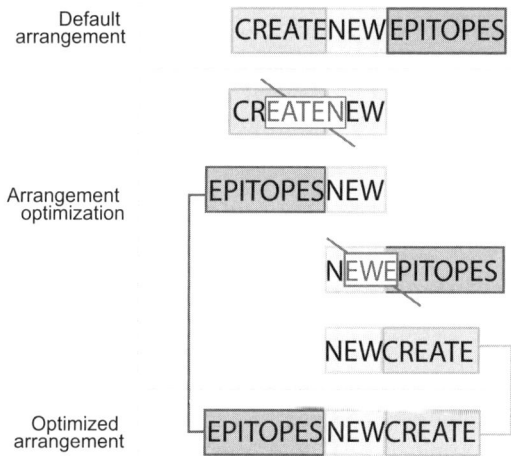

**Fig. 3** Vaccine-CAD, illustrated with three sample epitopes represented by the words 'create,' 'new,' and 'epitopes.' The default arrangement of the words results in unintended sequences, represented by the words 'eaten' and 'ewe,' at the junctions between the intended epitopes. Reiterative modifications in the arrangement of the epitopes results in the development of a sequence that has no 'pseudoepitopes' (new epitopes that were not intended) at the junctions of the juxtaposed epitopes

junctional epitopes, the insertion of spacers and breakers, the requirements for secretion or processing tags, and the evaluation of epitope strings for potential homologies to human protein fragments.

## 5.5 HLA Coverage for IDV

An important component in the epitope-driven vaccine process is the selection of epitopes from the regions of pathogens that are presented by MHC molecules for T cell recognition. 'MHC binding motifs,' first identified by Rötzschke and Falk, are the patterns of amino acids in peptides that are known to promote the binding of peptides containing these patterns to the MHC molecules on the surface of APCs (Falk et al. 1991; Rotzschke et al. 1991). Different MHC molecules have different binding motifs, limiting the set of MHC ligands that can be presented in the context of any given MHC.

While consideration of HLA alleles may lead to concern about the selection of epitopes for broad coverage of populations, Gulukota and DeLisi (1996) and Sette and Sidney (1998) have demonstrated that epitope-based vaccines that contain epitopes restricted by selected 'supertype' HLA can provide the broadest possible coverage of the human population. Furthermore, recent studies by Brander and Walker indicate that there may be even greater flexibility in the binding of epitopes to MHC than previously recognized; this is also consistent with the data recently presented by Frahm et al. (2004). The inclusion of 'promiscuous epitopes' – epitopes that are recognized in the context of more than one MHC (Paina-Bordignon et al. 1989; De Groot et al. 1991; Sette and Sidney 1998) in epitope-driven vaccines may therefore overcome the challenge of genetic restriction of immune response. In addition, the repertoire of possible MHC-restricted epitopes recognized by an individual's T cells has been shown to be quite variable, even between HLA-matched individuals (Jameson, Cruz and Ennis 1998; Gianfrani et al. 2000; Betts et al. 2001).

## 5.6 Aggregatrix: Aggregation of Epitopes into the Ideal IDV

Aggregatrix, a new algorithm that was recently developed at EpiVax, iteratively searches for the combination of epitopes that achieves maximal cross-clade representation. The authors performed this analysis for our HIV epitopes, as shown in Fig. 4, evaluating each HLA-A2-restricted HIV peptide individually and the set in aggregate for coverage of HIV-1 strains by year (1995–2005), by country of origin, and by clade. As can be seen in this figure, the set of highly conserved A2-restricted peptides we have tested and confirmed in ELISpot assays covered between 54% and 86% of strains in a given year, between 33 and 100% of strains in a given country, and between 0% and 100% of strains in a given clade. The HLA A2 peptides cover 85, 78, 78, and 80% of clades A, B, C, and D, respectively. This is a remarkable breadth of coverage for a limited set of

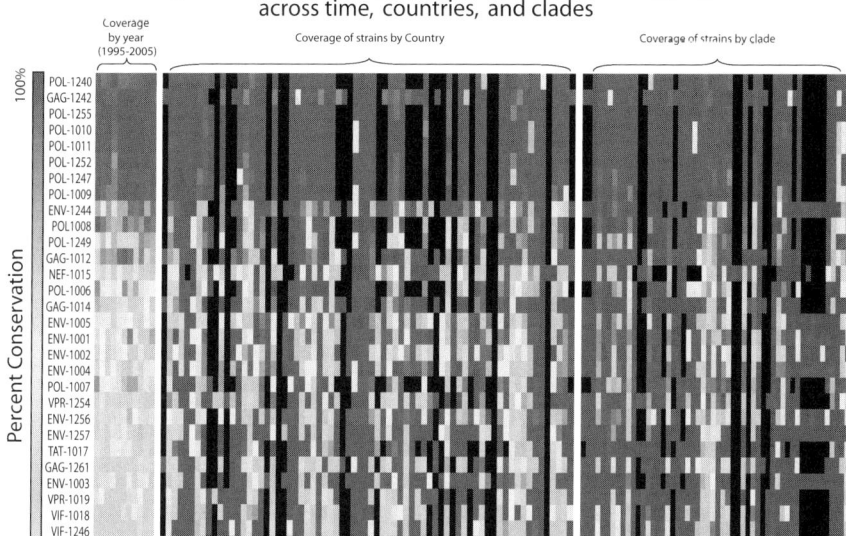

**Coverage of GAIA HIV A2 peptides--individually and in aggregate--across time, countries, and clades**

**Fig. 4** GAIA HIV A2 peptides — individually and in aggregate — percentage coverage of strains by year, country, and clade. Each row of the matrix denotes a specific peptide; the peptide's protein of origin is included within the peptide ID. Each column of the matrix denotes a specific year, country, or clade, grouped as indicated. The percentage coverage of strains is represented on a color gradient, with warm tones indicating values above 50% and cool tones indicating values below 50%. The *bottom* row of the matrix shows the percentage of each protein set that contains at least one peptide from our pool. *Black boxes* indicate that no isolates of the protein are available for that year, clade, or country. The *bottom* row represents the aggregate percent coverage for the epitope set. Each cell of the matrix represents the percentage coverage per peptide, except for the bottom row cells, which represent the aggregate percentage coverage for the peptide set. Column headers are listed here for space considerations: left to right, the year columns are 1995, 1996, 1997, 1998, 1999, 2000, 2001, 2002, 2003, 2004, and 2005; aggregate coverage of strains by year ranges from 47% (2000) to 63% (2004). The countries left to right are: Angola, Argentina, Australia, Belgium, Benin, Bolivia, Brazil, Botswana, Belarus, Canada, The democratic republic of the Congo (Zaire), Congo, Cote d'Ivoire, Chile, Cameroon, China, Colombia, Cuba, Cyprus, Germany, Djibouti, Dominican Republic, Ecuador, Estonia, Spain, Ethiopia, France, Gabon, United Kingdom, Georgia, Ghana, Gambia, Equatorial Guinea, Greece, Hong Kong, Israel, India, Italy, Japan, Kenya, Republic of Korea, Mali, Myanmar, Namibia, Niger, Nigeria, Netherlands, Norway, Portugal, Russian federation, Rwanda, Sweden, Senegal, Somalia, Chad, Thailand, Trinidad and Tobago, Taiwan, United Republic of Tanzania, Ukraine, Uganda, United States, Uruguay, Uzbekistan, Venezuela, Vietnam, Yemen, South Africa, Zambia, and Zimbabwe; aggregate coverage of strains by country ranges from 33% (Gabon) to 94% (Italy). The clade columns left to right are: 0206, 0209, 1819, A, AAD, AB, AC, ACD, ACDGKU, AD, ADF, ADGU, ADHK, AE, AF, AFG, AFU, AG, AGH, AGHU, AGJ, AGU, AHJU, AU, B, BC, BF, BG, C, CD, CGU, CK, CPX, CU, D, DF, DK, DO, DU, F, G, GHJKU, GK, GKU, GU, H, J, JKU, JU, K, N, O, U. The HLA-A2 peptides together cover 51, 51, 53, and 44% of clades A, B, C, and D, respectively. On average, the chimeric clade sequences were 50% covered by our HLA-A2 restricted epitopes (*See* Color Insert)

HLA A2 epitopes, given the well-known ability of HIV to mutate away from HLA (Nguyen et al. 2004; Iversen et al. 2006).

## 5.7 Individualized T Cell Epitope Measure (iTEM)

In studies of immune response to therapeutic proteins, the authors have observed that subject-to-subject variation in T cell response closely relates to (1) subject HLA type and (2) the number of protein-derived peptides that match the subjects HLA. To describe this relationship, the authors developed a metric that may be useful in the development of epitope-based vaccines or in the clinical assessment of immune response to vaccines, called the 'individualized T cell epitope measure' or iTEM. The iTEM can be calculated for each subject who is responding to a given epitope by summing the EpiMatrix Z-scores for each positive peptide for each HLA allele in a given subject's haplotype, thus:

iTEM score = [EpiMatrix score of peptide for HLA type 1]

+ [EpiMatrix score of peptide for HLA type 2] + . . .

The same calculation can be performed for larger peptides and proteins by summing all of the scores for the subject's HLA type. This calculated score allows for the individualized potential immunogenicity to be predicted, based on the number of putative epitopes contained in a protein that might be presented to their T cells, based on their HLA haplotype. Using this score, it is possible analyze the contribution of haplotype to the corresponding T cell response. In our prospective evaluations, significant correlations were found between the IFN-gamma response to a given antigen and the iTEM scores for individual subjects (for data published in Koren et al. (2007) $r = 0.69$, $p = 0.0134$). In addition, correlations between the iTEM score and patient HLA were also observed for their antibody titers. Further evaluations of this method for predicting individual responses to vaccines and therapeutic proteins are in progress.

## 5.8 Anticipating Processing and Presentation

Defining the immunome by identifying T cell epitopes and confirming their immunogenicity is but the first step of vaccine development. A number of conditions extrinsic to the MHC-ligand interaction may influence the final composition of the epitope ensemble. For example, whether or not a predicted epitope is confirmed is related to: (1) the quantitative expression of the source protein; (2) the number of different epitopes derived from this protein, which are presented on the surface of the APC (Wherry et al. 1999); (3) the amino acids that flank the epitope (Bergmann et al. 1994; Shastri, Serwold and Gonzalez 1995; Livingston et al. 2001); and (4) proper cleavage and trimming by proteolytic enzymes in the proteasome and processing pathways (Van Kaer et al. 1994; York et al. 1999; Chen et al. 2001; Toes et al. 2001). For example, it is very likely that end-to-end epitope presentation, such as was used in the design of the OXAVI HIV gag/epitope vaccine, impaired the presentation of the epitopes in immunogenicity studies.

Vaccine-CAD also takes into account the role of flanking residues: Studies conducted in murine models have demonstrated that residues flanking an MHC Class I epitope strongly influence the delivery of the intact epitope to TAP following proteasome degradation (Thomson et al. 1995; Hozhutter, Frommel and Kloetzel 1999; Mo et al. 1999). In addition, Livingston et al. (2001) have tested a standard spacer sequence (-GPGPG-) for vaccine constructs consisting of MHC-II-restricted, Th-cell epitopes; the use of this spacer disrupts junctional epitopes that might compete for degradation or for MHC binding (both G and P are unusual carboxy-terminal anchors for a peptide that binds to Class II MHC). This approach has been used for constructs with up to 20 epitopes, in assays where responses were detected to the majority of epitopes (Livingston et al. 2001).

# 6 Methods of Confirming IDV

## 6.1 Two Case Studies

### 6.1.1 Bacterial (Tularemia)

*Francisella tularensis* is a zoonotic bacterium. It is endemic to certain communities such as Martha's Vineyard, Massachusetts, USA, where it is known as Rabbit Fever. Tularemia represents a potentially dangerous biological weapon owing to its high degree of infectivity, ease of dissemination, and capacity to cause severe illness. Despite several decades of research, no vaccine for tularemia is licensed for public use. For a review of tularemia vaccines, see McMurry et al., 2007b.

We have been actively developing an epitope-based tularemia vaccine combining computational immunology with in vitro and in vivo validation (McMurry et al 2007a). The starting point of our vaccine was the fully annotated *F. tularensis* subsp. *tularensis* (SCHUS4) genome published in by Larsson et al. (2005). A prototype vaccine containing only Class II epitopes has been tested in challenge studies in HLA DRB1*0101 transgenic mice. For this vaccine, the EpiMatrix algorithm was utilized to identify highly promiscuous T cell epitopes within the tularemia genome. Twenty-five Class II-restricted epitopes were selected, synthesized, and screened in vitro using a recombinant soluble HLA Class II competition-binding assay (described above). Peptides that bound with high affinity were then tested ex vivo in ELISpot assays with blood obtained from *F. tularensis* subsp. *tularensis*-exposed individuals. Search-Light analysis was also performed on supernatants derived from human cell culture stimulated with peptide using a panel of nine cytokines. Forty-two percent of peptides bound to DRB*0101 and are likely to bind to several other alleles (in that they were predicted using the ClustiMer algorithm).

ELISpot assays showed positive IFN-gamma responses to 21/25 individual peptides and to peptide pools in nearly all of the 23 human study. The number of epitopes recognized per subject ranged from 1 to 17 and averaged 4 per subject; not every peptide was tested for every subject.

Peptides that elicit a robust memory response, as evaluated by these various assays, were incorporated into a vaccine construct and tested in challenge studies with HLA transgenic mice as a possible vaccine against tularemia. Immunogenicity studies in HLA transgenic DRB1*0101 mice were performed using the multi-Class II-epitope DNA constructs and/or peptides representing the epitopes, and T cell responses were evaluated in IFN-gamma ELISpots. HLA DR1 transgenic mice were challenged with five times the LD50 of *F. tularensis* LVS; 57% of vaccinated mice survived, all non-vaccinated mice died (manuscript in preparation). This result demonstrates the potential for a genome-derived epitope-based vaccine to protection from a Class A bioterror pathogen.

Importantly, the protection we observed is accounted for by only 10 of the 14 SCHUS4 epitopes in the vaccine, as they were conserved in the LVS challenge strain. The SCHUS4-specific epitopes that were found to be significantly immunogenic would further contribute to protection in a SCHUS4 challenge. This result is consistent with other findings that a limited set of epitopes may be sufficient to induce a protective immune response (Moutaftsi et al. 2006). Studies using SCHUS4 (the wild type tularemia) are planned. The results obtained to date appear to indicate that vaccine design that originates with the whole genome may lead to the development of a protective epitope-based vaccine.

### 6.1.2 Therapeutic HPV Vaccine

Cervical cancer is the second leading cause of death afflicting women worldwide; 40% of cervical patients develop persistent, recurrent, or widely metastatic disease. While a preventive vaccine now exists for HPV, there is a need for a therapeutic vaccine to treat existing cases of HPV, especially in resource-poor areas where access to the preventive vaccine is limited.

Cellular immune responses are believed to be critical for effective immune response to cancer; accordingly, EpiVax is pursuing the development of an 'immunotherapeutic vaccine' which would focus on the proteins primarily expressed either before or during carcinogenesis (E1 and E2, E6, and E7). In order to maximize immunogenicity across HPV subtypes, our strategy involves analyzing variant strains of HPV protein sequences, using (1) EpiMatrix to identify Class I and Class II HLA motif matches, (2) Conservatrix to identify those motif matches that are conserved, and (3) Epi-Assembler to weave together the conserved immunogenic sequences into a full immunogenic consensus sequence (ICS) antigen. A full ICS vaccine antigen would retain the fundamental structure of its naturally occurring counterparts; however, it will contain more and better epitopes than would occur in any such one counterpart. The authors have identified five conserved epitopes in E1 and E2, which

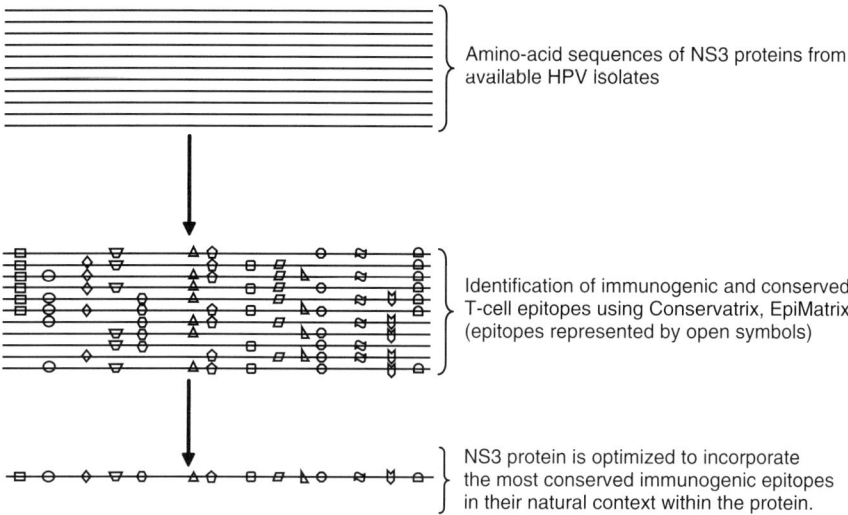

Amino-acid sequences of NS3 proteins from available HPV isolates

Identification of immunogenic and conserved T-cell epitopes using Conservatrix, EpiMatrix. (epitopes represented by open symbols)

NS3 protein is optimized to incorporate the most conserved immunogenic epitopes in their natural context within the protein.

**Fig. 5** Illustration of our novel strategy to generate an HPV vaccine candidate. The putative epitopes identified during this analysis will be combined to form several consensus sequences, which retain the fundamental structure of these HPV proteins but which also contain an unnaturally large number of conserved T cell epitopes

have stimulated significant responses in ELISpot IFN-g assays. In the proposed vaccine, these epitopes and others will be incorporated in their natural context within the proteins, which could be delivered as DNA, proteins, or a prime-boost combination. By preserving the natural flanking regions surrounding our epitopes, we hope to retain, in large part, the natural processes surrounding HPV protein degradation, transport, and presentation as they occur during natural infections. The ICS approach described here is the same as that illustrated in Fig. 5, but extending over the full natural length of the protein. For a more in-depth review of a similar approach being pursued for influenza, see McMurry et al. (2008).

The EpiVax HPV vaccine illustrates yet another aspect of vaccine design: 'megatope' proteins, re-engineered to increase the epitope content. This approach already had some success (Okazaki et al. 2006).

# 7 Advantages and Disadvantages of IDV

In the case of variable viruses such as HCV, influenza, and HIV, one limitation of conventional vaccination, and of natural infection, is that the immune system often focuses strongly on the most mutable immunogens. IDVs can be constructed from alternative antigens, which are more conserved or more protective, circumventing this problem (Russell and Liew 1979; Scherle and Gerhard 1986; Scherle and Gerhard 1988; Santra et al. 2002; Subbramanian et al. 2003).

In addition, broadening the T cell repertoire might make it possible to impair viral escape and decrease viral loads sufficiently to disrupt transmission.

Epitope-driven vaccines also offer distinct advantages over vaccines encoding whole protein antigens, since epitopes are safe and can be packaged into relatively small delivery vehicles. The epitope-driven approach offers platform independence: a delivery vehicle (peptide, DNA, multi-epitope construct) can be modified or selected midway into the development process. Multiple conserved epitopes, in addition to augmenting the efficacy of a preventive vaccine, could provide a broad and universal cellular immunity, known to be crucial for containment of infection, although perhaps ineffective for protection against infection.

Despite these advantages, there are a number of reasons that a given pathogen-directed, epitope-based vaccine might fail to reach clinical trials or protect humans: (1) the limited number of epitopes expressed by the vaccine (i.e., poor payload quantity); (2) limited conservation of epitopes (leading to limited coverage of variant clinical isolates) (3) the limited HLA population coverage (i.e., poor payload quality); (4) suboptimal vaccine delivery; and/or (5) the dearth of suitable animal models.

In addition, the concept of epitope-driven vaccines is relatively novel. Complete genome sequences have been available for only a little more than a decade now and the tools to process the data for vaccine design are only newer. Experimental validation needed to push forward these vaccines into clinical trials is now emerging and promises to enable epitope-based vaccines to claim a prominent place in the vaccine world.

## 8 Future Research

The technologies needed to identify immunostimulatory antigens and epitopes from pathogen genomes are already well developed. The principle focus of future research in this area will likely be in fine-tuning these technologies and expanding them to tailor immune responses in individuals. For example, development of epitope mapping algorithms for DQ and DP Class II HLA alleles will make it possible to completely characterize immunomes. This information will make it possible to generate comprehensive individual T cell epitope measures (iTEM) based on an individual's HLA genetic make-up and allow researchers to identify a priori clinically important epitopes and screen clinical cohorts for subjects that are more likely to develop targeted immune responses.

Furthermore, genome-mapping tools that are currently available are not yet useful for discovering B cell epitopes, whether from proteins or from non-protein components such as carbohydrates or lipid antigens. Immunoinformatics tools that are currently available cannot be used to accurately predict conformational (B cell) epitopes that interact with antibody, although such tools are being refined (Enshell-Seijffers et al. 2003). Thus, the immunogens identified using in silico approaches must be evaluated in vitro and also in appropriate challenge models, prior to progressing to vaccine trials.

Protective immune response probably also involves some engagement of the innate immune system; it has been impossible to differentiate between effective and non-protective epitopes. Cytokine milieu may affect the outcome of immunization; thus, a limited number of toll-receptor agonists (Imler and Hoffmann 2001) have been identified and these are under study in conjunction with IDVs. In the future, toll-receptor signaling 'pathogen-associated molecular patterns' (PAMPS) might also be modeled and selected using immunoinformatics tools.

Besides antigen identification, the success of IDVs relies heavily on delivery technologies. These areas continue to independently mature and provide important lessons to epitope-based vaccine design. The major areas of research to watch include biological macromolecule (including cytokines), lipopeptide, and polysaccharide adjuvants and particulate (liposomes, exosomes, virosomes, nanoparticles) and cell-based delivery systems.

# 9 Conclusion

The development of safe and effective vaccines against emerging infectious diseases such as influenza, both seasonal and pandemic, HIV, and TB, in addition to cancers associated with infectious pathogens such as HBV and HCV, is an urgent and achievable public health priority. In addition, vaccines for the prevention and treatment of cancer hold enormous promise for human health.

The threat of bioterrorism following the events of September 11, 2001, provided vaccinologists with a persuasive argument for more rapid development of vaccines against viral and bacterial pathogens that are now included on the NIH Category A-C Biopathogen list (http://www3.niaid.nih.gov/topics/BiodefenseRelated/Biodefense/research/CatA.htm). 'Emerging infectious diseases' were added to the vaccine wish list following the outbreak of Severe Acute Respiratory Syndrome (SARS) in Guangdong China in 2002. Indeed, only a few months following the publication of the SARS-Coronavirus (SARS-CoV) genome (Marra et al. 2003; Rota et al. 2003), researchers began to map vaccine components using new bioinformatics and immunoinformatics tools, coupled with improved immunology techniques and specialized animal models. New vaccines based on this approach are currently being evaluated in animal models, less than a year from the start of the epidemic.

Future vaccine approaches may need to move away from 'whole' protein vaccines for a wide range of reasons. Multiple antigen or epitope vaccinations such as the approach illustrated here could be one way to elicit the sort of strong TH1 response necessary to pathogens following infection, in the context of a therapeutic vaccine. This approach could also be useful for a wide range of pathogens for which genomes have been partially or completely mapped. As described in this chapter, our group is actively pursuing the development of epitope-driven vaccines for HIV (De Groot et al. 2005; Koita et al. 2006), *Franciscella tularensis*, *Helicobacter pylori*, and smallpox. We have progressed from genome-derived epitope mapping to challenge studies in less than one year for some of these vaccine development programs.

Epitope-based and whole antigen IDVs are now just beginning to enter clinical trials, but this relative disadvantage may be cured with the tincture of time. One reason for the relative paucity of IDVs in clinical development is that the immunoinformatics tools for developing these vaccines have really only evolved in the last 10 to 15 years. The average length of time to develop a vaccine may be 20 years or more. While immunoinformatics tools are useful for accelerating the discovery and pre-clinical stage of vaccine development, testing vaccines in animal models and developing clinical trials is a lengthy process. It is likely that IDV and epitope-based IDV will begin to enter clinical trials and emerge on the market in greater numbers in 5 to 10 years.

# References

Ahlers, J. D., Belyakov, I. M., Thomas, E. K., and Berzofsky, J. A. 2001. High-affinity T helper epitope induces complementary helper and APC polarization, increased CTL, and protection against viral infection. J Clin Invest. 108:1677–1685.

An, L. L., Rodriguez, F., Harkins, S., Zhang, J., and Whitton, J. L. 2000. Quantitative and qualitative analyses of the immune responses induced by a multivalent minigene DNA vaccine. Vaccine. 18:2132–2141.

An, L. L., Whitton, J. L. 1997. A multivalent minigene vaccine, containing B-cell, cytotoxic T-lymphocyte, and Th epitopes from several microbes, induces appropriate responses in vivo and confers protection against more than one pathogen. J Virol. 71:2292–2302.

Asjö, B., Stavang, H., Sørensen, B., Baksaas, I., Nyhus, J., and Langeland, N. 2002. Phase I trial of a therapeutic HIV type 1 vaccine, Vacc-4x, in HIV type 1-infected individuals with or without antiretroviral therapy. AIDS Res Hum Retroviruses. 18:1357–1365.

Barnea, E., Beer, I., Patoka, R., Ziv, T., Kessler, O., Tzehoval, E., Eisenbach, L., Zavazava, N., and Admon, A. 2002. Analysis of endogenous peptides bound by soluble MHC class I molecules: a novel approach for identifying tumor-specific antigens. Eur J Immunol. 32:213–222.

Bergmann, C. C., Tong, L., Cua, R., Sensintaffer, J., and Stohlman, S. 1994. Differential effects of flanking residues on presentation of epitopes from chimeric peptides. J Virol. 68:5306–5310.

Betts, M. R., Ambrozak, D. R., Douek, D. C., Bonhoeffer, S., Brenchley, J.M., Casazza, J.P., Koup, R. A., and Picker, L. J. 2001. Analysis of total human immunodeficiency virus (HIV)-specific CD4(+) and CD8(+) T cell responses: relationship to viral load in untreated HIV infection. J Virol. 75:11983–11991.

Bond, K. B., Sriwanthana, B., Hodge, T. W., De Groot, A. S., Mastro, T. D., Young, N. L., Promadej, N., Altman, J. D., Limpakarnjanarat, K., and McNicholl, J. M. 2001. An HLA-directed molecular and bioinformatics approach identifies new HLA-A11 HIV-1 subtype E cytotoxic T lymphocyte epitopes in HIV-1-infected Thais. AIDS Res Hum Retroviruses. 20:703–717.

Charo, J., Sundback M, Geluk, A., Ottenhoff, T., and Kiessling, R.. 2001. DNA immunization of HLA transgenic mice with a plasmid expressing mycobacterial heat shock protein 65 results in HLA class I- and II-restricted T cell responses that can be augmented by cytokines. Hum Gene Ther. 12:1797–1804.

Chen, W., Norbury, C. C., Cho, Y., Yewdell, J. W., and Bennink, J. R. 2001. Immuno-proteasomes shape immunodominance hierarchies of antiviral CD8( + ) T cells at the levels of T cell repertoire and presentation of viral antigens. J Exp Med. 193:1319–1326.

Crowe, S. R., Miller, S. C., Brown, D. M., Adams, P. S., Dutton, R. W., Harmsen, A. G., Lund, F. E., Randall, T. D., Swain, S.L., and Woodland, D. L. 2006. Uneven distribution of MHC class II epitopes within the influenza virus. Vaccine. 24:457–467.

De Groot, A. S. 2006. Immunomics: Discovering New Targets for Vaccine and Therapeutics. Drug Discov Today. 11:203–209.

De Groot, A. S. and Berzofsky, J. A. 2004. From Genome to Vaccine – New Immunoinfor-matics tools for vaccine design. Methods. 34:425–428.

De Groot, A. S., and Martin, W. 2003. From immunome to vaccine: epitope mapping and vaccine design tools. Novartis Found Symp. 254:57–72.

De Groot, A. S., Bishop, E., Khan, B., Lally, M., Marcon, L., Franco, J., Mayer, K., Carpenter, C., and Martin, W. 2004. Engineering immunogenic consensus T helper epitopes for a cross-clade HIV vaccine. Methods. 34:476–487.

De Groot, A. S., Bosma, A., Chinai, N., Frost, J., Jesdale, B. M., Gonzalez, B. M., Martin, W., and Saint-Aubin, C. 2001. From genome to vaccine: in silico predictions, ex vivo verifica-tion. Vaccine. 19:4385–4395.

De Groot, A. S., Clerici, M., Hosmalin, C. M., Hughes, A., Barnd, D., Hendrix, C. W., Houghten, R., Shearer, G. M. and Berzofsky, J. A. 1991. Human Immunodeficiency virus reverse transcriptase T helper epitopes identified in mice and humans: correlation with a cytotoxic T cell epitope. J Infect Dis. 164:1058–1065.

De Groot, A. S., Jesdale, B. M., Szu, E., Schafer, J. R., Chicz, R. M., and Deocampo, G. 1997. An interactive Web site providing major histocompatibility ligand predictions: application to HIV research. AIDS Res Hum Retroviruses. 13:529–531.

De Groot, A. S., Marcon, L., Bishop, E.A., Rivera, D., Kutzler, M., Weiner, D. B., and Martin, W. 2005. HIV vaccine development by computer assisted design: the GAIA vaccine. Vaccine. 23:2136–2148.

De Groot, A. S., Sbai, H., Aubin, C. S., McMurry, J., and Martin, W. 2002. Immuno-informatics: Mining genomes for vaccine components. Immunol Cell Biol. 80:225–269.

Depla, E., Van der Aa, A.,, Livingston, B. D., Crimi, C., Allosery, K., De Brabandere, V., Krakover, J., Murthy, S., Huang, M., Power, S., Babé, L., Dahlberg, C., McKinney, D., Sette, A., Southwood, S., Philip, R., Newman, M. J., and Meheus, L. 2008. Rational design of a multiepitope vaccine encoding T-lymphocyte epitopes for treatment of chronic hepatitis B virus infections. J Virol. 82:435–50.

Dong, Y., Demaria, S., Sun, X., Santori, F. R., Jesdale, B. M., De Groot, A. S., Rom, W. N., and Bushkin, Y. 2004. HLA-A2-restricted CD8 + -cytotoxic- T cell responses to novel epitopes in Mycobacterium tuberculosis superoxide dismutase, alanine dehydrogenase, and glutamine synthetase. Infect Immun. 72: 2412–2415.

Doytchinova, I. A., Taylor, P., and Flower, D. R. 2003. Proteomics in Vaccinology and Immunobiology: An Informatics Perspective of the Immunone. J Biomed Biotechnol. 2003:267–290.

Elliott, S. L., Suhrbier, A., Miles, J. J., Lawrence, G., Pye, S. J., Le, T. T., Rosenstengel, A., Nguyen, T., Allworth, A., Burrows, S. R., Cox, J., Pye, D., Moss, D. J., and Bharadwaj M. A. 2008. Phase I trial of a CD8 + T Cell Peptide Epitope-based Vaccine for Infectious Mono-nucleosis. J Virol. 82:1448–1457.

Enshell-Seijffers, D., Denisov, D., Groisman, B., Smelyanski, L., Meyuhas, R., Gross, G., Denisova, G., and Gershoni, J. M. 2003. The mapping and reconstitution of a conforma-tional discontinuous B-cell epitope of HIV-1. J Mol Biol. 334:87–101.

Falk, K., Rotzschke, O., Stevanovic, S., Jung, J., and Rammensee, H. G. 1991. Allele-specific motifs revealed by sequencing of self-peptides eluted from MHC molecules. Nature. 351:290–296.

Frahm, N., Korber, B. T., Adams, C. M., Szinger, J. J., Draenert, R., Addo, M. M., Feeney, M. E., Yusim, K., Sango, K., Brown, N. V., SenGupta, D., Piechocka-Trocha, A., Simonis, T., Marincola, F. M., Wurcel, A. G., Stone, D. R., Russell, C. J., Adolf, P., Cohen, D., Roach, T., StJohn, A., Khatri, A., Davis, K., Mullins, J., Goulder, P. J., Walker, B. D., and Brander, C. 2004. Consistent cytotoxic-T-lymphocyte targeting of immunodominant regions in human immunodeficiency virus across multiple ethnicities. J Virol. 78:2187–2200.

Franke, E. D., Sette, A., Sacci, J. Jr., Southwood, S., Corradin, G., and Hoffman, S. L. 2000. A subdominant CD8(+) cytotoxic T lymphocyte (CTL) epitope from the Plasmodium yoelii circumsporozoite protein induces CTLs that eliminate infected hepatocytes from culture. Infect Immun. 68:3403–3411.

Fraser, C. K., Diener, K. R., Brown, M. P., and Hayball, J. D. 2007. Improving vaccines by incorporating immunological coadjuvants. Expert Rev Vaccines. 6:559–578.

Gahery, H., Daniel, N., Charmeteau, B., Ourth, L., Jackson, A., Andrieu, M., Choppin, J., Salmon, D., Pialoux, G., and Guillet, J. G. 2006. New CD4+ and CD8+ T cell responses induced in chronically HIV type-1-infected patients after immunizations with an HIV type 1 lipopeptide vaccine. AIDS Res Hum Retroviruses. 22:684–94.

Ghosh, S., Pal, S., Das, S., Dasgupta, S. K., and Majumdar, S. 1998. Lipoarabinomannan induced cytotoxic effects in human mononuclear cells. FEMS Immunol Med Microbiol. 21:181–188.

Gianfrani, C., Oseroff, C., Sidney, J., Chesnut, R. W., and Sette, A. 2000. Human memory CTL response specific for influenza A virus is broad and multispecific. Hum Immunol. 61:438–452.

Godkin, A. J., Smith, K. J., Willis, A., Tejada-Simon, M. V., Zhang, J., Elliott, T., and Hill, A. V. 2001. Naturally processed HLA class II peptides reveal highly conserved immunogenic flanking region sequence preferences that reflect antigen processing rather than peptide-MHC interactions. J Immunol. 166:6720–6727.

Graham, C. M., Barnett, B. C., Hartlmayr, I., Burt, D. S., Faulkes, R., Skehel, J. J., and Thomas , D. B. 1989. The structural requirements for class II (I-Ad)-restricted T cell recognition of influenza hemaglglutinin: B cell epitopes define T cell epitopes. Eur J Immunol. 19:523.

Gulukota, K. and DeLisi, C. 1996. HLA allele selection for designing peptide vaccines. Genet Anal. 13:81–86.

Hernychova, L., Stulik, J., Halada, P., Macela, A., Kroca, M., Johansson, T., and Malina, M. 2001. Construction of a Francisella tularensis two-dimensional electrophoresis protein database. Proteomics. 1:508–515.

Hozhutter, H. G., Frommel, C., and Kloetzel., P. M. 1999. A theoretical approach towards the identification of cleavage determining amino acid motifs of the 20S proteasome. J. Mol Biol. 286:1251.

Imler, J. L., and Hoffmann, J. A. 2001. Toll receptors in innate immunity. Trends Cell Biol. 11:304–311.

Ishioka, G. Y., Fikes, J., Hermanson, G., Livingston, B., Crimi, C., Qin, M., del Guercio, M. F., Oseroff, C., Dahlberg, C., Alexander, J., Chesnut, R. W., and Sette, A.. 1999. Utilization of MHC class I transgenic mice for development of minigene DNA vaccines encoding multiple HLA-restricted CTL epitopes. J. Immunol. 162:3915–3925.

Itoh K. and Yamada, A. 2006. Personalized peptide vaccines: a new therapeutic modality for cancer.Cancer Sci. 97:970–976.

Iversen, A. K., Stewart-Jones, G., Learn, G. H., Christie, N., Sylvester-Hviid, C., Armitage, A. E., Kaul, R., Beattie, T., Lee, J. K., Li, Y., Chotiyarnwong, P., Dong, T., Xu, X., Luscher, M. A., MacDonald, K., Ullum, H., Klarlund-Pedersen, B., Skinhoj, P., Fugger, L., Buus, S., Mullins, J. I., Jones, E. Y., van der Merwe, P. A., and McMichael, A. J. 2006. Conflicting selective forces affect T cell receptor contacts in an immunodominant human immunodeficiency virus epitope. Nat Immunol. 7:179–189.

Jameson, J., Cruz, J., and Ennis, F.A. 1998. Human cytotoxic T-lymphocyte repertoire to influenza A viruses. J Virol. 72:8682–8689.

Johansson, B. E., Moran, T. M., and Kilbourne, E. D. 1987. Antigen-presenting B cells and helper T cells cooperatively mediate intravirionic antigenic competition between influenza A virus surface glycoproteins. Proc Natl Acad Sci U S A. 84:6869–6873.

Kaufmann, S. H., Gulle, H., Daugelat, S., and Schoel, B. 1992. Tuberculosis and leprosy: attempts to identify T-cell antigens of potential value for vaccine design. Scand J Immunol Suppl. 11:85–90.

Klitgaard, J. L., Coljee, V. W., Andersen, P. S., Rasmussen, L. K., Nielsen, L. S., Haurum, J. S., and Bregenholt S. 2006. Reduced susceptibility of recombinant polyclonal antibodies to inhibitory anti-variable domain antibody responses. J Immunol. 177:3782–3790.

Koibuchi, T., Allen, T. M., Lichterfeld, M., Mui, S. K., O'Sullivan, K. M., Trocha, A., Kalams, S. A., Johnson, R.P., and Walker, B. D. 2005. Limited sequence evolution within persistently targeted CD8 epitopes in chronic human immunodeficiency virus type 1 infection. J Virol. 79:8171–8181.

Koita, O. A., Dabitao, D., Mahamadou, I., Tall, M., Dao, S., Tounkara, A., Guiteye, H., Noumsi, C., Thiero, O., Kone, M., Rivera, D., McMurry, J. A., Martin, W., and De Groot, A. S. 2006. Confirmation of immunogenic consensus sequence HIV-1 T cell epitopes in Bamako, Mali and Providence, Rhode Island. Hum Vaccin. 2:119–128.

Koren, E., De Groot, A. S., Jawa, V., Beck, K. D., Boone, T., Rivera, D., Li, L., Mytych, D., Koscec, M., Weeraratne, D., Swanson, S., and Martin W. 2007. Clinical validation of the "in silico" prediction of immunogenicity of a human recombinant therapeutic protein. Clin Immunol. 124:26–32.

Kran, A. M., Sørensen, B., Nyhus, J., Sommerfelt, M. A., Baksaas, I., Bruun, J.N., and Kvale, D. 2004. HLA- and dose-dependent immunogenicity of a peptide-based HIV-1 immunotherapy candidate (Vacc-4x). AIDS. 18:1875–1883.

Krieg, A. M., Yi, A. K., Schorr, J., and Davis, H. L. 1998. The role of CpG dinucleotides in DNA vaccines. Trends Microbiol. 6:23–27.

Kuchroo, V. K., Das, M. P., Brown, J. A., Ranger, A. M., Zamvil, S. S., Sobel, R. A., Weiner, H. L., Nabavi, N., and Glimcher, L. H. 1995. B7-1 and B7-2 costimulatory molecule activate differentially the TH1/TH2 developmental pathways: application to autoimmune disease therapy. Cell. 80: 707–718.

Kulkarni-Kale, U., Bhosle, S., and Kolaskar, A. S. 2005. CEP: a conformational epitope prediction server. Nucleic Acids Res. 33:W168–171.

Larsson, P., Oyston, P. C., Chain, P., Chu, M. C., Duffield, M., Fuxelius, H. H., Garcia, E., Halltorp, G., Johansson, D., Isherwood, K. E., Karp, P. D., Larsson, E., Liu, Y., Michell, S., Prior, J., Prior, R., Malfatti, S., Sjostedt, A., Svensson, K., Thompson, N., Vergez, L., Wagg, J. K., Wren, B. W., Lindler, L. E., Andersson, S. G., Forsman, M., and Titball, R. 2005. W. The complete genome sequence of *Francisella tularensis*, the causative agent of tularemia. Nat Genet 37:153–159.

Lazarski, C. A., Chaves, F. A., Jenks, S. A., Wu, S., Richards, K. A., Weaver, J. M., and Sant, A. J. 2005. The kinetic stability of MHC class II peptide complexes is a key parameter that dictates immunodominance. Immunity. 23:29–40.

Leen, A. M., Christin, A., Khalil, M., Weiss, H., Gee, A. P., Brenner, M. K., Heslop, H. E., Rooney, C. M., and Bollard, C. M. 2008. Identification of hexon-specific CD4 and CD8 T-cell epitopes for vaccine and monotherapie. J Virol. 82:546–554.

Livingston, B. D., Newman, M., Crimi, C., McKinney, D., Chesnut, R., and Sette, A. 2001. Optimization of epitope processing enhances immunogenicity of multiepitope DNA vaccines. Vaccine. 19: 4652–4660.

Man, S., Newberg, M. H., Crotzer, V. L. Luckey, C. J., Williams, N. S., Chen, Y., Huczko, E. L. Ridge, J. P., and Engelhard V. H. 1995. Definition of a human T-cell epitope from influenza A non-structural protein 1 using HLA-A2.1 transgenic mice. Int. Immunol. 7:597–605.

Marra, M. A., Jones, S. J., Astell, C. R., Holt, R. A., Brooks-Wilson, A., Butterfield, Y. S., Khattra, J., Asano, J. K., Barber, S. A., Chan, S. Y., Cloutier, A., Coughlin, S. M., Freeman, D., Girn, N., Griffith, O. L., Leach, S. R., Mayo, M., McDonald, H., Montgomery, S. B., Pandoh, P. K., Petrescu, A. S., Robertson, A. G., Schein, J. E., Siddiqui, A.,

Smailus, D. E., Stott, J. M., Yang, G. S., Plummer, F., Andonov, A., Artsob, H., Bastien, N., Bernard, K., Booth, T. F., Bowness, D., Czub, M., Drebot, M., Fernando, L., Flick, R., Garbutt, M., Gray, M., Grolla, A., Jones, S., Feldmann, H., Meyers, A., Kabani, A., Li, Y., Normand, S., Stroher, U., Tipples, G. A., Tyler, S., Vogrig, R., Ward, D., Watson, B., Brunham, R. C., Krajden, M., Petric, M., Skowronski, D. M., Upton, C., Roper, R. L. 2003. The Genome Sequence of the SARS-Associated Coronavirus. Science 300:1399–1404.

Mathiassen, S., Lauemoller, S. L., Ruhwald, M., Claesson, M. H., and Buus, S. 2001. Tumor-associated antigens identified by mRNA expression profiling induce protective anti-tumor immunity. Eur J Immunol. 31:1239–1246.

McMurry, J. A., Gregory, S. H., Moise, L., Rivera, D., Buus, S., and De Groot, A. S. 2007a. Diversity of Francisella tularensis Schu4 antigens recognized by T lymphocytes after natural infections in humans: Identification of candidate epitopes for inclusion in a rationally designed tularemia vaccine. Vaccine. 25:3179–3191.

McMurry, J.A., Johansson, B.E., and De Groot, A.S. 2008. A call to cellular and humoral arms: enlisting cognate T cell help to develop broad-spectrum vaccines against influenza A. Hum Vaccin. 4(2):148–157.

McMurry, J. A., Moise, L., Gregory, S. H., and De Groot, A. S. 2007b. Tularemia vaccines – an overview. Med Health R I. 90:311–314.

McMurry, J., Sbai, H., Gennaro, M. L., Carter, E. J., Martin, W., and De Groot, A. S. 2005. Analyzing Mycobacterium tuberculosis proteomes for candidate vaccine epitopes. Tuberculosis (Edinb). 85:95–105.

Menne, S., Maschke, J., Tolle, T. K., Lu, M., and Roggendorf, M. Characterization of T-cell response to woodchuck hepatitis virus core protein and protection of woodchucks from infection by immunization with peptides containing a T-cell epitope. 1997. J Virol. 71:65–74.

Mo, X. Y., Cascio, P., Lemerise, K., Goldberg, A. L., and Rock, K. 1999. Distinct proteolytic processes generate the C and N termini of MHC class I-binding peptides. J Immunol 163:5851–5859.

Moutaftsi, M., Peters, B., Pasquetto, V., Tscharke, D. C., Sidney, J., Bui. H. H., Grey, H., and Sette, A. 2006. A consensus epitope prediction approach identifies the breadth of murine T(CD8 + )-cell responses to vaccinia virus. Nat Biotechnol. 24:817–819.

Nara, P. L. and Lin, G. 2005. HIV-1: the confounding variables of virus neutralization. Curr Drug Targets Infect Disord. 5:157–170.

Nardin, E. H., Oliveira, G. A., Calvo-Calle, J. M., Castro, Z. R., Nussenzweig, R. S., Schmeckpeper, B., Hall, B. F., Diggs, C., Bodison, S., and Edelman, R. 2000. Synthetic malaria peptide vaccine elicits high levels of antibodies in vaccines of defined HLA genotypes. J Infect Dis. 182:1486–1496.

Nguyen, L., Chaowanachan, T., Vanichseni, S., McNicholl, J. M., Mock, P. A., Nelson, R., Hodge, T. W., van Griensven, F., Choopanya, K., Mastro, T. D., Tappero, J. W., and Hu, D. J. 2004. Frequent human leukocyte antigen class I alleles are associated with higher viral load among HIV type 1 seroconverters in Thailand. J Acquir Immune Defic Syndr. 37:1318–1323.

Okazaki, T., Pendleton, C. D., Sarobe, P., Thomas, E. K., Iyengar, S., Harro, C., Schwartz, D., Berzofsky, J. A. 2006. Epitope enhancement of a CD4 HIV epitope toward the development of the next generation HIV vaccine. J Immunol. 176:3753–3759.

Olsen, A. W., Hansen, P. R., Holm, A., and Andersen, P. 2000. Efficient protection against Mycobacterium tuberculosis by vaccination with a single subdominant epitope from the ESAT-6 antigen. Eur J Immunol. 30:1724–1732.

Ostrowski, M., Galeota, J. A., Jar, A. M., Platt, K. B., Osorio, F. A., and Lopez, O. J. 2002. Identification of neutralizing and nonneutralizing epitopes in the porcine reproductive and respiratory syndrome virus GP5 ectodomain. J Virol. 76:4241–4250.

Paina-Bordignon, P., Tan, A., Termijtelen, A., Demotz, S., Corradin G., and Lanzavecchia, A. 1989. Universally immunogenic T cell epitopes: promiscuous recognition by T cells. Eur J Immunol. 19:2237–2242.

Pederson, T. 1999. The immunome. Mol Immunol. 36:1127–1128.

Petrovsky, N., and Brusic, V. 2002. Computational immunology: The coming of age. Immunol Cell Biology. 80:248–254.

Peyerl, F. W., Bazick, H. S., Newberg, M. H., Barouch, D. H., Sodroski, J., and Letvin, N. L. 2004. Fitness costs limit viral escape from cytotoxic T lymphocytes at a structurally constrained epitope. J. Virol. 78:13901–13910.

Pietersz, G. A., Pouniotis, D.S., and Apostolopoulos, V. 2006. Design of peptide-based vaccines for cancer. Curr Med Chem. 13:1591–1607.

Pizza, M., Scarlato, V., Masignani, V., Giuliani, M. M., Aricò, B., Comanducci, M., Jennings, G. T., Baldi, L., Bartolini, E., Capecchi, B., Galeotti, C. L., Luzzi, E., Manetti, R., Marchetti, E., Mora, M., Nuti, S., Ratti, G., Santini, L., Savino, S., Scarselli, M., Storni, E., Zuo, P., Broeker, M., Hundt, E., Knapp, B., Blair, E., Mason, T., Tettelin, H., Hood, D. W., Jeffries, A. C., Saunders, N. J., Granoff, D. M., Venter, J. C., Moxon, E. R., Grandi, G., and Rappuoli, R. 2000. Identification of vaccine candidates against serogroup B meningococcus by whole-genome sequencing. Science. 287:1816–1820.

Plotnicky, H., Cyblat-Chanal, D., Aubry, J. P., Derouet, F., Klinguer-Hamour, C., Beck, A., Bonnefoy, J. Y., and Corvaia, N. 2003. The immunodominant influenza matrix T cell epitope recognized in human induces influenza protection in HLA-A2/K(b) transgenic mice. Virology. 309:320–329.

Rajnavolgyi, E., Nagy, N., Thuresson, B., Dosztanyi, Z., Simon, A., Simon, I., Karr, R. W., Ernberg, I., Klein, E., and Falk, K.I. 1999. A repetitive sequence of Ebstein-Barr virus nuclear antigen 6 comprises overlapping T cell epitopes which induce HLA-DR restricted CD4+ T lymphocytes. Int Immunol. 12:281–293.

Rappuoli, R. and Covacci, A. 2003. Reverse vaccinology and genomics. Science. 302:602.

Rodriguez, F., Whitton, J. L. 2000. Enhancing DNA immunization. Virology. 268:233–238.

Rota, P. A., Oberste, M. S., Monroe, S. S., Nix, W. A., Campagnoli, R., Icenogle, J. P., Peñaranda, S., Bankamp, B., Maher, K., Chen, M. H., Tong, S., Tamin, A,. Lowe, L., Frace, M., DeRisi, J. L., Chen, Q., Wang, D., Erdman, D. D., Peret, T. C., Burns, C., Ksiazek, T. G., Rollin, P. E., Sanchez, A., Liffick, S., Holloway, B., Limor, J., McCaustland, K., Olsen-Rasmussen, M., Fouchier, R., Günther, S., Osterhaus, A. D., Drosten, C., Pallansch, M. A., Anderson, L. J., Bellini W. J. 2003. Characterization of a Novel Coronavirus Associated with Severe Acute Respiratory Syndrome. Science. 300:1394–1399.

Rotzschke, O., Falk, K., Stevanovic, S., Jung, J., Walden, P., and Rammensee, H. G.. 1991 Exact prediction of natural T cell epitope. Eur J Immunol. 21:2891–2894.

Russell, S. M. and Liew, F. Y. 1979. T cells primed by influenza virion internal components can cooperate in the antibody response to haemagglutinin. Nature. 280:147–148.

Santra, S., Barouch, D. H., Kuroda, M. J., Schmitz, J. E., Krivulka, G. R., Beaudry, K., Lord, C. I., Lifton, M. A., Wyatt, L. S., Moss, B., Hirsch, V. M., and Letvin, N. L. 2002. Prior vaccination increases the epitopic breadth of the cytotoxic T-lymphocyte response that evolves in rhesus monkeys following a simian-human immunodeficiency virus infection. J Virol. 76:6376–6381.

Schadeck, E. B., Partidos, C. D., Fooks, A. R., Obeid, O. E., Wilkinson, G. W., Stephenson, J. R., and Steward, M. W. 1999. CTL epitopes identified with a defective recombinant adenovirus expressing measles virus nucleoprotein and evaluation of their protective capacity in mice. Virus Res. 65:75–86.

Scherle P. A. and Gerhard, W. 1986. Functional analysis of influenza-specific helper T cell clones in vivo. T cells specific for internal viral proteins provide cognate help for B cell responses to hemagglutinin. J Exp Med. 164:1114–1128.

Scherle, P. A. and Gerhard, W. 1988. Differential ability of B cells specific for external vs. internal influenza virus proteins to respond to help from influenza virus-specific T-cell clones in vivo. Proc Natl Acad Sci U S A. 85:4446–4450.

Schreiber, A., Humbert, M., Benz, A., and Dietrich, U. 2005. 3D-Epitope-Explorer (3DEX): localization of conformational epitopes within three-dimensional structures of proteins. J Comput Chem. 26:879–887.

Sepkowitz, K. A. 2001. Tuberculosis control in the 21st century. Emerg Infect Dis. 7:259–262.

Sette, A. and Sidney, J. 1998. HLA supertypes and supermotifs: a functional perspective on HLA polymorphism. Curr Opin Immunol. 10:478–482.

Shahinian, A., Pfeffer, K., Lee, K. P., Kundig, T. M., Kishihara, T. M., Kishihara, A., Wakeham, A., Kawai, K., Ohashi, P. S., Thompson, C. B. and Mak, T. W. 1993. Differential T cell costimulatory requirements in CD28-deficient mice. Science. 261:609–612.

Shastri, N., Serwold, T., and Gonzalez, F. 1995. Presentation of endogenous peptide/MHC class I complexes is profoundly influenced by specific C-terminal flanking residues. J Immunol. 155:4339–4346.

Shirai, M., Arichi, T., Nishioka, M., Nomura, T., Ikeda, K., Kawanishi, K., Engelhard, V. H., Feinstone, S. M., and Berzofsky, J. A. 1995. CTL responses of HLA-A2.1-transgenic mice specific for hepatitis C viral peptides predict epitopes for CTL of humans carrying HLA-A2.1. J. Immunol. 154:2733–2742.

Sonnenberg, M. G., and Belisle, J. T. 1997. Definition of Mycobacterium tuberculosis culture filtrate proteins by two-dimensional polyacrylamide gel electrophoresis, N-terminal amino acid sequencing, and electrospray mass spectrometry. Infect Immun. 65:4515–4524.

Southwood, S., Sidney, J., Kondo, A., del Guercio, M, F., Appella, E., Hoffman, S., Kubo, R. T., Chesnut, R. W., Grey, H. M., and Sette, A. 1998. Several common HLA-DR types share largely overlapping peptide binding repertoires. J Immunol. 160:3363–3373.

Steere, A. C., Klitz, W., Drouin, E. E., Falk, B. A., Kwok, W. W., Nepom, G. T., and Baxter-Lowe, L. A. 2006. Antibiotic-refractory Lyme arthritis is associated with HLA-DR molecules that bind a Borrelia burgdorferi peptide. J Exp Med. 203:961–971.

Subbramanian, R. A., Kuroda, M. J., Charini, W. A., Barouch, D. H., Costantino, C., Santra, S., Schmitz, J. E., Martin, K. L., Lifton, M. A., Gorgone, D. A., Shiver, J. W., and Letvin, N. L. 2003. Magnitude and diversity of cytotoxic-T-lymphocyte responses elicited by multiepitope DNA vaccination in rhesus monkeys. J Virol. 77:10113–10118.

Thomson, S. A., Burrows, S. R,, Misko, I. S., Moss, D. J., Coupar, B. E., and Khanna, R. 1998. Targeting a polyepitope protein incorporating multiple class II-restricted viral epitopes to the secretory/endocytic pathway facilitates immune recognition by CD4+ cytotoxic T lymphocytes: a novel approach to vaccine design. J Virol. 72:2246–2252.

Thomson, S. A., Khanna, R., Gardner, J., Burrows, S. R., Coupar, B., Moss, D. J. and Suhrbier, A. 1995. Minimal epitoeps expressedin arecombinant polyepitope protein are processed and presented to CD8+ T cells : implications for vaccine design. Proc Natl Acad Sci U S A. 92:5845.

Tobery, T. W., Dubey, S. A., Anderson, K., Freed, D. C., Cox, K. S., Lin, S., Prokop, M. T., Sykes K. J., Mogg, R., Mehrotra, D. V., Fu, T. M., Casimiro, D. R., and Shiver, J. W. 2006. A comparison of standard immunogenicity assays for monitoring HIV type 1 gag-specific T cell responses in Ad5 HIV Type 1 gag vaccinated human subjects. AIDS Res HumRetroviruses. 22:1081–1090.

Toes, R. E., Nussbaum, A. K., Degermann, S., Schirle, M., Emmerich, N. P., Kraft, M., Laplace, C., Zwinderman, A., Dick, T. P., Muller, J., Schonfisch, B., Schmid, C., Fehling, H. J., Stevanovic, S., Rammensee, H. G. and Schild, H. 2001. Discrete cleavage motifs of constitutive and immunoproteasomes revealed by quantitative analysis of cleavage products. J Exp Med. 194:1–12.

Tomlinson, A. J., Jameson, S., and Naylor, S. 1996. Strategy for isolating and sequencing biologically derived MHC class I peptides. J Chromatogr A. 744:273–278.

Ueda, Y., Itoh, T., Nukaya, I., Kawashima, I., Okugawa, K., Yano, Y., Yamamoto, Y., Naitoh, K., Shimizu, K., Imura, K., Fuji, N., Fujiwara, H., Ochiai, T., Itoi, H., Sonoyama, T., Hagiwara, A., Takesako, K., Yamagishi, H. 2004. Dendritic cell-based immunotherapy of cancer with carcinoembryonic antigen-derived, HLA-A24-restricted CTL epitope: Clinical outcomes of 18 patients with metastatic gastrointestinal or lung adenocarcinomas. int J Oncol. 24:909–917.

Valmori, D., Dutoit, V., Ayyoub, M., Rimoldi, D., Guillaume, P., Lienard, D., Lejeune, F., Cerottini, J. C., Romero, P., and Speiser, D. E. 2003. Simultaneous CD8 + T cell responses to multiple tumor antigen epitopes in a multipeptide melanoma vaccine. Cancer Immun. 3:15.

Van Kaer, L., Ashton-Rickardt, P. G., Eichelberger, M., Gaczynska, M., Nagashima, K., Rock, K. L., Goldberg, A. L., Doherty, P. C., and Tonegawa, S. 1994. Altered peptidase and viral-specific T cell response in LMP2 mutant mice. Immunity 1:533–541.

Website for Category A-C pathogens: http://www3.niaid.nih.gov/topics/BiodefenseRelated/Biodefense/research/CatA.htm (last visited January 21, 2008).

Wherry, E. J., Puorro, K. A., Porgador, A., and Eisenlohr, L. C. 1999. The induction of virus-specific CTL as a function of increasing epitope expression: responses rise steadily until excessively high levels of epitope are attained. J Immunol. 163:3735–3745.

Willett, T. A., Meyer, A. L., Brown, E. L., and Huber, B. T. 2004. An effective second-generation outer surface protein A-derived Lyme vaccine that eliminates a potentially autoreactive T cell epitope. Proc Natl Acad Sci U S A. 101:1303–1308.

York, I. A., Goldberg, A. L., Mo, X. Y., and Rock, K. L. 1999. Proteolysis and class I major histocompatibility complex antigen presentation. Immunol Rev 172:49–66.

# Immunodeficiencies and Immunome: Diseases and Information Services

Mauno Vihinen

**Abstract** Essential human immunome is composed of about 900 genes and proteins. Primary immunodeficiencies (IDs) are a large and heterogenic group of inherited disorders of the immune system. Since defects in any part of the adaptive or innate immune system can cause disorders, numerous IDs have been detected. Immunodeficiency patients have increased susceptibility to recurrent and persistent, even life-threatening infections. Other symptoms vary greatly between IDs. To date, some 200 IDs and 150 affected genes have been identified. ID-related genes are distributed throughout the genome. Features of IDs and information sources for them are discussed. Genotype–phenotype correlations are rather rare for IDs. Those diseases in which mutation type has effect on disease phenotype are described. ImmunoDeficiency Resource (IDR) is a comprehensive knowledge base for all essential information about IDs. The service is freely available at http://bioinf.uta.fi/IDR. IDdiagnostics (http://bioinf.uta.fi/IDdiagnostics) is dedicated to health professionals looking for laboratories performing gene and clinical tests for IDs. ID mutation data can be accessed in locus-specific, patient-related mutation databases, IDbases (http://bioinf.uta.fi/IDbases). Mutations are described at DNA, mRNA, and protein levels, with links to reference sequences and reference articles. The mutation data have been collated into entries, along with some clinical information. Currently, we have databases for 123 ID genes with 5359 patient entries. Immunome genes and proteins, their evolution, mouse–human comparisons, phylogenetics trees and orthologs for Metazoan immunome entities are available in Immunome Database, ImmTree, and ImmunomeBase all of which can be accessed via Immunome Knowledge Base (IKB) (http://bioinf.uta.fi/IKB).

**Keywords** Immunodeficiency · Diagnostics · Registry · Data bank · Intravenous immunoglobulin

M. Vihinen
Institute of Medical Technology, University of Tampere; Research Unit,
Tampere University Hospital, FI-33520 Tampere, Finland
e-mail: mauno.vihinen@uta.fi

A. Falus (ed.), *Clinical Applications of Immunomics*,                                    71
DOI: 10.1007/978-0-387-79208-8_4, © Springer Science+Business Media, LLC 2009

Adaptive immune mechanisms recognize and neutralize foreign microorganisms or molecules in a specific manner. B and T cells respond selectively to thousands of non-self materials. While adaptive immunity mechanisms require time to mount the response, native (innate) immunity is able to respond almost immediately to infectious agents. The major components of innate immunity are the complement system, natural killer cells, and phagocytes. The innate immunity has only a limited specificity.

Primary immunodeficiencies (IDs) form a heterogenic group of mainly rare inherited disorders that affect the immune system. IDs are intrinsic defects of the immune system. They vary widely in regard to symptoms, genotype, phenotype, severity, or causative microorganisms, for example. This is because many types of cells and molecules are required for both natural and adaptive immunity. To all IDs is common the increased susceptibility to infections, but other symptoms vary and include autoimmune manifestations, malignancies, and allergies, for example.

Antibody deficiency disorders are defects in immunoglobulin-producing B cells. Either the development or the function of B-lymphocytes is impaired. About 70% of ID patients belong to this group. T cell deficiencies affect the capability to kill infected cells or help other immune cells. In combined B and T cell IDs, the most severe IDs, all adaptive immune functions are absent. Severe combined immunodeficiencies (SCIDs) can be fatal unless the immune system can be reconstituted by transplants of immunocompetent tissue, enzyme replacement, or in the future by gene therapy. Other IDs affect, for example, the complement system or phagocytic cells, impairing antimicrobial immunity. Infections in secondary IDs can be similar to primary IDs, but are associated with some other factors such as age, tumors, drugs, malnutrition, infections, including human immunodeficiency virus (HIV).

ID patients have recurrent, serious infections starting early after birth. The different immunological systems recognize and destroy different pathogens. Therefore, the symptoms vary depending on the component(s) of the immune system impaired in IDs. The immune system normally is not activated prior to birth. Maternal immunoglobulins protect newborn for 6–12 months in immunoglobulin deficiencies. IDs result in prenatal death only when the affected gene is crucial for organs outside the immune system.

Patients with antibody deficiencies are especially susceptible to pyogenic infections caused by encapsulated bacteria. Persons with T cell IDs and SCIDs have opportunistic infections caused by common environmental microorganisms, as well as an increased frequency of virus, parasitic, and fungal infections. In SCIDs life-threatening symptoms can arise already within the first few days of life. Individuals with natural killer (NK) cell disorders are mostly susceptible to viral infections, while patients with phagocyte deficiencies have mainly skin and oral bacterial infections and granuloma formation when the microorganisms spread to organs. Also fungal infections cause severe complication for affected individuals. Certain defects lead to an increased susceptibility to a few or even a single pathogenic agent(s). Some IDs result in primary autoimmune manifestations. In certain IDs, patients have an increased incidence of cancer such as Wiskott Aldrich syndrome (WAS) and ataxia telangiectasia (AT).

The management of IDs depends on the defective system. Prolonged treatment with high doses of antibiotics is needed to treat infections. Antibody deficiencies are treated with (intravenous) immunoglobulin substitution therapy. B and T cells are produced in stem cells in bone marrow. Therefore, in many IDs incuding SCIDs, bone marrow transplantation is the most effective treatment. In certain metabolic disorders (ADA and PNP deficiency), enzyme substitution therapy can be applied. Certain SCIDs are ideal diseases for the development of gene therapy methods because of selective pressure of transduced cells (Fischer et al., 2002). The first successful gene therapy was for common γ-chain SCID (Cavazzana-Calvo et al. 2000), although complications were developed in some patients. Subsequently, other IDs have also been cured with gene transfer.

The immune system is based on a large number of molecules and processes. A particular ID can originate from defects in any one of the molecules essential for a certain response. The incidence of IDs varies greatly from about 1:500 live births to only a few known cases in the most rare disorders. The most common IDs are antibody production defects, different forms of SCIDs, and defects in phagocyte system. ID-related proteins are involved in several cellular functions, including cell surface receptors, signal transduction, transcription factors, nucleotide metabolism, gene diversification, phagocytosis, etc. Mutation(s) can affect any part and function of the immune system.

Around 200 primary immunodeficiency diseases have been identified to date. The molecular bases of some 150 of them have been detected (for a review see Notarangelo et al. 2004, Ochs et al. 2006, Maródi and Notarangelo 2007). ID-related genes are distributed throughout the human genome, only from chromosomes 3, 18, and Y have not been identified. Far majority of the IDs are autosomal recessive (AR). The best known cases are X-linked forms: 8% of diseases, but 36% of cases investigated on sequence level (Table 1). There are also some autosomal dominant IDs. Consanguinity is common in families with autosomal recessive forms of IDs. Diseases with X-chromosomal origin have full penetrance because the single affected allele causes the phenotype. The rareness of dominant IDs can be because of a fatal outcome and death before the age of reproduction, which prevents the recognition of AD pedigrees. In Fig. 1 are shown the IDs for which the defective gene has been identified. The number of genes and disorders grow constantly. Mutation databases are available practically for all the IDs with known gene, either at the Institute of Medical Technology (IMT) Bioinformatics group or elsewhere.

**Table 1** Inheritance of the IDs

| Inheritance | Diseases (%) | Mutations (%) |
| --- | --- | --- |
| Autosomal recessive | 80.5 | 53 |
| Autosomal dominant | 9.0 | 8 |
| AD/AR | 2.5 | 3 |
| X-linked | 8.0 | 36 |

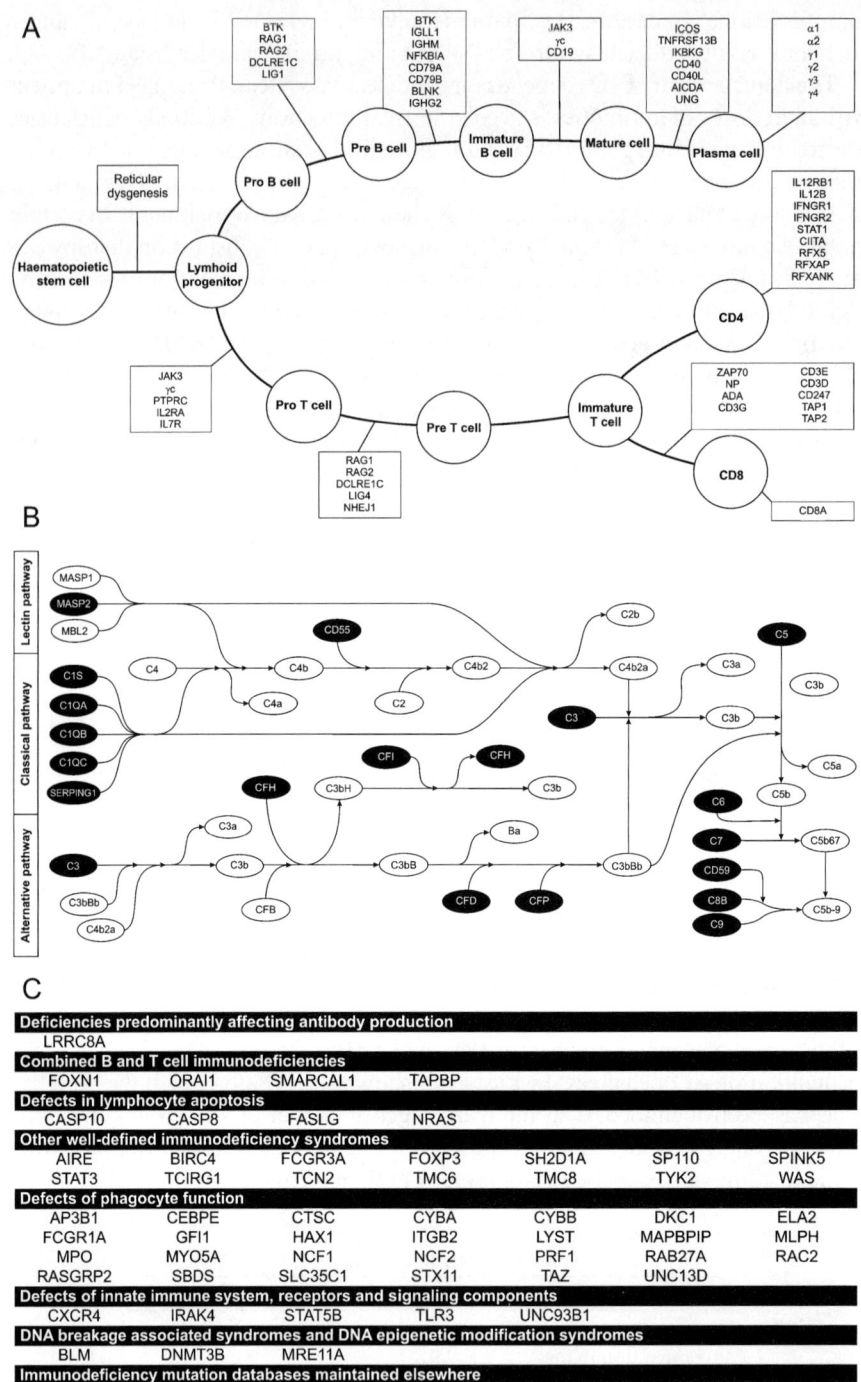

Some IDs have both autosomal dominant and autosomal recessive inheritance (see Piirilä et al. 2006) including interferon-γ receptor 1 deficiency. Both homozygous and single heterozygous mutations in tumor necrosis factor receptor superfamily member 13B (*TNFRSF13B*), which codes for transmembrane activator, and CAML interactor (TACI) cause common variable immunodeficiency (CVID) (Castigli et al. 2005, Salzer et al. 2005). The same kind of extreme variation can be seen with heterozygous mutations in tumor necrosis factor receptor superfamily member 6 (*TNFRSF6*) causing autoimmune lymphoproliferative syndrome (ALPS) (Infante et al. 1998).

Some IDs are monofactorial genetically heterogeneous diseases, i.e., defects in which similar phenotype can be caused by mutations in more than one gene. These conditions include hyper-immunoglobulin M (IgM) syndromes, which can be caused by mutations in five different genes: *AICDA*, *CD40LG*, *CD40*, *IKBKG*, and *UNG*. Other diseases to which several genes are linked are hemophagocytic lymphohistiocytosis (perforin, *PRF1* gene; unc-13 homolog D, *UNC13D*; and syntaxin 11, *STX11*) and Omenn syndrome (OS) caused by mutations in recombinase activating genes *RAG1* or *RAG2*, Artemis (*DLCRE1C*), or interleukin 7 receptor-α chain (*ILR7*). On the other hand, distinct mutations of the same gene can lead to different phenotypes, e.g., in elastase 2 (*ELA2*) mutations cause either severe congenital or cyclic neutropenia. For detailed discussion on these and other examples see Piirilä et al. (2006).

# 1 ImmunoDeficiency Resource (IDR)

There is plenty of information related to immunology and IDs available, but it is scattered in the literature and the Internet, which makes it difficult to find up-to-date and reliable data. To help in finding and using ID information, ImmunoDeficiency Resource (IDR) (http://bioinf.uta.fi/IDR) a knowledge base for the integration of the clinical, biochemical, genetic, genomic, proteomic, structural, and computational data of primary IDs, was developed (Väliaho et al. 2000, 2002, Samarghitean et al. 2007). It contains articles, instructional resources, analysis and visualization tools, and advanced search routines. The fact files, which form the core of the system, integrate biomedical knowledge from several sources (Väliaho et al. 2005). All information in IDR will be validated by expert curators.

**Fig. 1** Immunodeficiency diseases with known gene defects classified to those (**A**) affecting B and/or T cell development, (**B**) complement system, or (**C**) other parts of the immune system. The official gene names are given. In A genes involved in immunodeficiencies are indicated with their approximate location in the maturation pathway. Disease-related genes are highlighted in B with black background

**Fig. 2** Components of IDR system

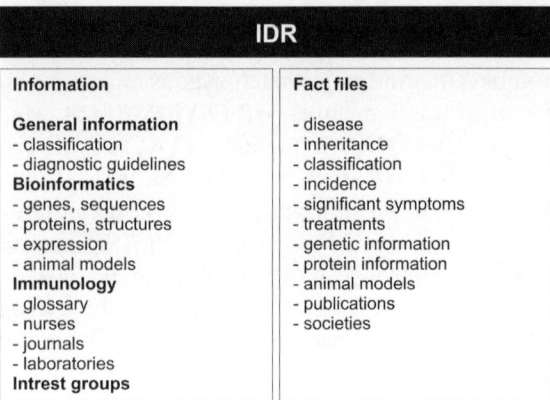

IDR pages are color-coded for different interest groups: researchers, physicians, nurses, patients, and families. By selecting the group of interest, the user can get specific pages produced and tailored for the particular group. This makes it easier for the user to find the most relevant and useful information.

The disease- and gene-specific information are stored in fact files that are eXtensible Markup Language (XML) based. Specific Inherited Disease Markup Language (IDML) was developed to distribute and collect information (Väliaho et al. 2005). IDR allows one to search and read all essential information about IDs at a glance, as well as to investigate numerous features more in detail. The major information categories in IDR are shown in Fig. 2.

## 2 Immunodeficiency Diagnostics Registry (IDdiagnostics)

Diagnosis of IDs can be very difficult, because several disorders can have similar symptoms and it is difficult to keep updated about the latest research results and recommendations. Numerous IDs are very rare. Early and reliable diagnosis is in many instances crucial for the efficient treatment, because delayed diagnosis and management can cause severe and irreversible complications, even the death of the patient.

The number of the laboratories analyzing the genetic defects of IDs is limited, sometimes just a single laboratory in the world. IDdiagnostics is a registry of laboratories performing tests for patients with IDs (Samarghitean et al. 2004) (http://bioinf.uta.fi/IDdiagnostics). It has two independent databases for laboratories performing genetic and clinical tests for IDs (Fig. 3). IDdiagnostics is intended for physicians, researchers, and other health care professionals involved with medical genetics.

Laboratories are included to the registry only on a voluntary basis, i.e., centers that want to have their information posted on the Internet. Those running either genetic or diagnostic test laboratories for IDs are requested to

**Fig. 3** Major data items for gene and clinical tests in IDdiagnostics

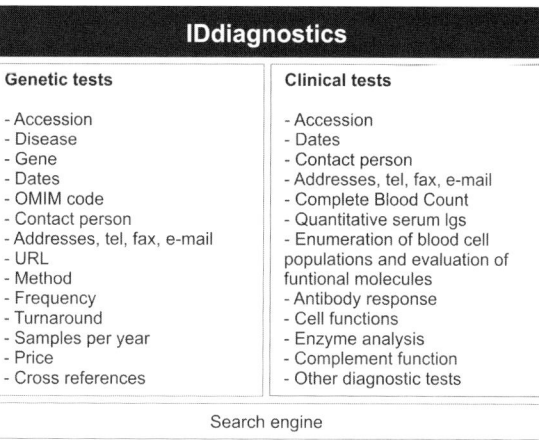

submit information to IDdiagnostics by using the user-friendly submission form available at the home page. Registration forms are available in both electronic and paper form. To update the IDdiagnostics data, laboratories are regularly contacted to verify the accuracy of their information. The data items and concept of the IDdiagnostiscs are depicted in Fig. 3.

From the database can be found, for example, the cost of the gene tests, the method used, the type of laboratory tests. The gene test laboratories provide information about the time required for a diagnosis and the turnaround time, how often the samples are run, how many samples are studied annually, and the cost of the analyses. Contact addresses for laboratories performing diagnosis are provided. Physicians have to contact the laboratory before sending in any samples.

## 3  PIDexpert

Diagnosis of an ID is often difficult, because similar symptoms characterize several IDs and mutations from the same gene can lead to distinct phenotypic consequences. We are currently developing a decision support system called PIDexpert (Samarghitean and Vihinen, 2008) to help in the diagnosis of IDs. PIDexpert is an online medical expert system based on artificial intelligence (AI), designed to give the diagnostic picture of IDs based on symptoms, signs, medical history, physical findings, and laboratory tests. The PIDexpert system will be available in the near future at http://bioinf.uta.fi/PIDexpert.

Medical expert systems (MESs) are computer programs that use a set of rules applied to the knowledge extracted from human experts (Samarghitean and Vihinen, 2008). MESs help in diagnostic processes and report generation, improve consistency in decisions, and increase timeliness in decision-making and productivity.

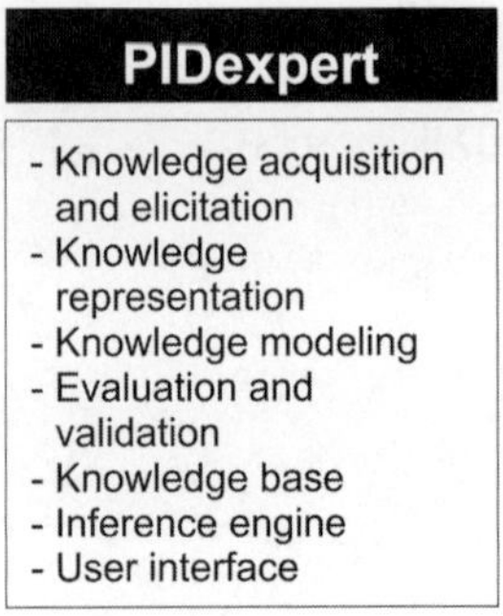

**Fig. 4** Components of
PIDexpert

| **PIDexpert** |
| --- |
| - Knowledge acquisition and elicitation<br>- Knowledge representation<br>- Knowledge modeling<br>- Evaluation and validation<br>- Knowledge base<br>- Inference engine<br>- User interface |

PIDexpert includes a knowledge acquisition system, a knowledge base, an inference engine, and a user interface (Fig. 4). Data to the knowledge base are collected from several sources including IDR, IDdiagnostics, literature, and medical experts. In addition, anonymous examples of differential diagnoses of patient cases will be used. The ESID/PAGID diagnostic guidelines (Conley et al. 1999) and practice parameters for the diagnosis and management of primary IDs established by the American College of Allergy, Asthma and Immunology (ACAAI), and the Joint Council of Allergy, Asthma and Immunology (Bonilla et al. 2005) are included in the knowledge base. These guidelines provide heuristics for possible/probable/definitive diagnosis for some IDs. The inference engine includes the rules or facts used for deduction. Various signs and symptoms are provided by the users on electronic input form. The system suggests, when possible, a diagnosis as well as identifies other conditions that may be associated with the disorder and how the diagnosis can be confirmed. If necessary, the program will also remind or suggest additional test(s). The PIDexpert is not intended to replace the physician, but to help in decision-making.

# 4 Immunodeficiency Mutation Databases, IDbases

Immunodeficiency-causing mutations have been identified from several genes. The Immunodeficiency mutation databases (IDbases) are available for 150 genes (Piirilä et al. 2006, http://bioinf.uta.fi/IDbases). The databases maintained at University of Tampere contain altogether more than 5000 entries. Mutation data are distributed along with patient-related clinical information. The first IDbase for X-linked agammaglobulinemia (XLA), BTKbase, was founded at 1994 (Vihinen et al. 1995, 1999, Väliaho et al. 2006).

IDbases at IMT are named according to the affected gene after systematic HUGO Gene Nomenclature (http://www.gene.ucl.ac.uk/nomenclature/) (Bruford et al. 2008). Mutations are named following the Human Genome Variation Society (HGVS) guidelines for mutation nomenclature (Antonarakis et al. 1998, den Dunnen and Antonarakis 2000) (http://www.hgvs.org/mutnomen/)

and their contents (Scriver et al. 1999) and curation (Cotton et al. in press) follow HGVS recommendations. Mutations are described at genomic, cDNA, and amino acid levels, and links to reference sequences are available in every entry. IDbases are linked to the University of California, Santa Cruz (UCSC) Genome Browser with PhenCode (Giardine et al. 2007).

Mutation data are collected mainly from literature (PubMed and Medline), and by direct submissions from the scientists analyzing mutations. Only mutations described unambiguously are accepted into IDbases. Scientists are requested to submit their mutation information either by contacting the curators or preferably via the Internet by a database-specific interactive submission form based on the MUTbase program (Riikonen and Vihinen, 1999) and available for each disorder.

The IDbases are locus-specific and patient-related databases. There is a separate entry for each patient. In addition to mutation data, many patient entries contain a variety of other information, including affected relatives, consanguinity, clinical symptoms, treatment, and other clinical information, if available. The entries have links to several central biological databases. The data items in IDbases are shown in Fig. 5.

**IDbases**

**Patient and mutation entry information**
ID
Accession
Systematic name
Original code
Description
Dates
PubMed links
References
Mutation description
  - DNA, RNA and protein level
  - variation(s)
  - reference sequences
Clinical parameters
  - symptoms
  - immunoglobulins
  - lymphocytes
  - other clinical signs
  - treatment
  - inheritance
**Bioinformatics**
  - statistics
  - IDR fact file
  - cross references
  - reference sequences
  - mutation browser
    - mutation checker
  - publications

**Fig. 5** Major data items for describing genetic and clinical parameters in IDbases

## 4.1 Using the IDbases

Currently, at IMT there are 123 gene-related immunodeficiency databases with 5359 public patient cases. The IDbases contain 2322 different unique mutations, 4489 mutations from unrelated families, and altogether 7547 mutations in affected alleles. 35% of unrelated cases are missense mutations, 19% nonsense mutations, 19% deletions, 14% splice site mutations, and 7% insertions. The remaining cases are undefined.

The IDbases can be used to discover many kinds of information. Patient-related mutation data with clinical features make it possible to find out statistically relevant trends from large data sets. The statistical studies of mutation sites indicate mutational hot spots, and reveal mutation mechanisms. The immunological data of the IDbases have been used to retrieve retrospective and prospective information on clinical presentation, immunological phenotype, long-term prognosis, or efficacy of available therapeutic options (Notarangelo et al. 2000, Notarangelo and Hayward, 2000). Genotype–phenotype correlations have been done for some IDs (Vihinen and Durandy, 2005). Structure–function relationships have been analyzed for several IDs such as Bruton tyrosine kinase (BTK) (e.g., Vihinen et al. 1994, 1999, Väliaho et al. 2006); for further details see Piirilä et al. 2006.

In addition to the IDbases, there are some immunodeficiency mutation databases maintained by others (see Fig. 1). At the moment we have links to 27 immunodeficiency mutation databases, and they contain altogether more than 3100 cases.

The European Society for Immunodeficiencies (ESID) patient registry (http://www.esid.org/) (Eades-Perner et al. 2007) collects patient information containing clinical data. The ESID patient registry and IDbases data collection are integrated. The collaboration facilitates direct submission both to the ESID registry and to the IDbases, and thus allows both systems to be updated by a single submission. A similar arrangement is under construction for American patient cases collected by US Immunodeficiency Network (USIDNET) (http://www.usidnet.org/).

Patient number variations between different IDs are very large. BTKbase, the largest and oldest IDbase, contains 1155 entries, and some of the registries have only one patient. As IDbases also have a public awareness function for IDs and rare diseases, databases are established as soon as a new gene and mutations are reported if not provided somewhere else. Detailed analysis of the numbers, types, and statistics of ID mutations have been published recently (Piirilä et al. 2006).

## 5 Genotype– Phenotype Correlations

Although large numbers of mutations have been identified, in certain IDs strict genotype–phenotype (GP) correlations are quite rare (Vihinen et al. 2001, Vihinen and Durandy, 2005). There are also some weaker correlations in certain disorders.

GP correlations are apparent in severe combined immunodeficiency (SCID) caused by mutations in the adenosine deaminase gene (ADA) and milder forms of ADA deficiency (Arredondo-Vega et al. 1998, Hershfield 2003), SCID caused by mutations in *RAG1* and *RAG2* (Villa et al. 2001), and in Wiskott-Aldrich syndrome (WAS) (Imai et al. 2003). Weaker correlations are apparent, e.g., in X-linked agammaglobulinemia (XLA) (Lindvall et al. 2005, Lopéz-Granados et al. 2005, Väliaho et al. 2006). The phenotype effects have been correlated to the consequences of mutations on protein structure in ADA, WAS, and XLA (Vihinen and Durandy, 2005, Lindvall et al. 2005, Väliaho et al. 2007). There are also some other IDs where GP correlations have been implicated including ELA2 in cyclic and congenital neutropenia (Dale et al. 2000), serpin (complement component 1) mutations in hereditary angioedema (HANE) (Verpy et al. 1996), and tafazzin in Barth syndrome (Johnston et al. 1997). Heterogeneity of IDs is becoming more clear as new genes and patients are identified. More cases, mutations, and clinical observations are needed for detailed statistical analyses to reveal further genotype–phenotype correlations.

Knowledge about GP correlations could help in the proper treatment and diagnosis, and provide means for understanding clinical features as well as support basic research on the function and structure of genes and proteins.

# 6 Immunome

ID genes are crucial for immune system as indicated by the associated diseases, but they form just a small fraction of all immunology-related genes. The human immune system is one of the most complex biological machineries known. Numerous genes and proteins are needed to work in concerted action to mount the necessary responses when the body is challenged by foreign molecules or organisms. We call for essential immunome the genes and their products, which are essential for immunological processes. Both innate and adaptive immunity are dependent on several processes, and therefore the genes and proteins vary greatly in their functions, subcellular localization, expression, tissue distribution, gene ontologies, and protein domain content, etc.

Identification of the essential immunome genes was made based on literature search from textbooks and publications, and database mining (Ortutay and Vihinen 2006,Ortutay et al. 2007a). Currently, the Immunome Database at http://bioinf.uta.fi/Immunome contains information for 847 entries. Information about the identified immunome genes was used to investigate the evolutionary history of immunological processes and protein domains and the functions essential for them. The human immunome consists of many genes that are distributed throughout almost all the chromosomes. The human–mouse immunome comparison was performed and found to indicate the conserved and variable genes and functions and the chromosome organization. Three types of emergence among immunome members were noted when independently considering evolutionary, domain, and functional studies. The evolutionary levels indicate ancient functions common to all taxa, constant growth in the number of functions, and one or just a few jumps in the number of assigned genes, domains, and functions.

Data for immunome were collected into a MySQL database and distributed in Internet. The web site at http://bioinf.uta.fi/Immunome/ contains search pages to present data about a particular gene or for gene groups. The immunome database can be a used for designing and interpreting results obtained with different experimental approaches. Immunome database integrates data from different sources and facilitates understanding immune system at higher level. It also serves as a catalog of essential immunome genes for systems biological studies.

For further studying the evolutionary history of the human immune system, we developed a database focused on the evolutionary relationships of immunity-related genes (Ortutay et al. 2007b). ImmTree at http://bioinf.uta.fi/ImmTree contains evolutionary trees for all the identified essential immunome genes. ImmTree provides also a data set for comparison of human–mouse ortholog pairs based on synonymous and non-synonymous mutation rates of the genes. There are data for orthologs of the human genes altogether in 80 species, including all the major model organisms.

The two previous services are centered on human immune system. Immunological molecules in Metazoan species are available in ImmunomeBase (http://bioinf.uta.fi/ImmunomeBase) (Rannikko et al. 2007). Immunome proteins and genes were identified from literature and databases and used as seed proteins for which orthologs were identified with reciprocal protein–protein Blast (Altschul et al. 1997) searches against proteins from the GenBank and RefSeq databases. Currently, the database contains 1811 metazoan seed genes and 10,333 non-seed orthologs. Different levels of ortholog groups were defined according to the order of reciprocal ortholog pairs among the seed immunity genes. ImmunomeBase contains all the immunity genes and their evidence of immune function, orthologs, and ortholog groups. The database contains information about the evidence of an immunity function for the seed genes in the form of journal citations and Gene Ontology (Ashburner et al. 2000) terms.

Immunome, ImmTree, ImmunomeBase, and additional information, e.g., about genetic variations in immunome genes and proteins, have been combined to a single server called Immune Knowledge Base (IKB) recently released at http://bioinf.uta.fi/IKB.

**Acknowledgments** Financial support from Finnish Academy and the Medical Research Fund of Tampere University Hospital is gratefully acknowledged.

# References

Ashburner, M., Ball, C. A., Blake, J. A., Botstein, D., Butler, H., Cherry, J. M., Davis, A. P., Dolinski, K., Dwight, S. S., Eppig, J. T., Harris, M. A., Hill, D. P., Issel-Tarver L., Kasarskis, A., Lewis, S., Matese, J. C., Richardson, J. E., Ringwald, M., Rubin, G. M., Sherlock, G. 2000. Gene ontology: tool for the unification of biology. The Gene Ontology Consortium. Nat. Genet. 25:25–29.

Altschul, S. F., Madden, T. L., Schäffer, A. A., Zhang, J., Zhang, Z., Miller, W., Lipman, D. J. 1997. Gapped BLAST and PSI-BLAST: a new generation of protein database search programs. Nucleic Acids Res. 25:3389–3402.

Antonarakis, S. E., Ashburner, M., Auerbach, A. D., Beaudet, A. L., Beckmann, J. S., Beutler, E., Cooper, D. N., Cotton, R. G. H., den Dunnen, J. T., Desnick, R. J., Eng, C., Fasman, K. H., Goldman, D., Hayashi, K., Hutchinson, F., Kazazian, H. H., Keen, J., King, M.-C., Lehväslaiho, H., McAlpine, P. J., McKusick, V., Motulski, A. G., Povey, S., Schorderet, D., Scriver, C. R., Shows, T. B., Superti-Furga, A., Tay, A. H. N., Tsui, L.-C., Valle, D., Vihinen, M. 1998. Recommendations for a nomenclature system for human gene mutations. Hum. Mut. 11: 1–3

Arredondo-Vega, F. X., Santisteban, I., Daniels, S., Toutain, S., Hershfield, M. S. 1998. Adenosine deaminase deficiency: genotype-phenotype correlations based on expressed activity of 29 mutant alleles. Am. J. Hum. Genet. 63:1049–1059.

Bonilla, F. A., Bernstein, I. L., Khan, D. A., Ballas, Z. K., Chinen, J., Frank, M. M., Kobrynski, L. J., Levinson, A. I., Mazer, B., Nelson, R. P. Jr., Orange, J. S., Routes, J. M., Shearer, W. T., Sorensen, R.U. 2005. Practice parameter for the diagnosis and management of primary immunodeficiency. Ann. Allergy Asthma Immunol. 94: S1–S63.

Bruford, E. A., Lush, M. J., Wright, M.W., Sneddon, T. P., Povey, S., Birney, E. 2008. The HGNC Database in 2008: a resource for the human genome. Nucleic Acids Res. 36:D445–448.

Cavazzana-Calvo, M., Hacein-Bey, S., de Saint Basile, G., Gross, F., Yvon, E., Nusbaum, P., Selz, F., Hue, C., Certain, S., Casanova, J. L., Bousso, P., Deist, F. L., Fischer A. 2000. Gene therapy of human severe combined immunodeficiency (SCID)-X1 disease. Science 288, 669–672.

Castigli E, Wilson, S. A., Garibyan, L., Rachid, R., Bonilla, F., Schneider, L., Geha, R. S. 2005. TACI is mutant in common variable immunodeficiency and IgA deficiency. Nat. Genet. 37:829–834.

Conley, M. E., Notarangelo, L. D., Etzioni, A. 1999. Diagnostic criteria for primary immunodeficiencies. Representing PAGID (Pan-American Group for Immunodeficiency) and ESID (European Society for Immunodeficiencies). Clin. Immunol. 93:190–197.

Cotton, R. G. H., Auerbach, A. D., Beckmann J. S., Blumenfeld, O. O., Brookes, A. J., Brown, A. F., Carrera, P., Cox, D. W., Gottlieb, B., Greenblatt, M. S., Hilbert, P., Lehvaslaiho, H., Liang, P., Marsh, S., Nebert, D. W., Povey, A., Rossetti, S., Scriver, C. R., Summar, M., Tolan, D. R., Verma, I. C., Vihinen, M., den Dunnen, J. T. 2008. Recommendations for locus specific databases and their curation. Hum. Mut. 29:2–5.

Dale, D. C., Person, R. E., Bolyard, A. A., Aprikyan, A. G., Bos, C., Bonilla, M. A., Boxer, L. A., Kannourakis, G., Zeidler, C., Welte, K., Benson, K. F., Horwitz, M. 2000. Mutations in the gene encoding neutrophil elastase in congenital and cyclic neutropenia. Blood 96:2317–2322.

den Dunnen, J. T., Antonarakis, S. E. 2000. Mutation nomenclature extensions and suggestions to describe complex mutations: A discussion. Hum. Mut. 15:7–12.

Eades-Perner, A. M., Gathmann, B., Knerr, V., Guzman, D., Veit, D., Kindle, G., Grimbacher, B. 2007. The European internet-based patient and research database for primary immunodeficiencies: results 2004–06. Clin Exp Immunol. 147:306–312.

Fischer, A., Hacein-Bey, S., Cavazzana-Calvo, M. 2002. Gene therapy of severe combined immunodeficiencies. Nat. Rev. Immunol. 8:615–621.

Giardine, B., Riemer, C., Hefferon, T., Thomas, D., Hsu, F., Zielenski, J., Sang, Y., Elnitski, L., Cutting, G., Trumbower, H., Kern, A., Kuhn, R., Patrinos, G. P., Hughes, J., Higgs, D., Chui, D., Scriver, C., Phommarinh, M., Patnaik, S. K., Blumenfeld, O., Gottlieb, B., Vihinen, M., Väliaho, J., Kent, J., Miller, W., Hardison, R. C. 2007. PhenCode: connecting ENCODE data with mutations and phenotype. Hum. Mut. 28:554–562.

Hershfield, M.S. 2003. Genotype is an important determinant of phenotype in adenosine deaminase deficiency. Curr. Opin. Immunol. 15:571–577.

Imai, K., Nonoyama, S., Ochs, H.D. 2003. WASP (Wiskott-Aldrich syndrome protein) gene mutations and phenotype. Curr. Opin. Allergy Clin. Immunol. 3:427–436.

Infante, A. J., Britton, H. A., DeNapoli, T., Middelton, L. A., Lenardo, M. J., Jackson, C. E., Wang, J., Fleisher, T., Straus, S. E., Puck, J. M. 1998. The clinical spectrum in a large kindred with autoimmune lymphoproliferative syndrome caused by a Fas mutation that impairs lymphocyte apoptosis. J. Pediatr. 133:629–633.

Johnston, J., Kelley, R. I., Feigenbaum, A., Cox, G. F., Iyer, G. S., Funanage, V. L., Proujansky, R. 1997. Am. J. Hum. Genet. 61:1053–1058.

Lindvall, J. M., Blomberg, K. E., Väliaho, J., Vargas, L., Heinonen, J. E., Berglof, A., Mohamed, A. J., Nore, B. F., Vihinen, M., Smith, C. I. E. 2005. Bruton's tyrosine kinase: cell biology, sequence conservation, mutation spectrum, siRNA modifications, and expression profiling. Immunol. Rev. 203:200–215.

Lopéz-Granados, E., Pérez de Diego, R., Ferreira, Cerdán. A., Fontán, Casariego. G., García, Rodríguez. M. C. 2005. A genotype-phenotype correlation study in a group of 54 patients with X-linked agammaglobulinemia. J. Allergy Clin. Immunol. 116:690–697.

Maródi, L., Notarangelo, L. D. 2007. Immunological and genetic bases of new primary immunodeficiencies. Nat. Rev. Immunol. 7:851–61.

Notarangelo, L., Casanova, J. L., Fischer, A., Puck, J., Rosen, F., Seger, R., Geha, R. 2004. Primary immunodeficiency diseases: an update. J. Allergy Clin. Immunol. 114:677–687.

Notarangelo, L. D., Hayward, A. R. 2000. X-linked immunodeficiency with hyper-IgM (XHIM). Clin. Exp. Immunol. 120:399–405.

Notarangelo, L. D., Giliani, S., Mazza, C., Mella, P., Savoldi, G., Rodriguez-Pérez, C., Mazzolari, E., Duse, M., Plebani, A., Ugazio, A. G., Vihinen, M., Candotti, F., Schumacher, R. F. 2000. Of genes and phenotypes: The immunological spectrum of combined immune deficiency. Defects of the γc-JAK3 signaling pathway as a model. Immunol. Rev. 178:39–49.

Ochs, H. H., Smith, C. I. E. and Puck, J. 2006. Primary immunodeficiency disaeses. A molecular and cellular approach. Oxford Univ. Press.

Ortutay, C. and Vihinen, M. 2006. Immunome: a reference set for system biology on the human immune system. Cell. Immunol. 244:87–89.

Ortutay, C., Siermala, M. and Vihinen, M. 2007a. Molecular characterization of the immune system: emergence of proteins, processes and domains. Immunogenet. 59:333–348.

Ortutay, C., Siermala, M. and Vihinen, M. 2007b. ImmTree: database of evolutionary relationships of genes and proteins in the human immune system. Immunome Res. 3:4.

Piirilä, H., Väliaho, J. and Vihinen, M. 2006. Immunodeficiency mutation databases (IDbases). Hum. Mut. 27:1200–1208.

Rannikko, K., Ortutay, C. and Vihinen, M. 2007. Immunity genes and their orthologs: a multi-species database. Int. Immunol. 19:1361–1370.

Riikonen, P., Vihinen M. 1999. MUTbase: maintenance and analysis of distributed mutation databases. Bioinformatics 15:852–859.

Salzer, U., Chapel, H. M., Webster, A. D., Pan-Hammarström, Q., Schmitt-Graeff, A., Schlesier, M., Peter, H. H., Rockstroh, J. K., Schneider, P., Schaffer, A. A., Hammarström, L., Grimbacher, B. 2005. Mutations in TNFRSF13B encoding TACI are associated with common variable immunodeficiency in humans. Nat. Genet. 37:820–828.

Samarghitean, C., Väliaho, J., Vihinen, M. 2004. Online registry of genetic and clinical immunodeficiency diagnostic laboratories, IDdiagnostics. J. Clin. Immunol. 24:53–61.

Samarghitean, C., Väliaho, J., Vihinen, M. 2007. IDR knowledgebase for primary immuno-deficiencies. Immunome Res. 3:6.

Samarghitean, C. and Vihinen, M. 2008. Medical expert systems. Curr. Bioinf. 3:56–65.

Scriver, C. R., Nowacki, P. M., Lehväslaiho, H. 1999. Guidelines and recommendations for content, structure and deployment of mutation databases. Hum. Mut. 13: 344–350.

Väliaho, J., Riikonen, P., Vihinen, M. 2000. Novel immunodeficiency data servers. Immunol. Rev. 178:177–185.

Väliaho, J., Pusa, M., Ylinen, T., Vihinen, M. 2002. IDR: the ImmunoDeficiency Resource. Nucleic Acids Res. 30:232–234.

Väliaho, J., Riikonen, P., Vihinen, M. 2005. Distribution of immunodeficiency fact files with XML – from Web to WAP. BMC Med. Inform. Decis. Mak. 5:21.

Väliaho, J., Smith, C. I. E., Vihinen, M. 2006. BTKbase: mutation database for X-linked agammaglobulinemia. Hum. Mut. 27:1209–1217.

Verpy, E., Biasotto, M., Brai, M., Misiano, G., Meo, T., Tosi, M. 1996. Exhaustive mutation scanning by fluorescence-assisted mismatch analysis discloses new genotype-phenotype correlations in angiodema. Am. J. Hum. Genet. 59:308–319.

Vihinen, M., Vetrie, D., Maniar, H. S., Ochs, H. D., Zhu, Q., Vorechovský, I., Webster, A. D., Notarangelo, L. D., Nilsson, L., Sowadski, J. M., Smith, C. I. E. 1994. Structural basis for chromosome X-linked agammaglobulinemia: a tyrosine kinase disease. Proc. Natl. Acad. Sci. USA. 91:12803–12807.

Vihinen, M., Cooper, M. D., de Saint Basile, G., Fischer, A., Good, R. A., Hendriks, R. W., Kinnon, C., Kwan, S. P., Litman, G. W., Notarangelo, L. D., Ochs, H. D., Rosen, F. S., Vetrie, D., Webster, A. D. B., Zegers, B. J. M, Smith, C. I. E. 1995. BTKbase: a database of XLA-causing mutations. Immunol. Today 16:460–465.

Vihinen, M., Durandy, A. 2005. Primary immunodeficiencies: genotype-phenotype correla-tions. Immunogenomics and human disease (ed. Falus, A.) John Wiley and Sons. pp. 443–460.

Vihinen, M., Kwan, S. P., Lester, T., Ochs, H. D., Resnick, I., Väliaho, J., Conley, M. E., Smith, C. I. E. 1999. Mutations of the human BTK gene coding for Bruton tyrosine kinase in X-linked agammaglobulinemia. Hum. Mutat. 13:280–285.

Vihinen, M., Arredondo-Vega, F. X., Casanova, J. L., Etzioni, A., Giliani, S., Hammarström, L., Hershfield, M. S., Heyworth, P. G., Hsu, A. P., Lähdesmäki, A., Lappalainen, I., Notarangelo, L. D., Puck, J. M., Reith, W., Roos, D., Schumacher, R. F., Schwarz, K., Vezzoni, P., Villa, A., Väliaho, J., Smith, C. I. E. 2001. Primary immunodeficiency mutation databases. Adv. Genet. 43:103–188.

Villa, A., Sobacchi, C., Notarangelo, L. D., Bozzi, F., Abinun, M., Abrahamsen, T. G., Arkwright, P. D., Baniyash, M., Brooks, E. G., Conley, M. E., Cortes, P., Duse, M., Fasth, A., Filipovich, A. M., Infante, A. J., Jones, A., Mazzolari, E., Muller, S. M., Pasic, S., Rechavi, G., Sacco, M. G., Santagata, S., Schroeder, M. L., Seger, R., Strina, D., Ugazio, A., Väliaho, J., Vihinen, M., Vogler, L. B., Ochs, H., Vezzoni, P., Friedrich, W., Schwarz, K. 2001. V(D)J recombination defects in lymphocytes due to RAG mutations: severe immunodeficiency with a spectrum of clinical presentations. Blood 97:81–88.

# Immunomics of Immune Rejection

Ena Wang, Marianna Sabatino, and Francesco M Marincola

**Abstract** There is overwhelming evidence that the innate and adaptive arms of the human immune system can interact with autologous tumor cells and can, in rare circumstances, clear the host of cancer. Yet, most often tumor cells strike a favorable balance with the host's immune response and continue their proliferation. The coexistence of immune responses with their targets is not a phenomenon restricted to cancer and it can be observed during therapeutically controlled chronic allograft rejection, during chronic viral diseases, or mild autoimmune reactions. Thus, systemic and peripheral immune reactions comprise an array of afferent cognitive loops and efferent effector mechanisms, whose balance dictates the final outcome of the disease. Recent surveys of the human tumor microenvironment using high-throughput scanning techniques suggest that autologous tumor rejection represents a distinct aspect of tissue-specific destruction similar to the clearance of pathogen, acute allograft rejection, or flares of autoimmunity. Identification of commonalities among the different immune pathologies may shed insights into the requirements for an effective immune response. We argue here that, due to the multiplicity of factors that may influence the balance of host/target interaction, a system biology approach based on high-throughput platforms may best suit the identification of this immunological constant of tissue rejection.

**Keywords** Immunization/Immunotherapy/ IL-2/melanoma · Epstein-Barr virus · Basal cell carcinoma · Renal cell carcinoma

**Abbreviations** CTL – Cytotoxic T lymphocytes; EBV – Epstein-Barr virus; HLA – Human Leukocyte Antigen; TIL – Tumor Infiltrating Lymphocytes.

F.M. Marincola
Infectious Disease and Immunogenetics Section, Department of Transfusion Medicine, Clinical Center, National Institutes of Health, Bethesda, MD, USA
e-mail: FMarincola@cc.nih.gov

A. Falus (ed.), *Clinical Applications of Immunomics*,
DOI: 10.1007/978-0-387-79208-8_5, © Springer Science+Business Media, LLC 2009

# 1 Introduction

There is overwhelming evidence that the innate and the adaptive arms of the human immune system can interact with autologous tumor cells and can, in rare circumstances, clear the host of cancer. Yet, most often tumor cells strike a favorable balance with the host's immune response and continue their proliferation in spite of a cognitive immune system. The coexistence of immune responses with their targets is not a phenomenon restricted to cancer and it can be observed during therapeutically controlled chronic allograft rejection, chronic unresolving viral diseases, or lingering autoimmune reactions. This is because the systemic and peripheral immune reactions comprise an array of afferent cognitive loops, efferent effector mechanisms, and regulatory reactions, whose balance dictates the final outcome of the disease. In this chapter, we argue that autologous tumor rejection represents a distinct aspect of immune-mediated tissue-specific destruction similar to the clearance of pathogen, acute allograft rejection, or to flares of autoimmunity, and it can occur only when sufficient immune stimulation is provided in the target organ to switch a chronic and lingering inflammatory process into an acute one; we came to this conclusion through direct observation in real-time of tumor responses to immune manipulation adopting high-throughput technologies, as well as comparing our results with the results derived from the literature in which similar approaches were applied to other disease models. Identification of commonalities among different disease models, various treatments inducing immune-mediated tissue destruction in the context of a heterogeneous genetic background may shed insights into the universal requirements for an effective immune response. We argue here that, due to the multiplicity of factors that may influence the balance of host/target interaction, a system biology approach based on high-throughput platforms may best suit the identification of this immunological constant of tissue rejection. Because humans are polymorphic, their diseases heterogeneous and, contrary to experimental models, such diversity cannot be controlled, a discovery-driven approach best suits the assessment of human pathology, particularly in the realm of immune interactions (Jin and Wang 2003; Jin et al. 2004). Yet, such heterogeneity, if approached with appropriate tools, may help in the identification of common characteristics associated with immune-mediated tumor rejection that must occur independently of the subject, disease, or treatment studied.

# 2 Relevance of Tumor-Specific Immune Responses in Humans

The recognition of human cancers by the immune system and, conversely, the mechanisms adopted by tumor cells to escape the host immune recognition are the result of different variables that could be grouped into two main categories: those related to systemic immune responses against cancer and their alterations in function, and those associated with local characteristics of tumor cells that may make them more or less susceptible to immune-mediated destruction by the immune-competent host (Marincola et al. 2000). Thus, the growth of cancer is the result of a fine balance between systemic immune responses and their ability to counteract the biological pursuit for survival of cancer cells. A classical example of such balance is the relationship between the host immune response to the Epstein-Barr virus (EBV) and the frequency of EBV-induced malignancies. EBV is a ubiquitous virus asymptomatically infecting the large majority of humans in their early years of life. Although in itself the virus has transforming potentials for lymphocytes as demonstrated by its ability to induce proliferation and expansion of B cells in vitro, EBV-induced lymphomas (lymphoproliferative disorders) occur only in immune-suppressed individuals (Ensoli and Sirianni 1998; Brander et al. 2001; Green 2001; Trofe et al. 2004). Withdrawal or reduction of immune suppression can reverse the natural course of post-transplant lymphoproliferative disorders (Green 2001). In addition, complete regression of post-transplant lymphoproliferative disorders can be mediated by adoptive transfer of Human Leukocyte Antigen (HLA)-matched EBV virus-specific cytotoxic T cells (CTL) (Heslop and Rooney 1997; Khanna et al. 1999; Haque et al. 2001), and their prophylactic administration can prevent the insurgence of malignancies (Rooney et al. 1998).

The control of EBV viral and oncogenic activity by a normal immune system may parallel other immune responses against other latent viral infections such as the immune response against cytomegalovirus. Flares of these infections occur with immune suppression, but tumors do not occur because this or other similar viruses lack strong oncogenic properties (Rinaldo and Torpey 1993; Moss and Khan 2004; Nicol et al. 2005). On the other hand, other viral systems appear to recapitulate the EBV-model-like, for instance, Kaposi's sarcoma and other Herpes Virus-related tumors (Ensoli and Sirianni 1998). The question arises, therefore, whether these virally induced tumors represent exceptions due to the high immunogenicity of the virus infecting the cancer cell while most non-virally induced solid tumors are invisible to a competent immune system. It has been observed that the prevalence of some solid tumors whose oncogenesis is not known to be related to viruses is also increased in immune-suppressed hosts (Trofe et al. 2004); this observation suggests that a continuum spectrum exists between the recognition of highly immunogenic virally induced tumors and other cancers, and opens the possibility that subtle alterations in immune recognition may affect the ability of the host to maintain an effective immune surveillance against most cancers if more subtle shifts in immune competence could occur, which could not be easily attributed to particular immune suppressive therapies or immune suppressive syndrome. What is then that may modulate the insurgence of cancers in borderline situations when an arguably competent immune system is insufficient to limit the growth of cancer? Among the various possibilities, it has been long hypothesized that differences in genetic backgrounds may be responsible for subtle phenotypic alterations of our immune function, which may result in a higher individual risk to develop cancer and control its growth.

## 3 Genetic Variables that May Affect the Ability of the Host to Control Cancer Growth

Monothematic observations have failed so far to identify specific immunogenetic characteristics of the host, which could convincingly explain the propensity of some individuals to be susceptible for the insurgence of cancer. Associations have been described between the incidence of nasopharyngeal cancer and HLA phenotype in Asian (Simons et al. 1974; Cui and Lin 1982; Ou et al. 1985; Zhang 1986; Hildesheim 2002) and northern African populations (Li et al. 2007). However, a convincing mechanistic explanation for this phenomenon is still lacking, and a predominant hypothesis (Simons 2003) suggests that this association is due to the presence of a predisposition locus in the short arm of chromosome 6 in strong linkage disequilibrium with the HLA class I genes rather than structural properties of distinct HLA alleles that may affect their ability to present antigen and, as a consequence, the ability of the host to recognize and control EBV function. Overall, the influence of the genetic background on immune responsiveness against cancer has been minimally explored. HLA genes have also been associated with various clinical parameters around the natural history of melanoma and other cancers (Marincola et al. 1996). However, correlates between individual alleles or extended haplotypes and treatment outcome or survival have been conflicting (Marincola et al. 1996; Rubin et al. 1991; Lee et al. 2004). Recently, associations were reported between cytokine promoter region polymorphism(s) and disease susceptibility. For instance, polymorphisms of the promoter region of IL-10 responsible for variable production of IL-10 identified phenotypes associated with decreased expression of this cytokine, which appear to predispose to increased incidence of melanoma and prostate cancer (Howell et al. 2001, 2002a,b). Others reported that polymorphism of the IFN-$\gamma$ gene may be responsible for higher likelihood of response to chemo-immunotherapy and improved survival of patients with melanoma (Liu et al. 2005). However, an overall understanding of the association of genomic variability and cancer predisposition will only be appreciated when high-throughput global approaches will be adopted, which could coordinately assess multiple variables at the same time that may be in concert responsible for the global immune phenotype of each individual (Jin and Wang 2003; Jin et al. 2004).

Some examples of successful system biology approaches for the identification of immune phenotypes at particular risk to develop cancer are emerging. Lee P et al. (Critchley-Thorne et al. 2007) observed that peripheral blood mononuclear cells (PBMC) of individuals with melanoma display a different transcriptional profile compared to non-tumor-bearing normal individuals. These differences could be partly attributed to a decreased responsiveness to IFN-$\alpha$ stimulation. This study, however, could not discriminate the cause of such striking differences between patients with melanoma and normal controls. In particular, it could not address whether the observed differences that affect the global transcriptional pattern of PBMC were due to primary genetic characteristics that predispose to the development of melanoma or represented a secondary effect of the tumor-bearing status that could affect, through secretion of immune regulatory factors, the function profile of PBMC. Studies from other human systems suggest that the first possibility cannot be a possibility; Xie et al (He et al. 2006) compared the transcriptional profile of PBMC stimulated with IFN-α in individuals with hepatitis C virus infection, which subsequently underwent systemic treatment with the same cytokine and observed that the responsiveness to IFN-α is genetically determined; in vitro testing of PBMC yielded signatures that were predictive of immune responsiveness in vivo; such signatures were more frequently observed in African American subjects with hepatitis C virus infection already known to be less responsive to IFN-α treatment, and were shared by individuals of other ethnic background who also were resistant to therapy. Thus, it is conceivable that this observation derived through a system biology approach may explain an individual's predisposition to establish a more or less effective immunological barrier toward the genesis of disease. Presently, the genetic basis of this phenotype remains unknown. Yet, much needs to be done to further this line of studies and, most importantly, to link functional genomics observations to genetic traits at the genome level to identify those polymorphisms responsible for distinct immune phenotypes.

## 4 Cancer Cell Biology as the Orchestrator of the Host Immune Response

While the control of EBV-induced lymphoproliferative disorders by a competent immune system suggests that, at least in these settings, CTLs play a major role in regulating tumor growth, other EBV-associated tumors such as Burkitt's lymphoma, Hodgkin's disease, and nasopharyngeal cancer occur in immune-competent individuals. Since they are characterized by a restricted expression of EBV proteins (Chua et al. 2001; Gottschalk et al. 2002, 2003; Straathof et al. 2005), they may be less immunogenic and at the same time this may explain why they are less susceptible to therapy with adoptively transferred EBV-specific CTL (Chua et al. 2001; Gottschalk et al. 2002; Straathof et al. 2005). Thus, the same virus, acting as a potent oncogene, may induce various types of cancers whose insurgence is predicated upon the immune status of the host. In some cases, insufficient systemic immune responses seem to allow the development of the neoplastic disorders in spite of a high immunogenic potential of the cancer cells (lymphoproliferative disorders), while in the other cases insufficient local (at the tumor site) immune responses may allow tumor growth in association with a lower immunogenic potential. The likely explanation, in the latter case, is that the tumor cells had undergone sufficient evolutionary adaptations to survive the natural response of the host against EBV and its by-products, primarily by decreasing its dependence upon EBV proteins to drive their ability to replicate.

It is reasonable to postulate that this compromise between host immune response and cancer cell immunogenic properties represents the rule for most human cancers. Indeed, adaptive humoral and cellular immune responses against tumors occur naturally in solid cancers even though these cancers do not necessarily express either foreign proteins or antigenic determinants resulting from mutations in the amino acid sequence of endogenous proteins (Boon et al. 1997; Old and Chen 1998). A good example is represented by the prototype human immune-sensitive cancer: metastatic melanoma, which is not only naturally immunogenic but also the immune responses toward it, whether humoral or cellular, are directed against endogenous, non-mutated self-proteins (Masramon et al. 2000). It is possible, therefore, that the immunogenic potential of a tumor is not related to the type of antigens (self vs non-self) expressed but by the concomitant microenvironment. Most solid tumors do not induce a pro-inflammatory microenvironment comparable to that induced by an even latent EBV infection of lymphocytes, and therefore tumor antigen–specific T cells are not exposed to the co-stimulation required for their activation at the tumor site (Monsurro' et al. 2003, 2004).

Thus, although several reasons could be responsible for reduced immune responses at the tumor site (Marincola et al. 2000) including insufficient localization of effector cells, their insufficient co-stimulation, activation-induced apoptosis of CTLs interacting with tumor cells, production of immune-suppressive factors by tumor cells, or the presence of T cells with suppressor activity referred to as regulatory T cells (Badoual et al. 2006; Curiel et al. 2004; Viguier et al. 2004; Woo et al. 2002; Yamaguchi and Sakaguchi 2005), it is likely that the simplest explanation is that tumors are physiologically least likely to represent immunological targets (Badoual et al. 2006; Curiel et al. 2004; Viguier et al. 2004). In addition, tumor cells could escape immune recognition by reducing the expression of target molecules such as HLA–epitope complexes: the latter being a component of a process generally referred to as immune escape (Marincola et al. 2000, 2003; Marincola 1997) and part of a complex process that molds tumor cell/host relationships in time referred to as immune editing (Dunn et al. 2002, 2004a,b)

The identification of the factors that regulate cancer recognition and destruction by the immune system, therefore, is limited when monothematic approaches are followed because it is unlikely that one factor, rather than the sum of several pro-inflammatory and regulatory factors, could be responsible for such an event. Thus, a system biology approach is more likely to provide the panoramic type of observations that may lead to the true mechanisms associated with the control of tumor growth by the host and its rejection in response to immune therapy. It is likely that, independent of possible genetic predispositions related to distinct immunogenetic properties of the host, tumors can dictate their own immune responsiveness in relation to their own production of immune regulatory factors that may affect the host systemically and/or within the tumor microenvironment (Marincola et al. 2003) We have previously observed that melanoma metastases can be categorized into two phenotypes of which one is characterized by high expression of interferon-stimulated genes (ISGs), and this expression is associated with that of several other immune regulatory cytokines as well as pro-angiogenic factors and other growth factors. In addition, Mandruzzato et al. (Mandruzzato et al. 2006) observed that immunological signatures had prognostic significance in melanoma. This is similar to the observations by others (Zhang et al. 2003; Pages et al. 2005; Galon et al. 2006) in the context of other solid tumors.

As the understanding of tumor/host interactions in humans has rapidly progressed after the identification of tumor antigens recognized by CTLs (Boon et al. 1994), studies have allowed a better characterization of tumor antigen–specific CTLs in vivo at the systemic level (Rivoltini et al. 1995; Marincola et al. 1996; Salgaller et al. 1995; D'Souza et al. 1998) and within the tumor microenvironment (Lee et al. 1998; Kammula et al. 1999; Panelli et al. 2000; Bioley et al. 2006). Further progress has been achieved through the study of the response of the host to the administration of antigen-specific and epitope-specific vaccines (Marincola 2005; Slingluff and Speiser 2005) or the adoptive transfer of tumor antigen–specific T cells (Yee 2005). Indeed, tumor antigen–specific immunization has allowed a precise enumeration of immune responses against cancer, their characterization, as well as the characterization of the target cells (Monsurro'et al. 2003, 2004, 2002). Through these studies, it has become apparent that the main reason for the lack of clearance of tumors by tumor antigen–specific T cells is not related to loss of tumor antigen–associated markers by cancer cells and, therefore, their escape from T cell recognition (Ohnmacht et al. 2001), but rather transcriptional profiling of vaccine-induced CTLs has shown that they lack the expression of genes associated with their complete activation and effector function (Monsurro' et al. 2004). Thus, tumor antigen–specific vaccines have clearly shown that while this strategy yields the reproducible systemic cellular responses against cancer cells, these are not sufficient because of the limited effectiveness of CTLs (Slingluff and Speiser 2005; Parmiani et al. 2002; Marincola and Ferrone 2003; Neidhardt-Berard et al. 2004; Paczesny et al. 2003; Slingluff et al. 2004). Thus, tissue specificity is not sufficient, and better results might be expected when a better understanding of the requirements for CTL activation will be achieved through the direct study of the tumor microenvironment at time points relevant to the natural history of the disease or its responsiveness to a given treatment.

For this purpose, the phenomenology of metastatic melanoma seems to adapt its self to this type of human investigation (Parmiani et al. 2002, 2003; Minev 2002; Speiser et al. 2003). Because of the accessibility of the subcutaneous metastases that is frequently present in patients with melanoma, these lesions can be more easily followed, biopsied, and studied; moreover, autologous tumor/ tumor infiltrating lymphocytes (TIL) pairs can be developed more easily (Wolfel et al. 1994). Melanoma has also a general tendency to regress upon systemic immune stimulation (Atkins et al. 1998). Finally, transcriptional analysis has clearly shown that melanoma bears signatures consistent with an immunologi-cally active microenvironment different from that of other solid tumors (Wang et al. 2004). As previously discussed, tumor antigen–specific immunization has lead to significant increases in the number of tumor antigen–specific CTLs, but this is not sufficient to induce cancer rejection (Marincola 2005; Slingluff and Speiser 2005; Lee et al. 1999). The understanding of the molecular basis of the events that may lead to an effective immune response of the host against melanoma points to a tissue-specific event similar to the selective rejection of allograft in transplantation, or the exquisite tissue-specificity of autoimmune responses or the accurate clearance of pathogen during acute infectious processes. As in allograft rejection and autoimmunity, the presence of antigen-specific immune responses seems to be a requisite for immune rejection, but it is not sufficient in itself and other triggering events are necessary, which lead to a switch from a chronic to an acute inflammatory process (Wang et al. 2002). Acute allograft rejection may suddenly follow long periods of chronic rejection well controlled with immune suppression, while autoimmune phenomena may wax and wane in between flares for reasons that are not totally understood (Lang et al. 2005; Ochsenbein et al. 1999; Sarwal et al. 2003). Possibly, the identity of the secondary stimuli required for the activation of immune-mediated, tissue-specific rejection are slowly unraveling through the study of various experimental and clinical models encompassing experimental diabetes (Lang et al. 2005), the immune-mediated rejection of basal cell carcinoma by toll receptor agonists (Urosevic et al. 2003; Panelli et al. 2006), and kidney allograft rejection (Sarwal et al. 2003). Gene profiling identified similarities among various rejection models; two patterns of transcriptional activation are frequently associated with immune-mediated tissue destruction that can be assigned to specific functional signatures: a first one, always present in response to immune stimulation but not sufficient alone to induce rejection is represented by the activation of ISGs related to the activity of both type I or type II interferons (Panelli et al. 2002; Stroncek et al. 2005). A second signature, more directly associated with immune rejection, appears to include activation of cellular effector mechanisms involving direct target killing (Monsurro' et al. 2004; Parmiani et al. 2003; Sarwal et al. 2003). This is in line with experimental observations, suggesting that factors other than T cell presence within the tumor-bearing host are required for tissue-specific rejection (Ochsenbein et al. 1999; Fuchs and Matzinger 1996; Matzinger 1998; Perez-Diez et al. 2007). Furthermore, these observations underlie the discrepancy between standard in vitro assays used to test the function of tumor antigen–specific CTLs and their actual in vivo function that can only be addressed by the utilization of novel assays capable to directly characterize T cell effector function ex vivo (Monsurro' et al. 2003, 2001; Shafer-Weaver et al. 2003; Malyguine et al. 2004; Rubio et al. 2003; Tomaru et al. 2003).

The level of CTL activation and differentiation in vivo remains to be elucidated (Marincola et al. 2003; Ohnmacht et al. 2001; McKee et al. 2005). Such characterization is important in view of the observation that tumor antigen–specific T cells can localize within tumors and interact with tumor cells by expressing IFN-$\gamma$. However, such interactions are not sufficient to induce cancer regression (Kammula et al. 1999; Panelli et al. 2000; Ohnmacht et al. 2001). Perhaps, the number of CTLs reaching the tumor site may be insufficient to clear large bulks of disease, or they are in a functional status incapable of killing efficiently. Studies performed in inbred mice suggest that the intensity of the immune response correlates with tumor clearance (Perez-Diez et al. 2002); however, in humans this correlation is difficult to demonstrate due to the extreme heterogeneity of tumor cells and of tumor-bearing individuals (Jin and Wang 2003; Marincola et al. 2000; Perez-Diez et al. 2002). In addition, T cell signaling may be altered in cancer patients (Zea et al. 1995; Lee et al. 1999) although the relevance of this may be controversial particularly in the context of immunization trials, as immunization-induced CTLs have been shown to express activation markers and can recognize cancer cells ex vivo displaying a classical cytokine profile (Monsurro' et al. 2002; Jin et al. 2006; Nielsen et al. 2000; Pittet et al. 2001a,b) on one hand, but lack other effector functions on the other hand (Kammula et al. 1999; Monsurro' et al. 2002, 2001; Nielsen et al. 2000; Pittet et al. 2001a,b; Hamann et al. 1997; Sallusto et al. 1999; Champagne et al. 2001; Speiser et al. 2001; Appay et al. 2002; Migueles et al. 2002; van Baarle et al. 2002).

## 5 The Study of the Requirements for Immune Rejection Within the Target Organ

Very little is known about the characteristics of the human microenvironment that may be permissive or resistant to immune-mediated rejection. This is because very little work has been applied to the study of human tissues biopsied at time points relevant to a particular event relevant to immune responsiveness. Ideally, the ability to predict immune responsiveness by obtaining pre-treatment biopsies could inform about the likelihood of predisposing factor that may influence tumor cell/host relationships. Biopsies obtained at salient time points during immunotherapy may inform about the actual mechanisms of action of the therapy and differentiate those that may impact outcome to those that are irrelevant. As a consequence of the limited study of the tumor microenvironment in humans, the requirements for the maintenance of tumor antigen–specific T cell function at the tumor site remain largely unknown. Transcriptional studies have shown that the tumor microenvironment is enriched in the expression of genes coding for soluble factors that could foster the growth of tumor cells through stimulation of angiogenesis and/or tissue repair (Marincola et al. 2003); these signatures are, indeed, very similar to those occurring during tissue repair when a short burst of inflammatory signals is rapidly quenched while tissue repair and remodeling quickly takes over (Deonarine et al. 2007). Yet, several of these factors have pleiotropic functions including potent immune modulatory activity exerting powerful effects locally and systemically. Importantly, the expression of such factors by individual tumors seems to be coordinated, making it extremely difficult to segregate at the individual gene level which of these genes may be most relevant in the determination of immune responsiveness or lack of. It could be hoped that, as a preliminary step, functional signatures could be identified predictive of response to immunotherapy as they have been described in association with the disease outcome in a small but significant number of patients (Mandruzzato et al. 2006).

Fine needle aspirates of melanoma metastases obtained before active specific immunization followed by systemic treatment with interleukin(IL)-2 suggested that tumors are pre-conditioned to respond to immunotherapy by displaying an already different immune microenvironment (Wang et al. 2002). Among the genes strongly upregulated during treatment in those metastases undergoing complete remission, interferon regulatory factor (IRF)-1 played a prominent role. This is an important observation because it suggests that IFN-γ activation most likely responsible for IRF-1 overexpression needs to be functional during immune responses. Interestingly, however, we have recently observed that in vitro administration of IL-2 to peripheral blood mononuclear cells or subset of lymphocytes also induces strong upregulation of IRF-1 although it is not known whether this is a primary effect of IL-2 or secondary to the release of IFN-γ by the IL-2-stimulated cells (Jin et al. 2006). Indeed, systemic IL-2 treatment appeared to be necessary to enhance the chances of immune rejection in the context of vaccination (Rosenberg et al. 1998), perhaps because (Boon et al. 1997; Rosenberg et al. 1998, 1989, 1997; Cormier et al. 1997) IL-2 may activate CTLs by inducing an acute inflammatory process within the tumor microenvironment mediated through a secondary release of cytokines (Panelli et al. 2004) as we previously observed by studying the transcriptional alterations induced by this cytokine in PBMC as well as in melanoma metastases (Panelli et al. 2002, 2004). This inflammatory process may lead to the recruitment of innate and/or adaptive immune components generating a pro-inflammatory milieu suitable for complete CTL activation (Panelli et al. 2003). In fact, immunization-induced CTLs, although displaying what is generally considered an effector phenotype (CD27 negative, CCR7 negative, CD45RA[high]) and being capable of secreting IFN-γ in response to antigen exposure, do not express perforin and cannot exert cytotoxic functions ex vivo (Monsurro' et al. 2002), nor display proliferative potential (Monsurro' et al. 2004). At the transcriptional level, this quiescent phenotype is characterized by a global downregulation of genes associated with T cell activation, proliferation, and effector function. However, this phenotype can be easily reversed in vitro by antigen recall and IL-2, suggesting that similar co-stimulation in vivo may lead to full CTL activation given appropriate immune stimulatory therapy (Monsurro' et al. 2003, 2004, 2002). Logically, within the tumor microenviron-ment, tumor antigen–specific CTLs, whether naturally occurring or immuniza-tion-induced, are likely to be exposed to tumor antigens but probably they are not exposed to sufficient co-stimulation (Fuchs and Matzinger, 1996). Under-standing how to safely provide such co-stimulation in the tumor microenviron-ment might be the key to successful implementation of tumor antigen–specific anti-cancer therapies. Thus, the limited studies so far performed within the tumor microenvironment strongly support the hypothesis that lack of tumor regression in the context of a cognitive immune system is related to a lack of sufficient activation of CTLs rather than loss of antigenic properties by tumor cells. Indeed, tumor cells progressively lose tumor antigen expression or presentation for recognition by CTL (Marincola et al. 2000, 2003); however, such losses are most likely to occur under immunologic pressure during an effective immune response, but are unlikely to be the primary cause of immune resistance. When the transcriptional profile of melanoma metastases was followed before and during therapy with a combination of active-specific immunization and IL-2, the level of expression of the tumor antigen targeted by the vaccine was not a predictor of responsiveness (Ohnmacht et al. 2001) consistent with a "silent" genetic profile of non-responding lesions noted by global gene expression analysis. On the contrary, loss of expression of the tumor antigen specifically targeted by the vaccination was consistently observed during therapy in lesions that ultimately underwent complete regression.

What are then the factors that may modulate tumor immune responsiveness? (Marincola et al. 2003) By transcriptional profiling, we have recently noted that metastatic melanomas are characterized by high expression of IL-16 messenger RNA compared with other less immunogenic tumors (Wang et al. 2004). Since IL-16 is the natural ligand for CD4 and it is a potent chemo attractant for CD4 positive T cells (Conti et al. 2002), it is possible that this cytokine may play an important role in anti-melanoma immune responses when biologically activated through the caspase cascade by attracting helper T cells. This may represent an example of how tumors may modulate the tumor microenvironment. More generally, the answer may lie beyond tumor immune biology and could be sought by the study of other immune pathologies characterized by different levels of immune responsiveness. Symptomatic acute hepatitis B and C virus infection results in viral clearance; however, asymptomatic infections result in chronic unresolving outcomes, suggesting that acute inflammation is necessary for pathogen clearance (Rehermann and Nascimbeni 2005). Similarly, allograft rejection follows for long periods an indolent course well controlled by mild immune suppression till sudden events trigger an uncontrollable acute rejection. Autoimmune diseases follow a waxing and waning course whose basis remains undefined. Could these patterns be explained by a common immunological requirement that determines the switch on of immune properties leading to tissue destruction turning a chronic into an acute inflammatory process?

As predicted by "danger" model, antigen exposure in the absence of co-stimulation is not sufficient to activate and sustain effective immune responses (Matzinger 1998, 2001, 2007) , and cancers lack the requisite dangers signals to activate sufficient immune responses (Fuchs and Matzinger 1996) while other infectious or autoimmune pathologies, or allograft conditions may vary in time, providing different levels of co-stimulatory activation. Global transcript analysis has provided a broad view of biological processes associated with immune-mediated tissue destruction and identified convergent characteristics. Neoplastic inflammation shares similarities with the inflammation seen in chronic hepatitis C virus (HCV) infection (Mantovani et al. 2008; Bowen and Walker 2005), and it is characterized by the expression of those ISGs previously described as commonly expressed by melanomas in natural conditions (Marincola et al. 2003) or in response to the systemic administration of IL-2 independently of clinical outcome (Panelli et al. 2002). Similar signatures can be identified in chronic allograft rejection controlled with standard immune suppression (Sarwal et al. 2003). Thus, it appears that ISGs are part of immunological processes associated with lingering unresolving inflammation. Genes associated with cytotoxic effector function, however, are rarely expressed in chronically inflamed tissues, but consistently appear when the inflammatory process causes destruction of a tumor (Sarwal et al. 2003; Panelli et al. 2006, 2002; Smith et al. 2003; Kim et al. 2005; Dinarello and Kim 2006). Transcriptional profiling suggested that genes associated with CTL and natural killer cell cytotoxic function are expressed by during response of melanoma to systemic IL-2 therapy (Panelli et al. 2002) or immune-mediated destruction of basal cell carcinomas by local application of toll-like receptor agonists (Panelli et al. 2006). Similar findings were reported by others who performed transcriptional analysis in renal biopsies obtained during acute allograft rejection and were compared with chronic rejection samples (Sarwal et al. 2003), during flares of Crohn's disease (Netea et al. 2005). These observations from disparate tissue rejection models suggest that immune-mediated tissue destruction comprises at least two components: a baseline expression of ISGs that may be required but it is not sufficient in itself to induce tissue rejection, and a less common signature associated with immune effector mechanism that is more tightly associated with rejection.

In conclusion, the mechanisms determining immune responsiveness of human tumors remain unclear; it is possible that a combination of immune genetic factors and characteristics of individual tumors may determine the natural history in each patient. Both of these possibilities need to be tackled simultaneously if a comprehensive picture is to be obtained. Although the phenomenon of immune-mediated cancer rejection is complex and multifactorial, modern technologies are suitable for genome-wide scanning and DNA and RNA level, and may provide insights on common patterns that may shed light not only on the mechanism leading to tumor rejection but also in general on the requirements for immune-mediated destruction on tissues in the context of other pathologies or in the context of allotransplantation.

# References

Appay V, Dunbar PR, Callan M, Klenerman P, Gillespie GMA, Papagno L et al.: Memory CD8 + T cells vary in differentiation phenotype in different persistent virus infections. *Nat Med* 2002, **8**: 379–385.
Atkins MB, Lotze MT, Dutcher JP, Fisher RI, Weiss G, Margolin K et al.: High-dose recombinant interleukin-2 therapy for patients with metastatic melanoma: analysis of 270 patients treated between 1985 and 1993. *J Clin Oncol* 1998, **17**: 2105–2116.

Badoual C, Hans S, Rodriguez J, Peyrard S, Klein C, Agueznay NH et al.: Prognostic value of tumor-infiltrating CD4 + T-cell subpopulations in head and neck cancers. *Clin Cancer Res* 2006, **12:** 465–472.

Bioley G, Jandus C, Tuyaerts S, Rimoldi D, Kwok WW, Speiser DE et al.: Melan-A/MART-1-specific CD4 T cells in melanoma patients: identification of new epitopes and ex vivo visualization of specific T cells by MHC class II tetramers. *J Immunol* 2006, **177:** 6769–6779.

Boon T, Cerottini J-C, Van den Eynde B, van der Bruggen P, Van Pel A: Tumor antigens recognized by T lymphocytes. *Annu Rev Immunol* 1994, **12:** 337–365.

Boon T, Coulie PG, Van den Eynde B: Tumor antigens recognized by T cells. *Immunol Today* 1997, **18:** 267–268.

Bowen DG, Walker CM: Adaptive immune responses in acute and chronic hepatitis C virus infection. *Nature* 2005, **436:** 946–952.

Brander C, O'Connor P, Suscovich T, Jones NG, Lee Y, Kedes D et al.: Definition of an optimal cytotoxic T lymphocyte epitope in the latently expressed Kaposi's sarcoma-associated herpesvirus kaposin protein. *J Infect Dis* 2001, **184:** 119–126.

Champagne P, Ogg GS, King A, Knabenhans C, Ellefsen K, Nobile M et al.: Skewed maturation of memory HIV-specific CD8 T lymphoctes. *Nature* 2001, **410:** 106–111.

Chua D, Huang J, Zheng B, Lau SY, Luk W, Kwong DL et al.: Adoptive transfer of autologous Epstein-Barr virus-specific cytotoxic T cells for nasopharyngeal carcinoma. *Int J Cancer* 2001, **94:** 73–80.

Conti P, Kempuraj D, Kandere K, Di Gioacchino M, Reale M, Barbacane RC et al.: Interleukin-16 network in inflammation and allergy. *Allergy Asthma Proc* 2002, **23:** 103–108.

Cormier JN, Salgaller ML, Prevette T, Barracchini KC, Rivoltini L, Restifo NP et al.: Enhancement of cellular immunity in melanoma patients immunized with a peptide from MART-1/Melan A [see comments]. *Cancer J Sci Am* 1997, **3:** 37–44.

Critchley-Thorne RJ, Yan N, Nacu S, Weber J, Holmes SP, Lee PP: Down-regulation of the interferon signaling pathway in T lymphocytes from patients with metastatic melanoma. *PLoS Med* 2007, **4:** e176.

Cui SH, Lin Y: Apparent correlation between nasopharyngeal carcinoma and HLA phenotype. *Zhong Hua Zhong Liu Zhi* 1982, **4:** 249–253.

Curiel TJ, Coukos G, Zou L, Alvarez X, Cheng P, Mottram P et al.: Specific recruitment of regulatory T cells in ovarian carcinoma fosters immune privilege and predicts reduced survival. *Nat Med* 2004, **10:** 942–949.

Deonarine K, Panelli MC, Stashower ME, Jin P, Smith K, Slade HB et al.: Gene expression profiling of cutaneous wound healing. *J Transl Med* 2007, **5:** 11.

Dinarello CA, Kim SH: IL-32, a novel cytokine with a possible role in disease. *Ann Rheum Dis* 2006, **65**(3): iii61–iii64.

D'Souza S, Rimoldi D, Lienard D, Lejeune F, Cerottini JC, Romero P: Circulating Melan-A/Mart-1 specific cytolytic T lymphocyte precursors in HLA-A2 + melanoma patients have a memory phenotype. *Int J Cancer* 1998, **78:** 699–706.

Dunn GP, Bruce AT, Ikeda H, Old LJ, Schreiber RD: Cancer immunoediting: from immunosurveillance to tumor escape. *Nature Immunol* 2002, **3:** 991–998.

Dunn GP, Old LJ, Schreiber RD: The three Es of cancer immunoediting. *Annu Rev Immunol* 2004a, **22:** 329–360.

Dunn GP, Old LJ, Schreiber RD: The immunobiology of cancer immunosurveillance and immunoediting. *Immunity* 2004b, **21:** 137–148.

Ensoli B, Sirianni MC: Kaposi's sarcoma pathogenesis: a link between immunology and tumor biology. *Crit Rev Oncog* 1998, **9:** 107–124.

Fuchs EJ, Matzinger P: Is cancer dangerous to the immune system? *Semin Immunol* 1996, **8:** 271–280.

Galon J, Costes A, Sanchez-Cabo F, Kirilovsky A, Mlecnik B, Lagorce-Pages C et al.: Type, density, and location of immune cells within human colorectal tumors predict clinical outcome. *Science* 2006, **313:** 1960–1964.

Gottschalk S, Edwards OL, Sili U, Huls MH, Goltsova T, Davis AR et al.: Generating CTLs against the subdominant Epstein-Barr virus LMP1 antigen for the adoptive immunotherapy of EBV-associated malignancies. *Blood* 2003, **101:** 1905–1912.

Gottschalk S, Heslop HE, Rooney CM: Treatment of Epstein-Barr virus-associated malignancies with specific T cells. *Adv Cancer Res* 2002, **84:** 175–201.

Green M: Management of Epstein-Barr virus-induced post-tranplant lymphoproliferative disease in recipients of solid organ transplanation. *Am J Transplant* 2001, **1:** 103–108.

Hamann D, Baars PA, Rep MHG, Hoolbrink B, Kerkhof-Garde SR, Klein MR et al.: Phenotype and functional separation of memory and effector human CD8+ T cells. *J Exp Med* 1997, **186:** 1407–1418.

Haque T, Taylor C, Wilkie GM, Murad P, Amlot PL, Beath S et al.: Complete regression of posttransplant lymphoproliferative disease using partially HLA-matched Epstein Barr virus-specific cytotoxic T cells. *Transplantation* 2001, **72:** 1399–1402.

He XS, Ji X, Hale MB, Cheung R, Ahmed A, Guo Y et al.: Global transcriptional response to interferon is a determinant of HCV treatment outcome and is modified by race. *Hepatology* 2006, **44:** 352–359.

Heslop HE, Rooney CM: Adoptive cellular immunotherapy for EBV lymphoproliferative disease. *Immunol Rev* 1997, **157:** 217–222.

Hildesheim A, Apple RJ, Chen C-J, Wang SS, Cheng Y-J, Klitz W et al.: Association of HLA class I and II alleles and extended haplotypes with nasopharyngeal carcinoma in Taiwan. *J Natl Cancer Inst* 2002, **94:** 1780–1789.

Howell WM, Bateman AC, Turner SJ, Collins A, Theaker JM: Influence of vascular endothelial growth factor single nucleotide polymorphisms on tumour development in cutaneous malignant melanoma. *Genes Immun* 2002a, **3:** 229–232.

Howell WM, Calder PC, Grimble RF: Gene polymorphisms, inflammatory diseases and cancer. *Proc Nutr Soc* 2002b, **61:** 447–456.

Howell WM, Turner SJ, Bateman AC, Theaker JM: IL-10 promoter polymorphisms influence tumour development in cutaneous malignant melanoma. *Genes Immun* 2001, **2:** 25–31.

Jin P, Panelli MC, Marincola FM, Wang E: Cytokine polymorphism and its possible impact on cancer. *Immunol Res* 2004, **30**(2): 181–190.

Jin P, Wang E: Polymorphism in clinical immunology. From HLA typing to immunogenetic profiling. *J Transl Med* 2003, **1:** 8.

Jin P, Wang E, Provenzano M, Deola S, Selleri S, Jiaqiang R et al.: Molecular signatures induced by interleukin-2 on peripheral blood mononuclear cells and T cell subsets. *J Transl Med* 2006, **4:** 26.

Kammula US, Lee K-H, Riker A, Wang E, Ohnmacht GA, Rosenberg SA et al.: Functional analysis of antigen-specific T lymphocytes by serial measurement of gene expression in peripheral blood mononuclear cells and tumor specimens. *J Immunol* 1999, **163:** 6867–6879.

Khanna R, Bell S, Sherritt M, Galbraith A, Burrows SR, Rafter L et al.: Activation and adoptive transfer of Epsten-Barr virus-specific cytotoxic T cells in solid organ transplant patients with posttransplant lymphoproliferative disease. *Proc Natl Acad Sci U S A* 1999, **96:** 10391–10396.

Kim SH, Han SY, Azam T, Yoon DY, Dinarello CA: Interleukin-32: a cytokine and inducer of TNFalpha. *Immunity* 2005, **22:** 131–142.

Lang KS, Recher M, Junt T, Navarini AA, Harris NL, Freigang S et al.: Toll-like receptor engagement converts T-cell autoreactivity into overt autoimmune disease. *Nat Med* 2005, **11:** 138–145.

Lee JE, Reveille JD, Ross MI, Platsoucas CD: HLA-DQB1*0301 association with increased cutaneous melanoma risk. *Int J Cancer* 1994, **59**(4): 510–3.

Lee K-H, Wang E, Nielsen M-B, Wunderlich J, Migueles.S., Connors M et al.: Increased vaccine-specific T cell frequency after peptide-based vaccination correlates with increased susceptibility to in vitro stimulation but does not lead to tumor regression. *J Immunol* 1999, **163:** 6292–6300.

Lee PP, Yee C, Savage PA, Fong L, Brockstedt D, Weber JS et al.: Characterization of circulating T cells specific for tumor-associated antigens in melanoma patients. *Nat Med* 1999, **5:** 677–685.

Li X, Ghandri N, Piancatelli D, Adams S, Chen D, Robbins FM et al.: Associations between HLA class I alleles and the prevalence of nasopharyngeal carcinoma (NPC) among Tunisians. *J Transl Med* 2007, **5:** 22.

Liu D, O'Day SJ, Yang D, Boasberg P, Milford R, Kristedja T et al.: Impact of gene polymorphisms on clinical outcome for stage IV melanoma patients treated with biochemotherapy: an exploratory study. *Clin Cancer Res* 2005, **11:** 1237–1246.

Malyguine A, Strobl S, Shafer-Weaver K, Ulderich T, Troke A, Baseler M et al.: A modified human ELISPOT assay to detect specific responses to primary tumor cell targets. *J Transl Med* 2004, **2:** 9.

Mandruzzato S, Callegaro A, Turcatel G, Francescato S, Montesco MC, Chiarion-Sileni V et al.: A gene expression signature associated with survival in metastatic melanoma. *J Transl Med* 2006, **4:** 50.

Marincola FM: A balanced review of the status of T cell-based therapy against cancer. *J Transl Med* 2005, **3:** 16.

Marincola FM, Ferrone S: Immunotherapy of melanoma: the good news, the bad news and what to do next. *Sem Cancer Biol* 2003, **13:** 387–389.

Marincola FM, Jaffe EM, Hicklin DJ, Ferrone S: Escape of human solid tumors from T cell recognition: molecular mechanisms and functional significance. *Adv Immunol* 2000, **74:** 181–273.

Marincola FM: The multiple ways to tumor tolerance [comment]. *J Immunother* 1997, **20:** 178–179.

Masramon L, Ribas M, Cifuentes P, Arribas R, Garcia F, Egozcue J et al.: Cytogenetic characterization of two colon cell lines by using conventional G-banding, comparative genomic hybridization, and whole chromosome painting. *Cancer Genet Cytogenet* 2000, **121:** 17–21.

Marincola FM, Rivoltini L, Salgaller ML, Player M, Rosenberg SA: Differential anti-MART-1/MelanA CTL activity in peripheral blood of HLA-A2 melanoma patients in comparison to healthy donors: evidence for in vivo priming by tumor cells. *J Immunother* 1996, **19:** 266–277.

Marincola FM, Shamamian P, Rivoltini L, Salgaller ML, Reid J, Restifo NP et al.: HLA associations in the anti-tumor response against malignant melanoma. *J Immunother* 1996, **18:** 242–252.

Marincola FM, Wang E, Herlyn M, Seliger B, Ferrone S: Tumors as elusive targets of T cell-based active immunotherapy. *Trends Immunol* 2003, **24:** 335–342.

Mantovani A, Romero P, Palucka AK, Marincola FM: Tumor immunity: effector response to tumor and role of the microenvironment. *Lancet* 2008, **371**(9614):711–783.

Matzinger P: An innate sense of danger. *Semin Immunol* 1998, **10:** 399–415.

Matzinger P: Danger model of immunity. *Scand J Immunol* 2001, **54:** 2–3.

Matzinger P: Friendly and dangerous signals: is the tissue in control? *Nat Immunol* 2007, **8:** 11–13.

McKee MD, Roszkowski JJ, Nishimura MI: T cell avidity and tumor recognition: implications and therapeutic strategies. *J Transl Med* 2005, **3:** 35.

Migueles SA, Laborico AC, Shupert WL, Sabbaghian MS, Rabin R, Hallahan CW et al.: HIV-specific CD8 + T cell proliferation is coupled to perforing expression and is maintained in nonprogressors. *Nature Immunol* 2002, **3:** 1061–1068.

Minev BR: Melanoma vaccines. *Semin Oncol* 2002, **29:** 479–493.

Monsurro' V, Nagorsen D, Wang E, Provenzano M, Dudley ME, Rosenberg SA et al.: Functional heterogeneity of vaccine-induced CD8 + T cells. *J Immunol* 2002, **168:** 5933–5942.

Monsurro' V, Nielsen M-B, Perez-Diez A, Dudley ME, Wang E, Rosenberg SA et al.: Kinetics of TCR use in response to repeated epitope-specific immunization. *J Immunol* 2001, **166:** 5817–5825.

Monsurro' V, Wang E, Panelli MC, Nagorsen D, Jin P, Smith K et al.: Active-specific immunization against melanoma: is the problem at the receiving end? *Sem Cancer Biol* 2003, **13**: 473–480.

Monsurro' V, Wang E, Yamano Y, Migueles SA, Panelli MC, Smith K et al.: Quiescent phenotype of tumor-specific CD8+ T cells following immunization. *Blood* 2004, **104**: 1970–1978.

Moss P, Khan N: CD8(+) T-cell immunity to cytomegalovirus. *Hum Immunol* 2004, **65**: 456–464.

Neidhardt-Berard EM, Berard F, Banchereau J, Palucka AK: Dendritic cells loaded with killed breast cancer cells induce differentiation of tumor-specific cytotoxic T lymphocytes. *Breast Cancer Res* 2004, **6**: R322–R328.

Netea MG, Azam T, Ferwerda G, Girardin SE, Walsh M, Park JS et al.: IL-32 synergizes with nucleotide oligomerization domain (NOD) 1 and NOD2 ligands for IL-1beta and IL-6 production through a caspase 1-dependent mechanism. *Proc Natl Acad Sci U S A* 2005, **102**: 16309–16314.

Nielsen M-B, Monsurro' V, Miguelse S, Wang E, Perez-Diez A, Lee K-H et al.: Status of activation of circulating vaccine-elicited CD8+ T cells. *J Immunol* 2000, **165**: 2287–2296.

Nicol AF, Fernandes AT, Bonecini-Almeida MG: Immune response in cervical dysplasia induced by human papillomavirus: the influence of human immunodeficiency virus-1 co-infection – review. *Mem Inst Oswaldo Cruz* 2005, **100**: 1–12.

Ochsenbein AF, Klenerman P, Karrer U, Ludewig B, Pericin M, Hengartner H et al.: Immune suerveillance against a solid tumor fails because of immunological ignorance. *Proc Natl Acad Sci U S A* 1999, **96**: 2233–2238.

Ohnmacht GA, Wang E, Mocellin S, Abati A, Filie A, Fetsch PA et al.: Short term kinetics of tumor antigen expression in response to vaccination. *J Immunol* 2001, **167**: 1809–1820.

Old LJ, Chen YT: New Paths in Human Cancer Serology. *J Exp Med* 1998, **187**: 1163–1167.

Ou B-X, Ruan H-Y, Fan Y: Study on association between HLA-A, B, C DR, and nasopharyngeal carcinoma in Guangzhou area. *Ai Zheng* 1985, **4**: 5–8.Rooney CM, Roskrow MA, Smith CA, Brenner MK, Heslop HE: Immunotherapy of Epstein-Barr virus-associated cancer. *J Natl Cancer Institute Monogr* 1998, **23**: 89–93.

Paczesny S, Ueno H, Fay J, Banchereau J, Palucka AK: Dendritic cells as vectors for immunotherapy of cancer. *Sem Cancer Biol* 2003, **13**: 439–447.

Pages F, Berger A, Camus M, Sanchez-Cabo F, Costes A, Molidor R et al.: Effector memory T cells, early metastasis, and survival in colorectal cancer. *N Engl J Med* 2005, **353**: 2654–2666.

Panelli MC, Martin B, Nagorsen D, Wang E, Smith K, Monsurro' V et al.: A genomic and proteomic-based hypothesis on the eclectic effects of systemic interleukin-2 administration in the context of melanoma-specific immunization. *Cells Tissues Organs* 2003, **177**: 124–131.

Panelli MC, Riker A, Kammula US, Lee K-H, Wang E, Rosenberg SA et al.: Expansion of Tumor-T cell pairs from Fine Needle Aspirates of Melanoma Metastases. *J Immunol* 2000, **164**: 495–504.

Panelli MC, Stashower M, Slade HB, Smith K, Norwood C, Abati A et al.: Sequential gene profiling of basal cell carcinomas treated with Imiquimod in a placebo-controlled study defines the requirements for tissue rejection. *Genome Biol* 2006, **8**: R8.

Panelli MC, Wang E, Phan G, Puhlman M, Miller L, Ohnmacht GA et al.: Genetic profiling of peripheral mononuclear cells and melanoma metastases in response to systemic interleukin-2 administration. *Genome Biol* 2002, **3**: RESEARCH0035.

Panelli MC, White RLJr, Foster M, Martin B, Wang E, Smith K et al.: Forecasting the cytokine storm following systemic interleukin-2 administration. *J Transl Med* 2004, **2**: 17.

Parmiani G, Castelli C, Dalerba P, Mortarini R, Rivoltini L, Marincola FM et al.: Cancer immunotherapy with peptide-based vaccines: what have we achieved? Where are we going? *J Natl Cancer Inst* 2002, **94**: 805–818.

Parmiani G, Castelli C, Rivoltini L, Casati C, Tully GA, Novellino L et al.: Immunotherapy of melanoma. *Sem Cancer Biol* 2003, **13**: 391–400.

Perez-Diez A, Joncker NT, Choi K, Chan WF, Anderson CC, Lantz O et al.: CD4 cells can be more efficient at tumor rejection than CD8 cells. *Blood* 2007.

Perez-Diez A, Spiess PJ, Restifo NP, Matzinger P, Marincola FM: Intensity of the vaccine-elicited immune response determines tumor clearence. *J Immunol* 2002, **168**: 338–347.

Pittet MJ, Speiser DE, Valmori D, Rimoldi D, Lienard D, Lejeune F et al.: Ex vivo analysis of tumor antigen specific CD8+ T cell responses using MHC/peptide tetramers in cancer patients. *Int Immunopharmacol* 2001a, **1**: 1235–1247.

Pittet MJ, Zippelius A, Speiser DE, Assenmacher M, Guillaume P, Valmori D et al.: Ex vivo IFN-γ secretion by circulating CD8 T lymphocytes: implications of a novel approach for T cell monitoring in infectious and malignant diseases. *J Immunol* 2001b, **166**: 7634–7640.

Rehermann B, Nascimbeni M: Immunology of hepatitis B virus and hepatitis C virus infection. *Nat Rev Immunol* 2005, **5**: 215–229.

Rinaldo CR, Jr., Torpey DJ, III: Cell-mediated immunity and immunosuppression in herpes simplex virus infection. *Immunodeficiency* 1993, **5**: 33–90.

Rivoltini L, Kawakami Y, Sakaguchi K, Southwood S, Sette A, Robbins PF et al.: Induction of tumor reactive CTL from peripheral blood and tumor infiltrating lymphocytes of melanoma patients by in vitro stimulation with an immunodominant peptide of the human melanoma antigen MART-1. *J Immunol* 1995, **154**: 2257–2265.

Rosenberg SA: Cancer vaccines based on the identification of genes encoding cancer regression antigens. *Immunol Today* 1997, **18**: 175–182.

Rosenberg SA, Lotze MT, Yang JC, Aebersold PM, Linehan WM, Seipp CA et al.: Experience with the use of high-dose interleukin-2 in the treatment of 652 cancer patients. *Ann Surg* 1989, **210**: 474–84; discus.

Rosenberg SA, Yang JC, Schwartzentruber D, Hwu P, Marincola FM, Topalian SL et al.: Immunologic and therapeutic evaluation of a synthetic tumor associated peptide vaccine for the treatment of patients with metastatic melanoma. *Nat Med* 1998, **4**: 321–327.

Rubin JT, Adams SD, Simonis T, Lotze MT: HLA polymorphism and response to IL-2 bases therapy in patients with melanoma. *Proc Soc Biol Ther Annu Meet* 1991, **1**: 18.

Rubio V, Stuge TB, Singh N, Betts MR, Weber JS, Roederer M et al.: Ex vivo identification, isolation and analysis of tumor-cytolytic T cells. *Nat Med* 2003, **9**: 1377–1382.

Salgaller ML, Afshar A, Marincola FM, Rivoltini L, Kawakami Y, Rosenberg SA: Recognition of multiple epitopes in the human melanoma antigen gp100 by peripheral blood lymphocytes stimulated in vitro with synthetic peptides. *Cancer Res* 1995, **55**: 4972–4979.

Sallusto F, Lenig D, Forster R, Lipp M, Lanzavecchia A: Two subsets of memory T lymphocytes with distinct homing potentials and effector functions. *Nature* 1999, **401**: 659–660.

Sarwal M, Chua MS, Kambham N, Hsieh SC, Satterwhite T, Masek M et al.: Molecular heterogeneity in acute renal allograft rejection identified by DNA microarray profiling. *N Engl J Med* 2003, **349**: 125–138.

Shafer-Weaver K, Sayers T, Strobl S, Derby E, Ulderich T, Baseler M et al.: The Granzyme B ELISPOT assay: an alternative to the 51Cr-release assay for monitoring cell-mediated cytotoxicity. *J Transl Med* 2003, **1**: 14.

Simons MJ, Day NE, Wee GB, Shanmugaratnam K, Ho HC, Wong SH et al.: Nasopharyngeal carcinoma V: immunogenetic studies of Southeast Asian ethnic groups with high and low risk for the tumor. *Cancer Res* 1974, **34**: 1192–1195.

Simons MJ: HLA and nasopharyngeal carcinoma: 30 years on. *ASHI Quarterly* 2003, **27:** 52–55.

Slingluff CL, Jr., Speiser DE: Progress and controversies in developing cancer vaccines. *J Transl Med* 2005, **3:** 18.

Slingluff CL, Jr., Petroni GR, Yamshchikov GV, Hibbitts S, Grosh WW, Chianese-Bullock KA et al.: Immunologic and clinical outcomes of vaccination with a multiepitope melanoma peptide vaccine plus low-dose interleukin-2 administered either concurrently or on a delayed schedule. *J Clin Oncol* 2004, **22:** 4474–4485.

Smith MW, Yue ZN, Korth MJ, Do HA, Boix L, Fausto N et al.: Hepatitis C virus and liver disease: global transcriptional profiling and identification of potential markers. *Hepatology* 2003, **38:** 1458–1467.

Speiser DE, Colonna M, Ayyoub M, Cella M, Pittet MJ, Batard P et al.: The activatory receptor 2B4 is expressed in vivo by human CD8+ effector alpha beta T cells. *J Immunol* 2001, **167:** 6165–6170.

Speiser DE, Pittet MJ, Rimoldi D, Guillame P, Luescher IF, Lienard D et al.: Evaluation of melanoma vaccines with molecularly defined antigens by ex vivo monitoring of tumor specific T cells. *Sem Cancer Biol* 2003, **13:** 461–472.

Straathof KC, Bollard CM, Popat U, Huls MH, Lopez T, Morriss MC et al.: Treatment of nasopharyngeal carcinoma with Epstein-Barr virus – specific T lymphocytes. *Blood* 2005, **105:** 1898–1904.

Stroncek DF, Basil C, Nagorsen D, Deola S, Arico E, Smith K et al.: Delayed Polarization of Mononuclear Phagocyte Transcriptional Program by Type I Interferon Isoforms. *J Transl Med* 2005, **3:** 24.

Tomaru U, Yamano Y, Nagai M, Maric D, Kaumaya PT, Biddison W et al.: Detection of virus-specific T cells and CD8+ T-cell epitopes by acquisition of peptide-HLA-GFP complexes: analysis of T-cell phenotype and function in chronic viral infections. *Nat Med* 2003, **9:** 469–476.

Trofe J, Beebe TM, Buell JF, Hanaway MJ, First MR, Alloway RR et al.: Posttransplant malignancy. *Prog Transplant* 2004, **14:** 193–200.

Urosevic M, Maier T, Benninghoff B, Slade H, Burg G, Dummer R: Mechanisms unerlying imiquimod-induced regression of basal cell carcinoma in vivo. *Arch Dermatol* 2003, **139:** 1325–1332.

van Baarle D, Kostense S, van Oers MHJ, Miedema F: Failing immune control as a result of impaired CD8+ T-cell maturation: CD27 might provide a clue. *Trends Immunol* 2002, **23:** 586–591.

Viguier M, Lemaitre F, Verola O, Cho MS, Gorochov G, Dubertret L et al.: Foxp3 expressing CD4+ CD25(high) regulatory T cells are overrepresented in human metastatic melanoma lymph nodes and inhibit the function of infiltrating T cells. *J Immunol* 2004, **173:** 1444–1453.

Wang E, Miller LD, Ohnmacht GA, Mocellin S, Petersen D, Zhao Y et al.: Prospective molecular profiling of subcutaneous melanoma metastases suggests classifiers of immune responsiveness. *Cancer Res* 2002, **62:** 3581–3586.

Wang E, Panelli MC, Zavaglia K, Mandruzzato S, Hu N, Taylor PR et al.: Melanoma-restricted genes. *J Transl Med* 2004, **2:** 34.

Wolfel T, Schneider J, Meyer zum Buschenfelde KH, Rammensee HG, Rotzschke O, Falk K: Isolation of naturally processed peptides recognized by cytolytic T lympho-cytes (CTL) on human melanoma cells in association with HLA-A2.1. *Int J Cancer* 1994, **57:** 413–418.

Woo EY, Yeh H, Chu CS, Schlienger K, Carroll RG, Riley JL et al.: Cutting edge: Regulatory T cells from lung cancer patients directly inhibit autologous T cell proliferation. *J Immunol* 2002, **168:** 4272–4276.

Yamaguchi T, Sakaguchi S: Regulatory T cells in immune surveillance and treatment of cancer. *Semin Cancer Biol* 2005, **2:** 115–123.

Yee C: Adoptive T cell therapy: addressing challanges in cancer immunotherapy. *J Transl Med* 2005, **3:** 17.

Zea AH, Curti BD, Longo DL, Alvord WG, Strobl SL, Mizoguchi H et al.: Alterations in T cell receptor and signal transduction molecules in melanoma patients. *Clin Cancer Res* 1995, **1:** 1327–1335.

Zhang L, Conejo-Garcia JR, Katsaros D, Gimotty PA, Massobrio M, Regnani G et al.: Intratumoral T cells, recurrence, and survival in epithelial ovarian cancer. *N Engl J Med* 2003, **348:** 203–213.

Zhang JZ: Correlation between nasopharyngeal carcinoma (NPC) and HLA in Hunan Province. *Zhong Hua Zhong Liu Zhi* 1986, **8:** 170–172.

# Spectrum, Function, and Value of Targets Expressed in Neoplastic Mast Cells

Peter Valent

**Abstract** A number of treatment concepts for myeloid neoplasms are based on molecular targets and respective targeted drugs. Systemic mastocytosis is a myeloid neoplasm that behaves as an indolent disease in most patients, but can also present as an aggressive disease or even as mast cell leukemia (MCL). In patients with aggressive mastocytosis or MCL, the response to conventional therapy is poor and the prognosis is grave. Therefore, a number of attempts have been made to identify novel targets and to develop targeted drugs for these malignancies. In the current paper, emerging new molecular targets expressed in neoplastic mast cells are discussed in light of novel therapeutic concepts, availability of drugs, and forthcoming clinical trials.

**Key words** Mastocytosis · Targets · KIT · Tyrosine kinase inhibitors

## 1 Pathogenesis of Mastocytosis and Classification

Mastocytosis is characterized by abnormal accumulation of mast cells (MC) in one or more organ systems (Lennert and Parwaresch 1979; Metcalfe 1991; Valent 1996). Cutaneous and systemic variants of the disease have been described (Lennert and Parwaresch 1979; Metcalfe 1991; Valent 1996; Travis et al. 1988; Horan and Austen 1991; Valent et al. 2003; Austen 1992). Cutaneous mastocytosis (CM) typically develops in childhood and is an indolent disease with frequent spontaneous regression. Systemic mastocytosis (SM) develops at any age and is characterized by involvement of one or more visceral organs, with or without skin involvement (Lennert and Parwaresch 1979; Metcalfe 1991; Valent 1996; Travis et al. 1988; Horan and Austen 1991; Valent et al. 2003; Austen 1992). In a high proportion of cases, the transforming *KIT* mutation D816V is detectable (Longley et al. 1996; Nagata et al. 1995; Longley et al. 1999; Feger et al. 2002; Akin et al.

P. Valent
Department of Internal Medicine I, Division of Hematology and Hemostaseology,
Medical University of Vienna, Waehringer Guertel 18-20, A-1090 Vienna, Austria
e-mail: peter.valent@meduniwien.ac.at

A. Falus (ed.), *Clinical Applications of Immunomics*,
DOI: 10.1007/978-0-387-79208-8_6, © Springer Science+Business Media, LLC 2008

2000). Sometimes, the mutation is also detectable in other myeloid cells (Akin et al. 2000; Yavuz et al. 2002). Thus, SM is a disease of multilineage hematopoietic progenitors, which is strongly supported by the notion that these patients may develop an associated clonal hematologic non-mast cell lineage disease (AHNMD) (Lawrence et al. 1991; Travis et al. 1988; Horny et al. 1990; Sperr et al. 2002; Valent et al. 2001). The WHO classification defines four categories of SM: indolent SM (ISM), SM with an AHNMD (SM-AHNMD), aggressive SM, and MCL (Table 1) (Valent et al. 2001; Valent et al. 2001). These entities differ greatly from each other in their clinical course and prognosis, and require different therapies. In fact, patients with ASM or MCL are candidates for cytoreductive drugs, whereas patients with ISM are treated with anti-mediator-type drugs, but not with cytoreductive agents (Table 1) (Austen 1992; Escribano et al. 2002; Valent et al. 2003).

So far, little is known about the pathogenesis of SM. Based on clinical and laboratory data, the *KIT* mutation D816V, which is detectable in most patients with SM, is considered to play an essential role as transforming event. This mutation leads to factor-independent tyrosine phosphorylation and activation of KIT, and thereby to autonomous growth of MC-progenitors (Feger et al. 2002; Furitsu et al. 1993; Piao and Bernstein 1996; Taylor and Metcalfe 2000). Interestingly, the mutation is found in all categories of SM including ISM, ASM, and MCL (Longley et al. 1996; Nagata et al. 1995; Longley et al. 1999; Feger et al. 2002; Akin et al. 2000; Sotlar et al. 2000; Valent et al. 2002 Metcalfe and Akin 2001). Therefore, apart from this mutation, also other factors apparently play a role in the clinical course (progression) of SM (Daley et al. 2001; Cervero et al. 1999). However, so far, no molecular defect specific for ASM or MCL, among SM categories, has been identified. Another important factor may be the involvement

**Table 1** Classification of mastocytosis and treatment options

| Disease variant | Treatment options |
|---|---|
| Cutaneous mastocytosis | Mediator-targeting drugs, PUVA* |
| Indolent systemic mastocytosis | Mediator-targeting drugs, PUVA* |
| Smouldering systemic mastocytosis | Mediator-targeting drugs, PUVA*, cytoreductive drugs in select cases (clear signs of disease progression) |
| Systemic mastocytosis with AHNMD | AHNMD is treated as if no SM was found and SM as if no AHNMD was diagnosed |
| Aggressive systemic mastocytosis | Interferon-alpha +/− glucocorticoids cladribine (2CdA), chemotherapy, cytoreductive drugs, splenectomy, tyrosine kinase inhibitors (trials) |
| Mast cell leukaemia | Chemotherapy +/− interferon-alpha, 2CdA, chemotherapy, splenectomy, stem cell transplantation (select cases), tyrosine kinase inhibitors (trials), palliative cytoreduction |
| Mast cell sarcoma | Chemotherapy plus radiation after surgery, palliative cytoreduction, tyrosine kinase inhibitors (trials) |
| Extracutaneous mastocytoma | Surgery |

* PUVA is recommended for severe cutaneous disease in CM and SM. AHNMD, associated hematologic clonal non-mast cell lineage disease

**Table 2** Pathogenetic factors and molecular mechanisms in systemic mastocytosis

| Pathogenetic factor | Mechanisms | Pathological correlate(s) |
|---|---|---|
| Gene polymorphisms, genetic predisposition | Susceptibility of stem cells to transformation, immune surveillance | Disease heterogeneity, indolent versus aggressive mast cell disease |
| *KIT* mutation D816V (somatic) | Abnormal signaling, enhanced MC survival, abnormal expression of critical molecules (adhesion- or survival-related) | Mast cell accumulation in various tissues, formation of clusters of mast cells, mastocytosis infiltrates |
| Disease evolution | Multiple genetic defects (genes unknown) | Progression of mastocytosis, occurrence of AHNMD |
| Involvement of immature myeloid stem cells | Stem cell susceptibility to transformation? clonal evolution | Multilineage involvement, smouldering mastocytosis, occurrence of AHNMD |
| Interactions between mast cells and the surrounding bone marrow stroma | Angiogenic cytokines, adhesion molecules, fibrogenic cytokines | Increased bone marrow angiogenesis, osteosclerosis, bone marrow fibrosis |

of different subsets of stem cells. Likewise, in patients with advanced SM (smouldering, aggressive, leukemic, AHNMD), the *KIT* mutation D816V is often detectable not only in MC but also in other myeloid cells, whereas in typical ISM the mutation is usually found in MC (Yavuz et al. 2002; Valent et al. 2002; Akin et al. 2001; Jordan et al. 2001; Hauswirth et al. 2002). Lastly, various interactions between MC and the microenvironment may contribute to disease progression. Likewise, neoplastic MC in SM express increased amounts of adhesion molecules (Escribano et al. 1998; Schernthaner et al. 2001; Escribano et al. 2001). LFA-2 (CD2) is of interest in this regard since this antigen is only expressed on MC in SM, but is not expressed on normal MC (Escribano et al. 1998; Schernthaner et al. 2001; Escribano et al. 2001) . Since MC also express CD58, the natural ligand of CD2, it has been hypothesized that abnormal expression of adhesion antigens on MC in SM is associated with the formation of MC clusters, typically found in ISM (Schernthaner et al. 2001). In addition, MC express an array of mediators and cytokines (VEGF, FGF, tryptases) that may influence the surrounding microenvironment (Li and Baek 2002; Wimazal et al. 2002; Schwartz et al. 1995; Horny et al.1998). Table 2 shows a summary of potential factors and mechanisms that may play a role in the evolution and progression of SM.

# 2 Current treatment options for patients with systemic mastocytosis

Patients suffering from mediator-related symptoms are usually treated with $H_1$ and $H_2$ anti-histamines or other "mediator-targeting" agents (Austen 1992; et al. 2002; Worobec 2000; Worobec and Metcalfe 2002; Godt et al. 1997; Wolff

**Table 3** Cytoreductive and targeted drugs currently used in patients with mastocytosis

| Drug | Patients/Indication | Response Rate (%) |
|------|---------------------|-------------------|
| Interferon-alpha | SM/osteoporosis | >50 |
| | ASM/organopathy | 10–20 |
| 2CdA | ASM/organopathy | ~50 |
| Glucocorticoids | SM/ascites | >50 |
| Polychemotherapy | ASM, MCL/organopathy | 10–20 |
| | SM-AML/organopathy | >50 (AML) |
| Imatinib | ASM (KIT D816V-) | ? |

SM, systemic mastocytosis; ASM, aggressive systemic mastocytosis; MCL, mast cell leukemia.

et al. 2001) (Table 1). However, so far, no effective therapy for uncontrolled growth of neoplastic MC in patients with ASM or MCL is available (Escribano et al. 2002; Valent et al. 2003; Travis et al. 1986). Thus, only a few drugs have an effect on malignant MC growth in such patients. These drugs include interferon alpha (IFN-α) and glucocorticoids (Table 3) (Valent et al. 2003; Kluin-Nelemans et al. 1992; Delaporte et al. 1995; Worobec et al. 1996; Weide et al. 1996; Butterfield 1998). Hence, treatment responses are variable and mostly transient, and only seen in a subgroup of patients (Valent et al. 2003; Kluin-Nelemans et al. 1992; Delaporte et al. 1995; Worobec et al. 1996; Weide et al. 1996; Butterfield 1998). Other drugs that have been used to treat ASM or MCL include cladribine (2CdA) and poly-chemotherapy (Table 3) (Escribano et al. 2002; Valent et al. 2003; Godt et al. 1997; Wolff et al. 2001; Travis et al. 1986; Tefferi et al. 2001; Kluin-Nelemans et al. 2003). Especially, 2CdA has been described to counteract malignant growth of MC in a group of patients with ASM (Tefferi et al. 2001; Kluin-Nelemans et al. 2003). Table 3 shows a summary of drugs currently used to treat ASM or MCL. However, the effects of most drugs may be short-lived without complete remission. Therefore, a number of attempts are currently made to identify suitable targets in neoplastic MC and to develop respective targeted drugs.

# 3 Expression of Molecular Targets on the Surface of Neoplastic Cells

MC derive from CD34+ multilineage progenitors and share several surface antigens with other myeloid cells (Kirshenbaum et al. 1999; Agis et al. 1993; Rottem et al. 1994; Valent et al. 1989; Valent and Bettelheim 1992; Agis et al. 1996). Among those are antigens that have already been employed as targets in clinical hematology (Stirewalt et al. 2003; Matthews et al. 1999; van Der Velden et al. 2001; Sievers et al. 2001; Frankel et al. 2002; Morris and Waldmann 2000; Valent et al. 2001; Saleh et al. 1998). Potential targets expressed on the surface of neoplastic MC include CD13 (aminopeptidase N), CD25 (IL-2Ra), CD33

**Table 4** Molecular targets expressed on the surface of neoplastic mast cells and their progenitors

| Target | CD | Normal MC | MC in SM | MC Progenitors | Targeted Drugs |
|---|---|---|---|---|---|
| Aminopeptidase N | CD13 | − | +/− | + | Betulinic acid (?) |
| IL-2R-alpha | CD25 | − | + | +/− | HAT, $^{90}$Y-HAT, DAB$_{389}$IL-2 |
| Siglec-3 | CD33 | + | + | +/− | Mylotarg, Gemtuzumab-Ozogamicin |
| Pgp1 | CD44 | + | + | + | CD44 antibodies |
| LCA | CD45 | + | + | + | $^{131}$I-CD45 |
| LAMP3 | CD63 | + | + | + | CD63 antibodies |
| uPA receptor | CD87 | + | + | +/− | DT(388)uPA |
| GM-CSF receptor | CD116 | − | − | −/+ | DT(388)GM-CSF |
| IL-3 receptor | CD123 | − | −/+ | +/− | DT(388)IL-3 |
| IFN receptors | − | +/− | nk | +/− | IFNα, IFNγ |

IL, interleukin; LCA, leukocyte common antigen; IFN, interferon

(Siglec-3), CD45 (CLA), CD46 (MCP), CD55 (DAF), CD59 (MACIF), CD63 (LAMP3), CD117 (KIT), and CD87 (uPAR) (Table 4) (Escribano et al. 1998; Schernthaner et al. 2001; Escribano et al. 2001; Valent et al. 2001). Of considerable interest is that neoplastic MC in SM, but not normal MC, express CD25 (Table 4) (Escribano et al. 1998, 2001). In addition, patients with SM exhibit elevated plasma levels of soluble CD25 (Akin et al. 2000; Akin and Metcalfe 2002). Whether CD25 can be employed as a target of therapy in patients with ASM or MCL is currently under investigation. Siglec-3 (CD33) is widely distributed on myeloid progenitor cells and in myeloid leukemias, as well as in neoplastic MC (Table 4) (Stirewalt et al. 2003; van Der Velden et al. 2001; Sievers et al. 2001). Based on this notion, CD33 targeted therapy has been established recently using a conjugate composed of a humanized CD33 antibody (P67.6; gemtuzumab) and the cytostatic drug calicheamicin (ozogamicine) (Stirewalt et al. 2003). This drug-conjugate (Mylotarg) has been described as a potent anti-leukemic agent in myeloid leukemias (Stirewalt et al. 2003; van Der Velden et al. 2001; Sievers et al. 2001). MC also express Siglec-3/CD33 (Table 4) (Valent et al. 1989; Valent and Bettelheim 1992; Agis et al. 1996). In addition, Mylotarg downregulates the growth of malignant MC without inducing mediator-secretion in mature MC (unpublished observation). However, the clinical value of Mylotarg in ASM or MCL remains to be determined. It is noteworthy in this regard that Mylotarg therapy is often associated with liver toxicity (Sievers et al. 2001). The same holds true for other (humanized) antibodies and ligand–toxin conjugates. Since patients with ASM or MCL often have significant liver involvement and abnormal liver function (Valent et al. 2003), the use of Mylotarg and other antibodies must be considered with great caution in these patients. Other potential targets of antibody therapy are CD87 and CD45 (Table 4). For both antigens, targeted therapies have been developed using ligand–toxin conjugates (Stirewalt et al. 2003; Matthews et al. 1999; Frankel et al. 2002).

However, in case of CD45, the approach is myeloablative (normal hemato-poietic stem cells express CD45) and therefore has to be combined with a stem cell transplantation strategy (Matthews et al. 1999).

Another probably less toxic strategy of targeting may be to generate an endogenous immune response against the target structure (Durrant and Spendlove 2001; Hafner et al. 2002; Popkov et al. 2000). One specific and elegant technique is the mimotope strategy in which a monoclonal and epitope-specific antibody is employed to define immunogenic peptides and to generate a specific, i.e., polyclonal, but still epitope-specific, immune response (Hafner et al. 2002; Popkov et al. 2000).

A number of different cell surface antigens (CD63, CD55, CD59, CAMs, others) have been considered as potential targets of passive immunotherapies for cancer patients (Durrant and Spendlove 2001; Hafner et al. 2002; Popkov et al. 2000). A recently proposed approach is to employ surface complement-regulatory (inhibitory) proteins such as CD46, CD55, or CD59 (Durrant and Spendlove 2001). The advantage of such strategy may be that "clonal selection" of CD-negative cells by targeted therapy would not occur because such "selected" cells are extremely susceptible to destruction by autologous complement (Durrant and Spendlove 2001). Considering this concept, it is noteworthy that normal and neoplastic MC express high levels of CD46, CD55, and CD59 (Valent and Bettelheim 1992; Agis et al. 1996; Valent et al. 2001; Füreder et al. 1995; Nunez-Lopez et al. 2003). In addition, MC express several CAMs and CD63, an activation-linked antigen that has recently been described as a potential immunogen (Smith et al. 1997).

When considering vaccination-type immunotherapies for patients with ASM or MCL, several important aspects have to be considered. First, such strategy may only work in patients with a reduced tumor burden after success-ful chemotherapy. However, in patients with ASM or MCL, most available therapies are not inducing complete remission (Valent et al. 2003; Travis et al. 1986; Kluin-Nelemans et al. 1992; Delaporte et al. 1995; Worobec et al. 1996; Weide et al. 1996; Butterfield 1998). Cladribine (2CdA) and the new KIT tyrosine kinase (TK) inhibitors may be exceptions in this regard and may significantly reduce the MC burden (mostly transiently) in patients with ASM or MCL (Tefferi et al. 2001; Kluin-Nelemans et al. 2003). Therefore, immuno-therapies may best be combined with preceding therapy with 2CdA, TK inhibitors, or polychemotherapy. A second important aspect is that in con-trast to active immunotherapy, vaccination is a "permanent" (long-lasting) maneuver. Therefore, it has to be clarified with certainty that the target is *not* expressed on normal stem cells. A novel promising target in this regard may be the ectoenzyme CD203c (E-NPP3) (Bühring et al. 1999). This enzyme is specifically expressed on MC and basophils as well as on basophil- and MC-committed CD34+ progenitors (which comprise only a small subset of all CD34+ cells), but is not expressed on immature CD34+ stem cells (Bühring et al. 1999).

# 4  Signal Transduction-Associated Targets in Neoplastic Mast Cells

Since the somatic *KIT* mutation D816V is a gene defect that leads to constitutive activation of the TK domain of the receptor and is involved in the pathogenesis of SM (Longley et al. 1996; Nagata et al. 1995; Longley et al. 1999; Feger et al. 2002; Furitsu et al. 1993; Piao and Bernstein 1996; Taylor and Metcalfe 2000), recent attempts have focused on agents capable of inhibiting the TK activity of KIT D816V, or of critical molecules that act downstream of the mutated receptor in oncogenic signaling networks.

Activation of wild type (wt) KIT by its ligand, stem cell factor (SCF), is associated with receptor-dimerization and a complex cascade of signal transduction events (Sattler et al. 1997; Chian et al. 2001; Blume-Jensen et al. 1994; Boissan et al. 2000; Linnekin 1999; Serve et al. 1995; Voseller et al. 1987; Okuda et al. 1992). Most of these pathways may also be activated by KIT D816V (Taylor and Metcalfe 2000; Chian et al. 2001; Boissan et al. 2000). However, some differences in signaling between wt KIT and KIT D816V have been described (Chian et al. 2001; Piao et al. 1996). Altered signaling through KIT D816V may also be associated with abnormal expression of important survival molecules. However, so far, only little is known about the nature of such KIT D816V-regulated effector molecules.

Because of its pathogenetic significance, increasing effort is currently undertaken to identify pharmacologic inhibitors of the KIT TK activity. A first promising observation has been that the TK inhibitor STI571 (imatinib) that is successfully applied in chronic myeloid leukemia (CML) is also capable of inhibiting the TK activity of wild type (wt) KIT (Buchdunger et al. 2000; Zermati et al. 2003). Moreover, it has been shown that imatinib downregulates (SCF-dependent) the in vitro growth of human MC (Akin et al. 2003). However, unfortunately, the D816V-mutated form of KIT is far less susceptible to the inhibitory action of imatinib compared to wt KIT (Akin et al. 2003; Ma et al. 2002). As a consequence, this drug can only be offered to patients with ASM or MCL in whom no activating *KIT* mutation is found (Table 5). However, these patients represent only a minority of SM patients requiring cytoreductive therapy. More recently, three novel KIT TK inhibitors that effectively counteract the TK activity of KIT D816V have been identified, namely PKC412 (midostaurin), BMS354825 (dasatinib), and AMN107 (nilotinib) (Gotlib et al. 2005; Gleixner et al. 2006; Shah et al. 2006; Fumo et al. 2003). These TK inhibitors reportedly counteract the growth of neoplastic MC harboring KIT D816V at pharmacologic concentrations (Table 5) (Gotlib et al. 2005; Gleixner et al. 2006; Shah et al. 2006) and are currently tested in clinical trials.

An alternative possibility to target KIT D816V is to influence its expression by antisense- or siRNA strategies or other specific pharmacologic approaches and drugs. One such drug is 17-allylaminogeldanamycin (17-AAG). This agent is considered to act by inhibiting the chaperone protein HSP90, which binds to

**Table 5** Signal transduction- and survival-related targets expressed in neoplastic mast cells

| Target | Targeted Drug |
|--------|---------------|
| Wild-type KIT tyrosine kinase | Imatinib, PKC412, AMN107, Dasatinib, ... |
| D816V KIT tyrosine kinase | PKC412, Dasatinib, AMN107, EXEL-0862 |
| RAS, farnesyltransferase (FT) | FT inhibitors (FTI), R115777, others |
| mTOR | Rapamycin, RAD001, CCI-779 |
| Hsp 90 chaperone | 17-AAG (17-allylaminogeldanamycin) |
| Proteasome | Bortezomib, MG-132, others |
| MITF | MITF antisense (?) |
| STAT5 | Sorafenib, Piceatannol (?) |
| Bcl-2 | Bcl-2 antisense, Genasense |
| Bcl-$x_L$ | Bcl-$x_L$ antisense, Bcl-$x_L$siRNA |
| Mcl-1 | Mcl-1 antisense, Mcl-1 siRNA, Sorafenib |
| VEGF | Bevacizumab, VEGF antisense or siRNA |
| SCF | SCF antisense or siRNA, glucocorticoids* |
| VEGF-R, KDR | SU5416, SU11248, SU6668, IMC-2C6, IMC-1121 |
| bFGF-R | SU6668 |
| PDGF-R beta | Imatinib, SU11248, SU6668 |
| Beta-tryptase | Beta-tryptase-antisense |

* Glucocorticoids suppodedly downregulate the expression of SCF in stroma cells

and stabilizes several cytokine receptors including wt KIT and KIT D816V (Fumo et al. 2003).

Another approach is to target KIT-dependent downstream signaling pathways in neoplastic cells using specific pharmacologic inhibitors. In this regard, it is important to be aware that D816V KIT-dependent signaling may not utilize exactly the same pathways compared to the wt KIT (Chian et al. 2001). Another important aspect is that KIT D816V activates multiple signaling pathways and that these pathways often cooperate through diverse cross-interactions. Therefore, the most promising approach may be to combine signal transduction targeting drugs in order to counteract the growth of neoplastic MC. Most promising agents in this regard may be Ras-targeting compounds (farnesyl transferase inhibitors, FTIs; farnesyl-thiosalicylic acid, FTS), PI3-kinase inhibitors, or inhibitors of mTOR, a PI3-kinase-downstream kinase that is involved in the regulation of cell cycle progression and is inhibited by rapamycin (Hidalgo and Rowinsky 2000; Gabillot-Carre et al. 2006). Another important question is whether resistance of KIT D816V against imatinib can be overcome by the addition of other drugs. Finally, such targeted drugs may be used in combination with standard therapy such as cladribine (2CdA), interferon-alpha (IFN-$\alpha$), or chemotherapy (chemosensitization) (Gleixner et al. 2006).

An alternative approach that may also target multiple signaling cascades is the use of proteasome inhibitors (Guzman et al. 2002). In fact, such inhibitors have recently been described to counteract the growth of neoplastic hematopoietic stem cells (Guzman et al. 2002). The effect of these agents on neoplastic MC and MC-stem cells is currently under investigation.

Table 5 shows a summary of targeted drugs that may be employed to define new treatment strategies in patients with ASM or MCL.

# 5  Molecular Targets that Play a Role in Growth or Survival of Mast Cells

A number of different "survival factors" may play a role in the accumulation and growth of normal and neoplastic MC. As mentioned above, KIT and its ligand, SCF, are well-recognized survival factors for MC. Other factors appear to act downstream of the (mutated) KIT receptor. The biologic role of such survival factors is best documented in deficiency models. Thus, disruption of genes encoding survival factors in normal mice is associated with MC deficiency. Likewise, mice lacking a functionally active *SCF* gene (Sl/Sl$^d$), *c-kit* gene (W/W$^v$), or *MITF* gene (mi/mi) are virtually deficient in MC (Kitamura et al. 1978; Kitamura and Go 1979; Ebi et al. 1990; Kitamura et al. 2002). In Bcl-x$_L$-deficient mice, the entire myeloid compartment (including MC) is depressed (Motoyama et al. 1995). More recent data suggest that STAT5 knock-out mice are MC-deficient (Shelburne et al. 2003). It is tempting to speculate that such MC-survival factors may also represent suitable targets for therapy (Table 5). However, it has to be emphasized that such deficiency models are not investigating neoplastic MC carrying mutated KIT, which may utilize different pathways of signaling and different survival factors compared to wt KIT (Chian et al. 2001). Therefore, the suitability of each target has to be reconfirmed for the growth and survival of neoplastic human MC.

Among the potential survival factors that may play a role in MC are certain transcription factors (e.g., STAT5, MITF) and members of the Bcl-2 family (Bcl-x$_L$, Bcl-2, Mcl-1) (Kitamura et al. 1978; Kitamura and Go 1979; Ebi et al. 1990; Kitamura et al. 2002; Motoyama et al. 1995; Shelburne et al. 2003). At least for some of these survival factors, it has been suggested that KIT D816V upregulates their expression or function (Mayerhofer et al. 2008). In line with this notion, neoplastic bm MC in patients with SM exhibit higher levels or/and a constitutively activated form of certain survival factors. An interesting molecule in this regard is Bcl-2. Notably, it has been shown that Bcl-2 is overexpressed in MC in patients with MCL compared to normal MC or MC in patients with ISM (Cervero et al. 1996). Since neoplastic MC in patients with MCL are particularly resistant against antineoplastic drugs, Bcl-2 (or other Bcl-2 members) may be attractive targets for chemosensitization.

A number of different approaches to target intracellular survival factors have been proposed. One example is the use of antisense oligonucleotides (Jansen and Zangemeister-Wittke 2002; Jansen et al. 2000). Likewise, the application of an antisense against Bcl-2 has been reported to lead to chemosensitization in patients with malignant melanoma (Jansen and Zangemeister-Wittke 2002; Jansen et al. 2000). This antisense approach may also be considered for other gene products such as MITF, Mcl-1, or KIT. However, the problem with antisense strategies is that it remains unclear how much of the construct is entering the cytoplasmic compartment in a given tumor cell. A more recently developed strategy is the use of small interfering RNA (siRNA) (Scherr et al. 2003; Wohlbold et al. 2003). However, so far, the effects of antisense probes or siRNA on the growth of neoplastic MC in vivo remain unknown.

## 6 Mast Cell-Derived Effector Molecules as Targets of Therapy

A number of mediators and cytokines, derived from MC, may be involved in disease evolution. Likewise, MC express several angiogenic and fibrogenic growth factors such as vascular endothelial growth factor (VEGF), tryptases, or basic fibroblast growth factor (bFGF) (Li and Baek 2002; Wimazal et al. 2002; Schwartz et al. 1995). These molecules may play a role in SM-related angiogenesis and fibrosis. Other MC-derived cytokines may act as autocrine or paracrine growth regulators. Likewise, activated MC reportedly produce and secrete several interleukins (Gordon et al. 1990; Galli et al. 1991). Furthermore, MC may produce and secrete SCF under certain conditions (de Paulis et al. 1999; Akin et al. 2002). In addition, SCF may be produced by stromal cells in various tissues, and may act as a paracrine growth regulator for neoplastic MC (Flanagan et al. 1991; Kiener et al. 200; Finotto et al. 1987). Notably, MC also express tumor necrosis factor (TNF) alpha (Gordon et al. 1990; Galli et al. 1991; Gordon and Galli 1990), a cytokine that can induce the expression of SCF and other cytokines in stromal cells (Kiener et al. 2000).

A number of attempts have been made to identify the drugs that would interfere with the production or action of MC-derived growth regulators (Table 5). In case of SCF, glucocorticoids may act as inhibitors of production of this cytokine in stromal cells (Finotto et al. 1987). Thus, in some patients with aggressive SM, glucocorticoids counteract malignant MC growth, probably through downregulation of expression of SCF. In case of VEGF, targeted drugs have also been developed (Table 5). One such drug is bevacizumab, a monoclonal anti-VEGF antibody. However, no data on the effects of bevacizumab on MC are available. An interesting observation is that the immunosuppressive agent rapamycin, which acts through the inhibition of mTOR, counteracts not only VEGF production in neoplastic cells and thus neoangiogenesis but also the growth of malignant (mast) cells (Hidalgo and Rowinsky 2000; Gabillot-Carre et al. 2006; Mayerhofer et al. 2002; Guba et al. 2002). Other drugs that interfere with angiogenesis in myeloid neoplasms are thalidomide, revlimid, and several tyrosine kinase inhibitors targeting VEGF receptors. However, the effects of such inhibitors on neoplastic MC remain unknown.

# 7 Targeting of Mast Cell Progenitors and Neoplastic Stem Cells in Mastocytosis

In common with all other myeloid cells, normal MC derive from uncommitted myeloid progenitors (Kirshenbaum et al. 1999; Agis et al.1993). In patients with SM, MC also derive from CD34+ progenitors (Rottem et al. 1994). Some of these cells may exhibit unlimited self-renewal capacity comparable to normal stem cells, and thus may be termed "SM stem cells.". A similar concept has recently been proposed for clonal stem cells in myeloid leukemias, and may be applicable to neoplastic disorders in general (Bonnet and Dick 1997; Sutherland et al. 1996; Jordan et al. 2000). By definition, these neoplastic stem cells are the source of persisting disease and of disease relapses, and therefore represent a most critical target in curative therapies (Bonnet and Dick 1997; Sutherland et al. 1996; Jordan et al. 2000). However, due to their low frequency and the difficulties of their detection (requiring a NOD/SCID mouse model and high capacity sort facilities), little is known so far about neoplastic stem cells in hematopoietic malignancies. In myeloid leukemias, neoplastic stem cells reportedly are CD34+/CD38– cells and often express CD123, whereas the levels of KIT/CD117 vary depending on the type of disorder (Bonnet and Dick 1997; Sutherland et al. 1996; Jordan et al. 2000). The phenotype of the SM stem cell is so far not known, although the phenotype may be related to the marker profile of neoplastic stem cells in AML (Florian et al. 2006).

In myeloid leukemias, a number of curative treatment strategies have recently been developed on the basis of expression of molecular target antigens in leukemic stem cells (Guzman et al. 2002; Black et al. 2003; Feuring-Buske et al. 2002). Likewise, expression of CD123 on leukemic stem cells in patients with myeloid leukemias has prompted several investigators to develop CD123-targeted therapies (Black et al. 2003; Feuring-Buske et al. 2002). Leukemic stem cells also express CD33, the target receptor for Mylotarg (Florian et al. 2006). Both antigens, i.e., CD33 and CD123, are not expressed on normal stem cells, so that the treatment approach is non-myeloablative and therefore can be performed without bone marrow transplantation (Black et al. 2003; Feuring-Buske et al. 2002). In other instances, however, normal hematopoietic stem cells also express the target (e.g., CD45), so that targeted therapy has to be combined with a stem cell transplantation approach. An interesting observation had been that the CD34+/CD38– cells in SM also express CD33, CD44, CD45, and CD123 (Florian et al. 2006).

At a more advanced stage of stem cell differentiation, the progenitor cells loose their ability to regenerate neoplastic cells for a long time. Targeting of such more mature progenitors without stem cell targeting is a non-curative approach. However, in the absence of any available curative therapy in malignant MC disorders, it may still be desirable to target these more mature progenitors. One approach of progenitor cell targeting is the use of IFN-α (Fiorani et al. 1999). In fact, at the CFU-level, progenitor cells supposedly express receptors for IFN-α.

Moreover, IFN-α inhibits colony (CFU) growth in myeloid progenitor cells (Broxmeyer et al. 1985; Broxmeyer 1992). In this regard, it has to be pointed out, however, that contrasting other cell lineages, MC development from such progenitors takes a long time (several weeks to months) (Födinger et al. 1994). Since mature MC are also long-lived cells and do not undergo apoptosis when exposed to IFN-α, the clinical effect of IFN-α is usually seen only after several months (Schernthaner et al. 2000; Hauswirth et al. 2004). By contrast, the effects of IFN-α on granulocyte hyperplasia in CML takes only a few weeks as the differentiation of CFU into granulocytes takes only 2–3 weeks and the life span of a mature granulocyte is short. Another approach to counteract the growth of neoplastic progenitor cells is the application of chemotherapy using cell cycle–specific drugs. Although such therapy may not target all neoplastic stem cells, these therapies may cause marked aplasia as a side effect. Such treatment may not be regarded as a "curative approach" in SM, as most neoplastic MC-progenitors may be dormant (non-cycling) cells and therefore escape therapy. Thus, it is generally assumed that such chemotherapies work better in malignancies with a high turnover of cells. Therefore, patients with rapidly progressing ASM or MCL may be more suitable candidates for polychemotherapy. Patients who show a response to such therapy should then be considered for further consolidation, immunotherapy, or hematopoietic stem cell transplantation.

## 8 Concluding Remarks

Targeted drug therapy of myeloid neoplasms is becoming an increasingly important issue in clinical hematology. Since advanced MC disorders have a grave prognosis, it seems desirable to search also for targets and targeted drugs in these malignancies. Notably, a number of potential targets are expressed in neoplastic mast cells and their progenitors in SM, and some of these targets may even be expressed in neoplastic stem cells. Whether these targets and targeted drugs can be employed to counteract the growth of neoplastic cells in patients with ASM/MCL will be determined in clinical trials in the near future.

## References

Agis H, Füreder W, Bankl HC, Kundi M, Sperr WR, Willheim M, et al. Comparative immunophenotypic analysis of human mast cells, blood basophils, and monocytes. *Immunology* 1996;**87**:535–543.

Agis H, Willheim M, Sperr WR, Wilfing A, Krömer E, Kabrna E, et al. Monocytes do not make mast cells when cultured in the presence of SCF. Characterization of the circulating mast cell progenitor as a c-kit +, CD34 +, Ly-, CD14-, CD17-, colony forming cell. *J Immunol* 1993;**151**:4221–4227.

Akin C, Brockow K, D'Ambrosio C, Kirshenbaum A, Ma Y, Longley JB, Metcalfe DD. Effects of tyrosine kinase inhibitor STI571 on human mast cells bearing wild-type or mutated forms of c-kit. *Exp Hematol* 2003;**31**:686–692.

Akin C, Jaffe ES, Raffeld M, Kirshenbaum AS, Daley T, Noel P, Metcalfe DD. An immunohistochemical study of the bone marrow lesions of systemic mastocytosis: expression of stem cell factor by lesional mast cells. *Am J Clin Pathol* 2002;**118**:242–247.

Akin C, Kirshenbaum AS, Semere T, Worobec AS, Scott LM, Metcalfe DD. Analysis of the surface expression of c-kit and occurrence of the c-kit Asp816Val activating mutation in T cells, B cells, and myelomonocytic cells in patients with mastocytosis. *Exp Hematol* 2000;**28**:140–147.

Akin C, Metcalfe DD. Surrogate markers of disease in mastocytosis. *Int Arch Allergy Immunol* 2002;**127**:133–136.

Akin C, Schwartz LB, Kitoh T, Obayashi H, Worobec AS, Scott LM, Metcalfe DD. Soluble stem cell factor receptor (CD117) and IL-2 receptor alpha chain (CD25) levels in the plasma of patients with mastocytosis: relationships to disease severity and bone marrow pathology. *Blood* 2000;**96**:1267–1273.

Akin C, Scott LM, Metcalfe DD. Slowly progressive systemic mastocytosis with high mast cell burden and no evidence of a non-mast cell hematologic disorder. An example of a smoldering case ? *Leuk Res* 2001;**25**:635–638.

Austen KF. Systemic Mastocytosis. *N Engl J Med* 1992;**326**:639–640.

Black JH, McCubrey JA, Willingham MC, Ramage J, Hogge DE, Frankel AE. Diphtheria toxin-interleukin-3 fusion protein (DT(388)IL3) prolongs disease-free survival of leukemic immunocompromised mice. *Leukemia* 2003;**17**:155–159.

Blume-Jensen P, Ronnstrand L, Gout I, Waterfield MD, Heldin CH. Modulation of Kit/stem cell factor receptor-induced signaling by protein kinase C. *J Biol Chem* 1994;**269**: 21793–21802.

Boissan M, Feger F, Guillosson JJ, Arock M. c-Kit and c-kit mutations in mastocytosis and other hematological diseases. *J Leukoc Biol* 2000;**67**:135–148.

Bonnet D, Dick JE. Human acute myeloid leukemia is organized as a hierarchy that originates from a primitive hematopoietic cell. *Nat Med* 1997;**3**:730–737.

Broxmeyer HE. Suppressor cytokines and regulation of myelopoiesis. Biology and possible clinical uses. *Am J Pediatr Hematol Oncol* 1992;**14**:22–30.

Broxmeyer HE, Cooper S, Rubin BY, Taylor MW. The synergistic influence of human interferon-gamma and interferon-alpha on suppression of hematopoietic progenitor cells is additive with the enhanced sensitivity of these cells to inhibition by interferons at low oxygen tension in vitro. *J Immunol* 1985;**135**:2502–2506.

Buchdunger E, Cioffi CL, Law N, Stover D, Ohno-Jones S, Druker BJ, Lydon NB. Abl proteintyrosine kinase inhibitor STI571 inhibits in vitro signal transduction mediated by c-kit and platelet-derived growth factor receptors. *J Pharmacol Exp Ther* 2000;**295**:139–145.

Bühring HJ, Simmons PJ, Pudney M, Müller R, Jarrossay D, vanAgthoven A, et al. The monoclonal antibody 97A6 defines a novel surface antigen expressed on human basophils and their multi- and unipotent progenitors. *Blood* 1999;**94**:2343–2356.

Butterfield JH. Response of severe systemic mastocytosis to interferon alpha. *Br J Dermatol* 1998;**138**:489–495.

Cervero C, Escribano L, San Miguel JF, Diaz-Agustin B, Bravo P, Villarubia J, et al. Expression of Bcl-2 by human mast cells and its overexpression in mast cell leukemia. *Am J Hematol* 1999;**60**:191–195.

Chian R, Young S, Danilkovitch-Miagkova A, Rönnstrand L, Leonard E, Ferrao P, et al. Phosphatidylinositol 3 kinase contributes to the transformation of hematopoietic cells by the D816V c-Kit mutant. *Blood* 2001;**98**:1365–1373.

Daley T, Metcalfe DD, Akin C. Association of the Q576R polymorphism in the interleukin-4 receptor alpha chain with indolent mastocytosis limited to the skin. *Blood* 2001;**98**: 880–882.

de Paulis A, Minopoli G, Arbustini E, de Crescenzo G, Dal Piaz F, Pucci P, et al. Stem cell factor is localized in, released from, and cleaved by human mast cells.*J Immunol* 1999; **163**:2799–2808.

Delaporte E, Pierard E, Wolters BG, Desreumaux P, Janin A, Cortot A, et al. Interferon-alpha in combination with corticosteroids improves systemic mast cell disease. *Br J Dermatol* 1995;**132**:479–482.

Durrant LG, Spendlove I. Immunization against tumor cell surface complement-regulatory proteins. *Curr Opin Investig Drugs* 2001;**2**:959–966.

Ebi Y, Kasugai T, Seino Y, Onoue H, Kanemoto T, Kitamura Y. Mechanism of mast cell deficiency in mutant mice of mi/mi genotype: an analysis by co-culture of mast cells and fibroblasts. *Blood* 1990;**75**:1247–1251.

Escribano L, Akin C, Castells M, Orfao A, Metcalfe DD. Mastocytosis: current concepts in diagnosis and treatment. *Ann Hematol* 2002;**81**:677–690.

Escribano L, Díaz-Agustín B, Bellas C, Navalón R, Nuñez R, Sperr WR, et al. Utility of flow cytometric analysis of mast cells in the diagnosis and classification of adult mastocytosis. *Leuk Res* 2001;**25**:563–570.

Escribano L, Orfao A, Diaz-Agustin B, Villarrubia J, Cervero C, Lopez A, et al. Indolent systemic mast cell disease in adults: immunophenotypic characterization of bone marrow mast cells and its diagnostic implication. *Blood* 1998;**91**:2731–2736.

Feger F, Ribadeau Dumas A, Leriche L, Valent P, Arock M. Kit and *c-kit* mutations in mastocytosis: a short overview with special reference to novel molecular and diagnostic concepts. *Int Arch Allergy Immunol* 2002;**127**:110–114.

Feuring-Buske M, Frankel AE, Alexander RL, Gerhard B, Hogge DE A diphtheria toxin-interleukin 3 fusion protein is cytotoxic to primitive acute myeloid leukemia progenitors but spares normal progenitors. *Cancer Res* 2002;**62**:1730–1736.

Finotto S, Mekori YA, Metcalfe DD. Glucocorticoids decrease tissue mast cell number by reducing the production of the *c-kit* ligand, stem cell factor, by resident cells: *in vitro* and *in vivo* evidence in murine systems. *J Clin Invest* 1987;**99**:1721–1728.

Fiorani C, Tonelli S, Casolari B, Sacchi S. The role of interferon-alpha in the treatment of myeloproliferative disorders. *Curr Pharm Des* 1999;**5**:987–1013.

Flanagan JG, Chan DC, Leder P. Transmembrane form of kit ligand growth factor is determined by alternative splicing and is missing in the Sld mutant. *Cell* 1991;**64**:1025–1035.

Florian S, Sonneck K, Hauswirth AW, Krauth MT, Schernthaner GH, Sperr WR, Valent P. Detection of molecular targets on the surface of CD34+/CD38– stem cells in various myeloid malignancies. *Leuk Lymphoma* 2006;**47**:207–22.

Födinger M, Fritsch G, Winkler K, Emminger W, Mitterbauer G, Gadner H, et al. Origin of human mast cells: Development from transplanted hemopoietic stem cells after allogeneic bone marrow transplantation. *Blood* 1994;**84**:2954–2959.

Frankel AE, Beran M, Hogge DE, Powell BL, Thorburn A, Chen YQ, et al. Malignant progenitors from patients with CD87+ acute myelogenous leukemia are sensitive to a diphtheria toxin-urokinase fusion protein.*Exp Hematol* 2002;**30**:1316–1323.

Fumo G, Akin C, Metcalfe DD, Neckers L. 17-allylamino-17-demethoxy-geldanamycin (17-AAG) is effective in down-regulating mutated, constitutively activated KIT protein in human mast cells. *Blood* 2003;**103**:1078–1084.

Füreder W, Agis H, Willheim M, Bankl HC, Maier U, Kishi K, et al. Differential expression of complement receptors on human mast cells and basophils: Evidence for mast cell heterogeneity and C5aR/CD88 expression on skin mast cells. *J Immunol* 1995;**155**:3152–3160.

Furitsu T, Tsujimura T, Tono T, Ikeda H, Kitayama H, Koshimizu U, et al. Identification of mutations in the coding sequence of the proto-oncogene c-kit in a human mast cell leukemia cell line causing ligand-independent activation of the c-kit product. *J Clin Invest* 1993;**92**:1736–1744.

Gabillot-Carre M, Lepelletier Y, Humbert M, et al. Rapamycin inhibits growth and survival of D816V-mutated c-kit mast cells. *Blood* 2006;**108**:1065–1072.

Galli SJ, Gordon JR, Wershil BK. Cytokine production by mast cells and basophils. *Curr Opin Immunol* 1991;**3**:865–872.

Gleixner KV, Mayerhofer M, Aichberger KJ, et al. The tyrosine kinase-targeting drug PKC412 inhibits in vitro growth of neoplastic human mast cells expressing the D816V-mutated variant of kit: comparison with AMN107, imatinib, and cladribine (2CdA), and evaluation of cooperative drug effects. *Blood* 2006;**107**:752–759.

Godt O, Proksch E, Streit V, Christophers E. Short- and long-term effectiveness of oral and bath PUVA therapy in urticaria pigmentosa and systemic mastocytosis. *Dermatology* 1997;**195**:35–39.

Gordon JR, Burd PR, Galli SJ. Mast cells as a source of multifunctional cytokines. *Immunol Today* 1990;**11**:458–464.

Gordon JR, Galli SJ. Mast cells as a source of both preformed and immunologically inducible TNF-alpha/cachectin. *Nature* 1990;**346**:274–276.

Gotlib J, Berube C, Growney JD, et al. Activity of the tyrosine kinase inhibitor PKC412 in a patient with mast cell leukemia with the D816V KIT mutation. *Blood* 2005;**106**:2865–2870.

Growney JD, Clark JJ, Adelsperger J, et al. Activation mutations of human c-KIT resistant to imatinib are sensitive to the tyrosine kinase inhibitor PKC412. *Blood* 2005;**106**: 721–724.

Guba M, von Breitenbuch P, Steinbauer M, Koehl G, Flegel S, Hornung M, Bruns CJ, et al. Rapamycin inhibits primary and metastatic tumor growth by antiangiogenesis: involvement of vascular endothelial growth factor. *Nat Med* 2002;**8**: 128–135.

Guzman ML, Swiderski CF, Howard DS, Grimes BA, Rossi RM, Szilvassy SJ, Jordan CT. Preferential induction of apoptosis for primary human leukemic stem cells. *Proc Natl Acad Sci (USA)* 2002;**99**:16220–16225.

Hafner C, Samwald U, Wagner S, Felici F, Heere-Ress E, Jensen-Jarolim E, et al. Selection of mimotopes of the cell surface adhesion molecule Mel-CAM from a random pVIII-28aa phage peptide library. *J Ivest Dermatol* 2002;**119**:865–869.

Hauswirth AW, Simonitsch-Klupp I, Uffmann M, Koller E, Sperr WR, Lechner K, Valent P: Response to therapy with interferon alpha-2b and prednisolone in aggressive systemic mastocytosis: report of five cases and review of the literature. *Leuk Res* 2004;**28**:249–257.

Hauswirth A, Sperr WR, Jordan JH, Ghannadan M, Schernthaner GH, Lechner K, et al. A case of smouldering mastocytosis with lymphadenopathy and eosinophilia. *Leuk Res* 2002;**26**:601–606.

Hidalgo M, Rowinsky EK. The rapamycin-sensitive signal transduction pathway as a target for cancer therapy. *Oncogene* 2000;**19**:6680–6686.

Horan RF, Austen KF. Systemic mastocytosis: retrospective review of a decadés clinical experience at the Brigham and Womeńs Hospital. *J Invest Dermatol* 1991;**96**: 5S–13S.

Horny H-P, Ruck M, Wehrmann M, Kaiserling, E. Blood findings in generalized mastocytosis: evidence of frequent simultaneous occurrence of myeloproliferative disorders. *Br J Haematol* 1990;**76**:186–193.

Horny H-P, Sillaber C, Menke D, Kaiserling E, Wehrmann M, Stehberger B, et al. Diagnostic value of immunostaining for tryptase in patients with mastocytosis. *Am J Surg Pathol* 1998;**22**:1132–1140.

Jansen B, Wacheck V, Heere-Ress E, Schlagbauer-Wadl H, Höller C, Lucas T, et al. Chemosensitisation of malignant melanoma by BCL2 antisense therapy. *Lancet* 2000,**356**. 1/28–1/33.

Jansen B, Zangemeister-Wittke U. Antisense therapy for cancer-the time of truth. *Lancet Oncol* 2002;**3**:672–683.

Jordan JH, Fritsche-Polanz R, Sperr WR, Mitterbauer G, Födinger M, Schernthaner GH, et al. A case of smouldering mastocytosis with high mast cell burden, monoclonal myeloid cells, and *C-KIT* mutation Asp-816-Val. *Leuk. Res.* 2001; **25**:627–634.

Jordan CT, Upchurch D, Szilvassy SJ, Guzman ML, Howard DS, Pettigrew AL, et al. The interleukin-3 receptor alpha chain is a unique marker for human acute myelogenous leukemia stem cells. *Leukemia* 2000;**14**:1777–1784.

Kiener HP, Hofbauer R, Tohidast-Akrad M, Walchshofer S, Redlich K, Bitzan P, et al. Tumor necrosis factor alpha promotes the expression of stem cell factor in synovial fibroblasts and their capacity to induce mast cell chemotaxis. *Arthritis Rheum* 2000;**43**: 164–174.

Kirshenbaum AS, Goff JP, Semere T, Foster B, Scott LM, Metcalfe DD. Demonstration that human mast cells arise from a progenitor cell population that is CD34+, c-kit+, and expresses aminopeptidase N (CD13). *Blood* 1999;**94**:2333–2342.

Kluin-Nelemans HC, Oldhoff JM, Van Doormaal JJ, Van 't Wout JW, Verhoef G, Gerrits WB, Van Dobbenburgh OA, Pasmans SG, Fijnheer R. Cladribine therapy for systemic mastocytosis. *Blood* 2003;**102**:4270–4276.

Kitamura Y, Go S. Decreased production of mast cells in *Sl/Sld* anemic mice. *Blood* 1979;**53**:492–497.

Kitamura Y, Go S, Hatanaka K. Decrease of mast cells in *W/Wv* mice and their increase by bone marrow transplantation. *Blood* 1978;**52**:447–452.

Kitamura Y, Morii E, Jippo T, Ito A. Regulation of mast cell phenotype by MITF. *Int Arch Allergy Immunol* 2002;**127**:106–109.

Kluin-Nelemans HC, Jansen JH, Breukelman H, Wolthers BG, Kluin PM, et al. Response to interferon alfa-2b in a patient with systemic mastocytosis. *N Engl J Med* 1992;**326**: 619–623.

Kreitman RJ, Bailon P, Chaudhary VK, FitzGerald DJP, Pastan I. Recombinant immuno-toxins containing anti-Tac(Fv) and derivatives of *Pseudomonas* exotoxin produce complete regression in mice of an interleukin-2 receptor-expressing human carcinoma. *Blood* 1994;**83**:426–434.

Lawrence JB, Friedman BS, Travis WD, Chinchilli VM, Metcalfe DD, et al. Hematologic manifestations of systemic mast cell disease: a prospective study of laboratory and morphologic features and their relation to prognosis. *Am J Med* 1991;**91**:612–624.

Lennert K, Parwaresch MR. Mast cells and mast cell neoplasia: a review. *Histopathology* 1979;**3**:349–365.

Li CY, Baek JY. Mastocytosis and fibrosis: role of cytokines. *Int Arch Allergy Immunol* 2002;**127**:123–126.

Liao AT, Chien MB, Shenoy N, Mendel DB, McMahon G, Cherrington JM, London CA. Inhibition of constitutively active forms of mutant kit by multitargeted indolinine tyrosine kinase inhibitors. *Blood* 2002;**100**:585–593.

Linnekin D. Early signaling pathways activated by c-Kit in hematopoietic cells. *Int J Biochem Cell Biol* 1999;**31**:1053–1074.

Longley BJ, Metcalfe DD, Tharp M, Wang X, Tyrrell L, Lu SZ, et al. Activating and dominant inactivating c-KIT catalytic domain mutations in distinct clinical forms of human mastocytosis. *Proc Natl Acad Sci (USA)* 1999;**96**:1609–1614.

Longley BJ, Tyrrell L, Lu SZ, Ma YS, Langley K, Ding TG, et al. Somatic c-KIT activating mutation in urticaria pigmentosa and aggressive mastocytosis: establishment of clonality in a human mast cell neoplasm. *Nat Genet* 1996;**12**:312–314.

Ma Y, Zeng S, Metcalfe DD, Akin C, Dimitrijevic S, Butterfield JH, et al. The c-KIT mutation causing human mastocytosis is resistant to STI571 and other KIT kinase inhibitors; kinases with enzymatic site mutations show different inhibitor sensitivity profiles than wild-type kinases and those with regulatory type mutations. *Blood* 2002;**99**:1741–1744.

Matthews DC, Appelbaum FR, Eary JF, Fisher DR, Durack LD, Hui TE, et al. Phase I study of [131]I-anti-CD45 antibody plus cyclophosphamide and total body irradiation for advanced acute leukemia and myelodysplastic syndrome. *Blood* 1999;**94**:1237–1247.

Mayerhofer M, Gleixner KV, Hoelbl A, Florian S, Hoermann G, Aichberger KJ, et al. Unique effects of KIT D816V in BaF3 cells: induction of cluster formation, histamine synthesis, and early mast cell differentiation antigens. *J Immunol* 2008;**180**:5466-5476.

Mayerhofer M, Valent P, Sperr WR, Griffin JD, Sillaber C. BCR/ABL induces expression of vascular endothelial growth factor and its transcriptional activator, hypoxia inducible factor-1alpha, through a pathway involving phospho-inositide 3-kinase and the mammalian target of rapamycin. *Blood* 2002;**100**:3767–3775.

Metcalfe DD. Classification and diagnosis of mastocytosis: current status. *J Invest Dermatol* 1991;**96**:2S–4S.

Metcalfe DD, Akin C. Mastocytosis: molecular mechanisms and clinical disease heterogeneity.*Leuk Res* 2001;**25**:577–582.

Morris JC, Waldmann TA. Advances in interleukin-2 receptor targeted treatment. *Ann Rheumatol* 2000;**59**:i109–i114.

Motoyama N, Wang F, Roth KA, Sawa H, Nakayama K, Nakayama K, et al. Massive cell death of immature hematopoietic cells and neurons in Bcl-x-deficient mice. *Science* 1995;**267**:1506–1510.

Nagata H, Worobec AS, Oh CK, Chowdhury BA, Tannenbaum S, Suzuki Y, Metcalfe DD. Identification of a point mutation in the catalytic domain of the protooncogene c-kit in peripheral blood mononuclear cells of patients who have mastocytosis with an associated hematologic disorder. *Proc Natl Acad Sci (USA)* 1995;**92**:10560–10564.

Nunez-Lopez R, Escribano L, Schernthaner GH, Prados A, Rodriguez-Gonzalez R, Diaz-Agustin B, Lopez A, Hauswirth A, Valent P, Almeida J, Bravo P, Orfao A. Overexpression of complement receptors and related antigens on the surface of bone marrow mast cells in patients with systemic mastocytosis. *Br J Haematol* 2003;**120**:257–265.

Okuda K, Sanghera JS, Pelech SL, Kanakura Y, Hallek M, Griffin JD, et al. Granulocyte-macrophage colony-stimulating factor, interleukin-3, and steel factor induce rapid tyrosine phosphorylation of p42 and p44 MAP kinase. *Blood* 1992;**79**:2880–2887.

Piao X, Bernstein A. A point mutation in the catalytic domain of c-kit induces growth factor independence, tumorigenicity, and differentiation of mast cells. *Blood* 1996;**87**:3117–3123.

Piao X, Paulson R, van der Geer P, Pawson T, Bernstein A. Oncogenic mutation in the Kit receptor tyrosine kinase alters substrate specificity and induces degradation of the protein tyrosine phosphatase SHP-1. *Proc Natl Acad Sci (USA)* 1996;**93**:14665–14669.

Popkov M, Sidrac-Ghali S, Alakhov V, Mandeville R. Epitope-specific antibody response to HT-1080 fibrosarcoma cells by mimotope immunization. *Clin Cancer Res* 2000;**6**: 3629–3635.

Rottem M, Okada T, Goff JP, Metcalfe DD. Mast cells cultured from peripheral blood of normal donors and patients with mastocytosis originate from a CD34 + /Fc(RI- cell population. *Blood* 1994;**84**:2489–2496.

Saleh MN, LeMaistre CF, Kuzel TM, Foss F, Platanais LC, Schwartz G, et al. Antitumor activity of DAB389IL-2 fusion toxin in mycosis fungoides. *J Am Acad Dermatol* 1998;**39**:63–73.

Sattler M, Salgia R, Shrikhande G, Verma S, Pisick E, Prasad KV, Griffin JD. Steel factor induces tyrosine phosphorylation of CRKL and binding of CRKL to a complex containing c-kit, phosphatidylinositol 3-kinase, and p120(CBL). *J Biol Chem* 1997;**272**:10248–10253.

Schernthaner GH, Jordan JH, Ghannadan M, Agis H, Bevec D, Nuñez R, et al. Expression, epitope analysis, and functional role of the LFA-2 antigen detectable on neoplastic mast cells. *Blood* 2001;**98**:3784–3792.

Schernthaner GH, Spanblöchl E, Sperr WR, Sillaber C, Semper H, Jurecka W, et al. Effects of interferon-α-2b treatment on ex vivo differentiation of mast cells from circulating progenitors in a patient with systemic mastocytosis. *Ann Hematol* 2000;**79**: 660–666.

Scherr M, Battmer K, Winkler T, Heidenreich O, Ganser A, Eder M. Specific inhibition of bcr-abl gene expression by small interfering RNA. *Blood* 2003;**101**:1566–1569.

Schwartz LB, Sakai K, Bradford TR, Ren S, Zweiman B, Worobec AS, Metcalfe, DD. The alpha form of human tryptase is the predominant type present in blood at baseline in normal subjects and is elevated in those with systemic mastocytosis. *J Clin Invest* 1995; **96**:2702–2710.

Serve H, Yee NS, Stella G, Sepp-Lorenzino L, Tan JC, Besmer P. Differential roles of PI3-kinase and Kit tyrosine kinase 821 in Kit-receptor mediated proliferation, survival and cell adhesion in mast cells. *EMBO J* 1995;**14**:473–483.

Shah NP, Lee FY, Luo R, Jiang Y, Donker M, Akin C. Dasatinib (BMS-354825) inhibits KITD816V, an imatinib-resistant activating mutation that triggers neoplastic growth in most patients with systemic mastocytosis. *Blood* 2006;**108**:286–291.

Shelburne CP, McCoy ME, Piekorz R, Sexl V, Roh KH, Jacobs-Helber SM, et al. Stat5 expression is critical for mast cell development and survival. Blood 2003; **102**:1290–1297.

Sievers EL, Larson RA, Stadtmauer EA, Estey E, Löwenberg B, Dombret H, et al. Efficacy and safety of gemtuzumab ozogamicin in patients with CD33-positive acute myeloid leukemia in first relapse. *J Clin Oncol* 2001;**19**:3244–3254.

Smith M, Bleijs R, Radford K, Hersey P. Immunogenicity of CD63 in a patient with melanoma.*Melanoma Res* 1997;7:S163–S170.

Sotlar K, Marafioti T, Griesser H, Theil J, Aepinus C, Jaussi R, et al. Detection of c-kit mutation Asp 816 to Val in microdissected bone marrow infiltrates in a case of systemic mastocytosis associated with chronic myelomonocytic leukaemia. *Mol Pathol* 2000;**53**:188–193.

Sperr WR, Horny HP, Valent P. Spectrum of associated clonal hematologic non-mast cell lineage disorders occurring in patients with systemic mastocytosis. *Int Arch Allergy Immunol* 2002;**127**:140–142.

Stirewalt DL, Meshinchi S, Radich JP. Molecular targets in acute myelogenous leukemia. *Blood Rev* 2003;**17**:15–23.

Sutherland HJ, Blair A, Zapf RW. Characterization of a hierarchy of human acute myeloid leukemia progenitor cells. *Blood* 1996;**87**:4754–4761.

Taylor ML, Metcalfe DD. Kit signal transduction. *Hematol Oncol North Am* 2000;**14**:517–535.

Tefferi A, Li CY, Butterfield JH, Hoagland HC. Treatment of systemic mast-cell disease with cladribine. *N Engl J Med* 2001;**344**:307–309.

Travis WD, Li CY, Bergstrahl EJ, Yam LT, Swee RG. Systemic mast cell disease. Analysis of 58 cases and literature review. *Medicine* 1988;**67**:345–368.

Travis WD, Li CY, Yam LT, Bergstralh EJ, Swee RG. Significance of systemic mast cell disease with associated hematologic disorders. *Cancer* 1988;**62**:965–972.

Travis WD, Li CY, Hoagland HC, Travis LB, Banks PM. Mast cell leukemia: report of a case and review of the literature. *Mayo Clin Proc* 1986;**61**:957–966.

Valent, P. Biology, classification and treatment of human mastocytosis. *Wien Klin Wochenschr* 1996;**108**:385–397.

Valent P, Akin C, Sperr WR, Escribano L, Arock M, Horny H-P, et al. Aggressive systemic mastocytosis and related mast cell disorders: current treatment options and proposed response criteria. *Leuk Res* 2003;**27**:635–641.

Valent P, Akin C, Sperr WR, Horny HP, Arock M, Lechner K, et al. Diagnosis and treatment of systemic mastocytosis: state of the art. *Br J Haematol* 2003;**122**:695–717.

Valent P, Akin C, Sperr WR, Horny H-P, Metcalfe DD. Smouldering mastocytosis: a new type of systemic mastocytosis with slow progression. *Int Arch Allergy Immunol* 2002;**127**: 137–139.

Valent P, Ashman LK, Hinterberger W, Eckersberger F, Majdic O, Lechner K, Bettelheim P. Mast cell typing: demonstration of a distinct hemopoietic cell type and evidence for immunophenotypic relationship to mononuclear phagocytes. *Blood* 1989; **73**:1778–1785.

Valent P, Bettelheim P. Cell surface structures on human basophils and mast cells: biochemical and functional characterization. *Adv Immunol* 1992;**52**:333–423.

Valent P, Horny H-P, Escribano L, Longley BJ, Li CY, Schwartz LB, et al. Diagnostic criteria and classification of mastocytosis: a consensus proposal. *Leuk Res,* 2001;**25**: 603–625.

Valent P, Horny H-P, Li CY, Longley JB, Metcalfe DD, Parwaresch RM, Bennett JM. Mastocytosis (Mast cell disease). In: *World Health Organization (WHO) Classification of Tumours. Pathology & Genetics. Tumours of Haematopoietic and Lymphoid Tissues.* Eds: Jaffe ES, Harris NL, Stein H, Vardiman JW. pp 291–302. IARC Press, Lyon, 2001.

Valent P, Schernthaner GH, Sperr WR, Fritsch G, Agis H, Willheim M, et al. Variable expression of activation-linked surface antigens on human mast cells in health and disease. *Immunol Rev* 2001;**179**:74–81.

Valent P, Sperr WR, Samorapoompichit P, Geissler K, Lechner K, Horny H-P, Bennett JM. Myelomastocytic overlap syndromes: Biology, criteria, and relationship to mastocytosis. *Leuk Res* 2001;**25**:595–602.

van Der Velden VH, te Marvelde JG, Hoogeveen PG, Bernstein ID, Houtsmuller AB, Berger MS, et al. Targeting of the CD33-calicheamicin immunoconjugate Mylotarg (CMA-676) in acute myeloid leukemia: in vivo and in vitro saturation and internalization by leukemic and normal myeloid cells. *Blood* 2001;**97**:3197–3204.

Voseller K, Stella G, Yee NS, Besmer P. c-kit receptor signaling through its phosphatidyli-nositol-3-kinase-binding site and protein kinase C: role in mast cell anhancement of degranulation, adhesion, and membrane ruffling. *Mol Biol Cell* 1987;**8**:909–922.

Weide R, Ehlenz K, Lorenz W, Walthers E, Klausmann M, Pfluger KH. Successful treatment of osteoporosis in systemic mastocytosis with interferon alpha-2b.*Ann Hematol* 1996;**72**:41–43.

Wimazal F, Jordan JH, Sperr WR, Chott A, Dabbass S, Lechner K, et al. Increased angio-genesis in the bone marrow of patients with systemic mastocytosis. *Am J Pathol* 2002;**160**:1639–1645.

Wohlbold L, Van Der Kuip H, Miething C, Vornlocher HP, Knabbe C, Duyster J, et al. Inhibition of bcr-abl gene expression by small interfering RNA sensitizes for imatinib mesylate (STI571). *Blood* 2003;**102**:2236–2239.

Wolff K, Komar M, Petzelbauer P. Clinical and histopathological aspects of cutaneous mastocytosis. *Leuk Res* 2001;**25**:519–528.

Worobec AS. Treatment of systemic mast cell disorders. *Hematol Oncol Clin North Am* 2000;**14**:659–687.

Worobec AS, Kirshenbaum AS, Schwartz LB, Metcalfe DD. Treatment of Three Patients with Systemic Mastocytosis with Interferon Alpha-2b. *Leuk Lymphoma* 1996;**22**: 501–508.

Worobec AS, Metcalfe DD. Mastocytosis: current treatment concepts.*Int Arch Allergy Immunol* 2002;**127**:153–155.

Yavuz AS, Lipsky PE, Yavuz S, Metcalfe DD, Akin C. Evidence for the involvement of an hematopoietic progenitor cell in systemic mastocytosis from single cell analysis of muta-tions in the *c-kit* gene.*Blood* 2002;**100**:661–665.

Zermati Y, De Sepulveda P, Feger F, Letard S, Kersual J, Casteran N, et al. Effect of tyrosine kinase inhibitor STI571 on the kinase activity of wild-type and various forms of mutated receptors found in mast cell neoplasms. *Oncogene* 2003;**22**:660–664.

# Structure, Allergenicity, and Cross-Reactivity of Plant Allergens

Christian Radauer and Heimo Breiteneder

**Abstract** Within the last two decades, hundreds of allergenic proteins from diverse sources have been identified. The availability of numerous allergen sequences provided the basis for classifying allergens into families of evolutionary and structurally related proteins. This chapter gives an introduction into the protein family classification of allergens of plant origin. Our analysis showed that the majority of plant allergens could be grouped into a small number of protein families. The most important families of plant allergens are the prolamin superfamily, which contains several groups of important food and respiratory allergens, the profilins and the Bet v 1-related proteins, two families of allergens responsible for cross-reactivity between pollen and plant foods, and the seed storage globulins, major allergenic components of legumes, tree nuts, and other dicotyledonous seeds. The molecular classification of allergens can be used to establish a correlation between sequence and structural similarity and cross-reactivity among homologous allergens, delineate common properties of allergens, and deduce possible factors that make proteins allergenic.

**Keywords** Food allergy · Pollen allergy · Allergens · Protein families

**Abbreviations** HMM – hidden Markov model; IgE – immunoglobulin; nsLTP – non-specific lipid transfer protein; PR-10 – pathogenesis-related proteins family 10.

## 1 Introduction

Identification and characterization of allergens is one of the major research areas in molecular allergology. Allergens are molecules that are capable of eliciting an allergic immune response in humans, which is characterized by an excess of

C. Radauer
Department of Pathophysiology, Center of Physiology and Pathophysiology, Medical University of Vienna, Vienna, Austria
e-mail: Christian.Radauer@meduniwien.ac.at

A. Falus (ed.), *Clinical Applications of Immunomics*,
DOI: 10.1007/978-0-387-79208-8_7, © Springer Science+Business Media, LLC 2009

T helper 2 cells that trigger the differentiation of B cells into immunoglobulin E (IgE) producing plasma cells. Most allergens are proteins. During the last two decades, hundreds of allergenic proteins from all types of sources (pollen, mites, molds, animal dander, insect venoms, natural rubber latex, and foods) have been identified and their sequences, physicochemical properties, and clinical data made available in various allergen databases (Mari 2005) such as the Official List of Allergens (http://www.allergen.org) and the Allergome (http://www.allergome.org). Most of these databases classify allergens mainly by source. However, the growing number of available allergen sequences and structures made it possible to establish a more natural classification scheme that identifies the protein families of allergens based on sequence and structural similarity.

This chapter provides the application of protein family classification to plant allergens. Our classification scheme provides a framework for discussing the common molecular properties of allergens and assessing the connection between sequence and structural similarity and cross-reactivity.

## 2 Protein Families of Plant Allergens

Since the 1970s, a large number of protein family databases, which use different methods for grouping proteins, have been developed (Redfern, Grant, Maibaum, and Orengo 2005). They classify proteins either by common structures (SCOP, CATH) or based on similar sequences using conserved motifs (PROSITE, PRINTS), position-specific scoring matrices (ProDom), or profile hidden Markov models (Pfam, SMART) for aligning new sequences with existing protein families.

The Pfam database (http://Pfam.sanger.ac.uk) is a semi-automatic collection of multiple sequence alignments and profile hidden Markov models (HMMs), which contains in its version 20.0, released in May 2006, 8296 families that cover 74% of all protein sequences in the Uniprot database (Finn, Mistry, Schuster-Bockler, Griffiths-Jones, Hollich, Lassmann, Moxon, Marshall, Khanna, Durbin, Eddy, Sonnhammer, and Bateman 2006).

We chose Pfam as a resource for the classification of allergens for several reasons: (i) sequence alignments serving as starting points for HMM calculations are manually compiled and hence more reliable than automatically generated family definitions, (ii) protein families in Pfam are defined based on sequence rather than structural similarity, which makes this database quite comprehensive, and (iii) sequence comparison using profile HMMs is more sensitive than classical sequence alignment algorithms, thus enabling the grouping of proteins sharing minimal portions of their sequences.

The search for plant allergen sequences in the Allergome database, one of the currently most comprehensive collections of allergen data (Mari 2005), yielded 321 complete or almost complete sequences (excluding variant and allelic sequences from the same source). These sequences were compared to the Pfam database to generate a list of Pfam families and Pfam architectures of plant allergens. The Pfam architecture represents the domain organization of a protein and was chosen as the criterion for classifying allergens in order to avoid ambiguities concerning multi-domain proteins. Of the 321 plant allergen sequences, 315 were distributed among 62 Pfam architectures (Table 1) and 66 Pfam families; these are only 0.8% of all protein families in the Pfam database. In comparison, all seed plant proteins in the Uniprot database were classified into 2,615 Pfam families (Radauer and Breiteneder 2006a). The narrow distribution of plant allergens in the protein family space is highlighted by the fact that the 17 most populated allergen families contain 75% of all allergen sequences (Table 1). In comparison, the 17 most abundant families of seed plant proteins comprise only 25% of all seed plant proteins.

In addition, the protein family distribution of plant allergens does not match the distribution of protein families among all plant proteins. The prolamin superfamily, the family containing the greatest number of plant allergens, on 23rd position in the ranking of all plant protein families, is the only allergen-containing protein family ranked among the top 25 (Radauer and Breiteneder 2006a).

The two most important types of sources of plant-derived allergens are pollen and plant foods. The distribution of allergens among protein families shows significant differences between these types of sources (Fig. 1). While some allergen families are confined to certain sources such as seed storage globulins in plant foods and polcalcins in pollen, other families are found in

**Table 1** The most abundant protein families of plant allergens

| Protein family name | Pfam architecture | Number |
| --- | --- | --- |
| Prolamin superfamily | PF00234 | 52 |
| Profilin | PF00235 | 42 |
| Bet v 1-related | PF00407 | 21 |
| Seed storage globulin | PF00190-PF00190 | 20 |
| Group 2 grass pollen allergen | PF01357 | 11 |
| Thaumatin-like | PF00314 | 11 |
| Polcalcin | PF00036-PF00036 | 11 |
| Expansin | PF03330-PF01357 | 10 |
| Pectate lyase | PF00544 | 9 |
| Group 5 grass pollen allergen | PF01620 | 9 |
| Ole e 1-related | PF01190 | 8 |
| Polygalacturonase | PF00295 | 7 |
| Papain-like protease | PF08246-PF00112 | 6 |
| Berberine bridge enzyme | PF01565-PF08031 | 6 |
| Class I chitinase | PF00187-PF00182 | 6 |
| Ribosome inactivating protein | PF00161 | 6 |
| Beta-1,3-glucanase | PF00332 | 5 |
| Members of 45 allergen families with less than 5 members | | 75 |
| Members of uncharacterized families | | 6 |

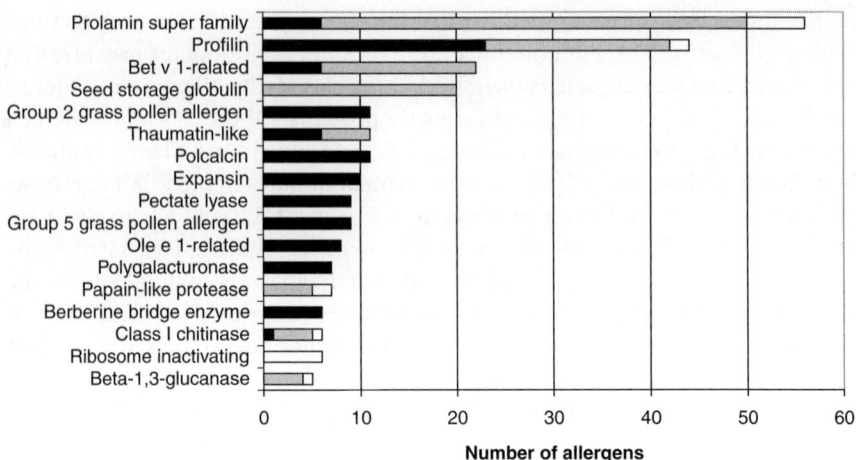

**Fig. 1** Source distribution of plant allergens. *Black*: pollen, *grey*: food, *white*: other (latex and non-pollen inhalative allergens)

all types of allergen sources. Cross-reactivity among members of some of these ubiquitous allergen families, for example, profilins and Bet v 1-related allergens, are responsible for the well-known pollen and latex-associated food allergy syndromes (Vieths, Scheurer, and Ballmer-Weber 2002).

In the following sections, we describe the biochemical properties, species distributions, and allergological significance of some of the most important plant allergen families.

## 2.1  The Prolamin Superfamily

The existence of the prolamin superfamily was proposed as early as 1985 (Kreis, Forde, Rahman, Miflin, and Shewry 1985), and it derives its name from the alcohol-soluble proline and glutamine-rich storage proteins of cereals such as wheat, barley, and rye, but not rice. Although allergic reactions to prolamins from wheat have been reported (Sandiford, Tatham, Fido, Welch, Jones, Tee, Shewry, and Newman Taylor 1997), the major allergens that are thought to sensitize atopic individuals through the gastrointestinal tract belong to one of the three other related protein families of this superfamily. These are the 2S albumins, the non-specific lipid transfer proteins (nsLTPs), and the cereal inhibitors of alpha-amylase and/or trypsin (Mills, Jenkins, Alcocer, and Shewry 2004). Like all members of the prolamin superfamily, these proteins are characterized by the presence of a conserved pattern of six or eight cysteine residues that form three or four intramolecular disulfide bonds. Disulfide bonds are a molecular feature that conveys stability to heating and digestion to these molecules, an important characteristic of true food allergens (Breiteneder and

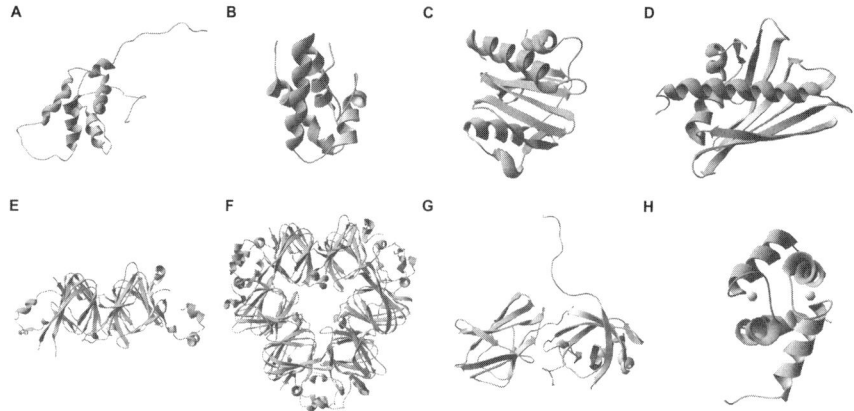

**Fig. 2** Structures of the members of important plant allergen families. A: Ara h 6, a 2S albumin-related allergen from peanut (PDB accession number 1W2Q), B: Pru p 3, the non-specific lipid transfer protein from peach (2ALG), C: *A. thaliana* pollen profilin (1A0K), D: Bet v 1 from birch pollen (1BV1), E, F: single subunit and trimer of soybean β-conglycinin (1FXZ and 1OD5), G: Phl p 1, a β-expansin from timothy grass pollen (1N10), H: Bet v 4, a polcalcin from birch pollen (1H4B)

Mills 2005c). All of these low-molecular-weight proteins have a related architecture that consists of bundles of four α-helices stabilized by their intra-chain disulfide bonds (Fig. 2A, B). Interestingly, the order of the disulfide bonds in the nsLTPs is different from the 2S albumins or the bifunctional inhibitors (Mills et al. 2004).

The 2S albumins are a major group of seed storage proteins from a botanically diverse range of dicotyledonous plants. These include not only economically important temperate crops such as oilseed rape, mustard, sunflower, lupine, and pea, but also crops from warmer climates such as sesame and peanut. Apart from their function as storage proteins, activities as inhibitors of proteinases have been reported for 2S albumins from the brassicas *Sinapis arvensis* (Svendsen, Nicolova, Goshev, and Genov 1994) and *Brassica nigra* (Genov, Goshev, Nikolova, Georgieva, Filippi, and Svendsen 1997), and an antifungal activity has been described for a 2S albumin from radish (*Raphanus sativa*) (Terras, Schoofs, De Bolle, Van Leuven, Rees, Vanderleyden, Cammue, and Broekaert 1992).

Many of the seed and tree nut allergens belong to the 2S albumins. The best characterized allergenic 2S albumin is probably the one from yellow mustard, Sin a 1 (Menendez-Arias, Moneo, Dominguez, and Rodriguez 1988). Other allergenic 2S albumins are Ber e 1 from Brazil nut (Alcocer, Murtagh, Bailey, Dumoulin, Meseguer, Parker, and Archer 2002), Jug r 1 from English walnut (Teuber, Dandekar, Peterson, and Sellers 1998), and Ana 3 from cashew nut (Teuber, Sathe, Peterson, and Roux 2002). The allergens Ara h 2 and Ara h 6 from peanut are members of the conglutin family, which is closely related to the 2S albumin family (Kleber-Janke, Crameri, Appenzeller, Schlaak, and Becker

1999). Recently, a high-resolution structure of Ara h 6 (Fig. 2A) became available (Lehmann, Schweimer, Reese, Randow, Suhr, Becker, Vieths, and Rosch 2006). Additional allergenic 2S albumins have been identified in chickpea (Vioque, Sanchez-Vioque, Clemente, Pedroche, Bautista, and Millan 1999), sunflower (Kelly and Hefle 2000), almond (Poltronieri, Cappello, Dohmae, Conti, Fortunato, Pastorello, Ortolani, and Zacheo 2002), and sesame seed (Pastorello, Varin, Farioli, Pravettoni, Ortolani, Trambaioli, Fortunato, Giuffrida, Rivolta, Robino, Calamari, Lacava, and Conti 2001).

The family of nsLTPs is not restricted to seed tissues as the 2S albumins, but is frequently found in the epidermal tissues of fruits. Comparisons of their sequences show that the nsLTPs are closely homologous basic proteins. All nsLTPs possess eight conserved cysteines and are stabilized by four intra-chain disulfide bonds. nsLTPs have been well defined at a structural level, with the first structure being defined for corn (Shin, Lee, Hwang, Kim, and Suh 1995) and the most recent one for peach (Pasquato, Berni, Folli, Folloni, Cianci, Pantano, Helliwell, and Zanotti 2006), shown in Fig. 2B. All nsLTPs are characterized by a related fold consisting of bundles of four α-helices stabilized by disulfide bonds with a central hydrophobic tunnel into which lipophilic molecules can bind (Breiteneder and Mills 2005b). nsLTPs have been suggested to mediate the transfer of phospholipids between vesicles and membranes (Guerbette, Grosbois, Jolliot-Croquin, Kader, and Zachowski 1999). However, it is now becoming clear that plants have used the three-dimensional scaffold of the nsLTPs in a promiscuous fashion and that many nsLTPs are not able to transfer lipids, and indeed in one instance the hydrophobic tunnel has been lost completely (Tassin, Broekaert, Marion, Acland, Ptak, Vovelle, and Sodano 1998). Instead, lipid transfer proteins may play an important role in plant defense against fungi and bacteria, the exact mechanism of action still being unknown (Cheng, Samuel, Liu, Shyu, Lai, Lin, and Lyu 2004).

Allergenic nsLTPs are widely distributed, with sequences available from fruits, nuts, seeds, vegetables, pollen, and *Hevea brasiliensis* latex. The peach nsLTP, Pru p 3, was the first to be characterized (Pastorello, Farioli, Pravettoni, Ortolani, Ispano, Monza, Baroglio, Scibola, Ansaloni, Incorvaia, and Conti 1999; Sanchez-Monge, Lombardero, Garcia-Selles, Barber, and Salcedo 1999). Recent reviews have summarized the current clinical and molecular knowledge of allergenic plant food (Pastorello and Robino 2004; Salcedo, Sanchez-Monge, Diaz-Perales, Garcia-Casado, and Barber 2004) and pollen nsLTPs (Breiteneder and Mills 2005a; Weber 2005).

Seeds of wheat, barley, rye, corn, and rice contain a family of inhibitors of trypsin and non-plant alpha amylases. These inhibitors interfere with the digestion of plant starches and proteins by impeding insect gut enzymes. They possess eight, nine, or ten cysteine residues that form four or five intra-chain disulfide bridges. Allergens belonging to the cereal inhibitor family are involved in both respiratory and food allergies (James, Sixbey, Helm, Bannon, and Burks 1997; Pastorello et al. 1999). Members of this family of proteins exist as monomers, dimers, or tetramers, with the most active containing glycosylated subunits in a monomeric or tetrameric form. In barley these subunits were designated BMAI-1 (Mena, Sanchez-Monge, Gomez, Salcedo, and Carbonero 1992), now Hor v 15, and BTAI-CMb (Sanchez-Monge, Gomez, Barber, Lopez-Otin, Armentia, and Salcedo 1992), and in wheat WTAI-CM16 (Sanchez-Monge et al. 1992). The other well-characterized allergenic cereal inhibitor family is from rice. RDAI-1 and RDAI-3 are rice dimeric alpha amylase inhibitors (Nakase, Adachi, Urisu, Miyashita, Alvarez, Nagasaka, Aoki, Nakamura, and Matsuda 1996; Salcedo, Diaz-Perales, and Sanchez-Monge 2001).

## 2.2 Profilins

Profilins are cytosolic proteins of 12–15 kDa in size, which are found in all eukaryotic cells. Profilins bind to monomeric actin (G actin) and a plethora of other proteins, thus regulating the dynamics of actin polymerization during processes such as cell movement, cytokinesis, and signaling (Witke 2004).

Profilins from higher plants constitute a family of highly conserved proteins showing sequence identities of at least 75% even between members from distantly related organisms (Radauer, Willerroider, Fuchs, Hoffmann-Sommergruber, Thalhamer, Ferreira, Scheiner, and Breiteneder 2006b). Plant profilins were originally discovered as cross-reactive pollen allergens eliciting IgE responses in 10–20% of pollen-allergic patients. Later, they were also described as allergens in plant foods and *Hevea* latex (reviewed in Radauer and Hoffmann-Sommergruber 2004). Since profilin-specific IgE usually cross-reacts with homologs from virtually every plant source (Radauer et al. 2006b), sensitization to these allergens has been considered a risk factor for allergic reactions to multiple pollen sources (Mari 2001) and for pollen-associated food allergy (Asero, Mistrello, Roncarolo, Amato, Zanoni, Barocci, and Caldironi 2003). However, the clinical relevance of profilin-specific IgE has frequently been questioned (Wensing, Akkerdaas, van Leeuwen, Stapel, Bruijnzeel-Koomen, Aalberse, Bast, Knulst, and van Ree 2002).

Structures of three plant profilins have been published: those from *Arabidopsis thaliana* pollen (Thorn, Christensen, Shigeta, Huddler, Shalaby, Lindberg, Chua, and Schutt 1997), birch pollen (Fedorov, Ball, Mahoney, Valenta, and Almo 1997), and *H. brasiliensis* latex (PDB accession number 1G5U). Profilins fold into globular structures consisting of a central seven-stranded anti-parallel β-sheet and two α-helices each on either side (Fig. 2C). Despite sequence identities of below 30%, plant profilin structures are highly similar to the structures of profilins from mammals, fungi, and amoeba. This explains the finding that IgE directed to plant profilins weakly binds to the human homolog as well (Valenta, Duchene, Pettenburger, Sillaber, Valent, Bettelheim, Breitenbach, Rumpold, Kraft, and Scheiner 1991). However, no profilins from sources other than plants have been shown to elicit allergic reactions.

## 2.3  Bet v 1-Related Allergens

The identification of the major birch pollen allergen Bet v 1 was published in 1989 (Breiteneder, Pettenburger, Bito, Valenta, Kraft, Rumpold, Scheiner, and Breitenbach 1989). Since then, this allergen has become the most intensely studied allergen worldwide. Bet v 1 is a PR-10 (pathogenesis-related proteins family 10) like protein. Apart from pathogenesis-related proteins and Bet v 1-related allergens, the PR-10 family contains the family of major latex proteins, named after a major component of the latex of opium poppy, as well as several enzymes involved in secondary plant metabolism (Liscombe, MacLeod, Loukanina, Nandi, and Facchini 2005). PR-10 proteins fold into structures consisting of a seven-stranded anti-parallel β-sheet wrapped around a long C-terminal α-helix and two small N-terminal helices (Fig. 2D). The crystal structure of the naturally occurring isoform Bet v 1 1 was recently determined in complex with deoxycholate (Markovic-Housley, Degano, Lamba, von Roepenack-Lahaye, Clemens, Susani, Ferreira, Scheiner, and Breiteneder 2003). In addition, ligand-binding studies with Bet v 1 1 (Markovic-Housley et al. 2003), Bet v 1.2801 (Mogensen, Wimmer, Larsen, Spangfort, and Otzen 2002), and with the cherry homolog Pru av 1 (Neudecker, Schweimer, Nerkamp, Scheurer, Vieths, Sticht, and Rösch 2001) showed interaction with various phytosteroids. Therefore, it was suggested that Bet v 1 and related PR-10 proteins might function as plant steroid carriers (Markovic-Housley et al. 2003).

Bet v 1-related tree pollen allergens are limited to the botanical order of the Fagales; they include Cor a 1 from hazel, Aln g 1 from alder, Car b 1 from hornbeam and major allergens from other Fagales trees. On the other hand, cross-reactive Bet v 1 homologs have been identified from fruits of the Rosaceae family, e.g., Mal d 1 from apple (Vanek-Krebitz, Hoffmann-Sommergruber, Laimer da Camara Machado, Susani, Ebner, Kraft, Scheiner, and Breiteneder 1995), Pru av 1 from cherry (Scheurer, Metzner, Haustein, and Vieths 1997), and Pyr c 1 from pear (Karamloo, Scheurer, Wangorsch, May, Haustein, and Vieths 2001). Furthermore, Bet v 1-related allergens are found in vegetables from the Apiaceae family, such as Api g 1 from celery (Breiteneder, Hoffmann-Sommergruber, O'Riordain, Susani, Ahorn, Ebner, Kraft, and Scheiner 1995) and Dau c 1 from carrot (Hoffmann-Sommergruber, O'Riordain, Ahorn, Ebner, Laimer Da Camara Machado, Pühringer, Scheiner, and Breiteneder 1999), as well as from the legume seeds peanut (Mittag, Akkerdaas, Ballmer-Weber, Vogel, Wensing, Becker, Koppelman, Knulst, Helbling, Hefle, Van Ree, and Vieths 2004), soybean (Kleine-Tebbe, Vogel, Crowell, Haustein, and Vieths 2002), and mung bean (Mittag, Vieths, Vogel, Wagner-Loew, Starke, Hunziker, Becker, and Ballmer-Weber 2005). The presence of high concentrations of Bet v 1-related allergens in these foods is the explanation for the high frequency of allergic reactions to plant foods among birch pollen-allergic patients (Vieths et al. 2002).

## 2.4  Seed Storage Globulins

Seed storage globulins are classified into two groups based on their sedimentation coefficients. The 7/8 S globulins occur in a range of monocot and dicot plant species and are often referred to as vicilins. The 11S-12S globulins are the most widespread group and are present in monocot and dicot seeds, particularly legume seeds, hence they are often called legumins. Both vicilins and legumins belong to the cupin superfamily. The cupins are a large and functionally immensely diverse superfamily of proteins (Dunwell, Purvis, and Khuri 2004) that have a common origin and whose evolution can be followed from bacteria to eukaryotes including animals and higher plants (Dunwell 1998; Dunwell, Khuri, and Gane 2000). These proteins share a beta-barrel structural core domain to which the term cupin (derived from the Latin word *cupa* for barrel) was given (Fig. 2E). Cupins can be divided into single-domain cupins and two-domain bicupins, which are thought to have evolved from the duplication of a microbial sequence. Bicupins include the globulin seed storage proteins that constitute major components of the human diet. Globulins are highly abundant in nuts and seeds, comprising up to 70% of the seed protein.

Mature 7/8S globulins are homotrimeric proteins of about 150–190 kDa. Presently, the three-dimensional structures of four 7/8S globulins have been solved, namely canavalin from jack bean (Ko, Day, and McPherson 2000; Ko, Kuznetsov, Malkin, Day, and McPherson 2001), phaseolin from French bean (Lawrence, Izard, Beuchat, Blagrove, and Colman 1994), the beta subunit of beta-conglycinin from soybean (Maruyama, Adachi, Takahashi, Yagasaki, Kohno, Takenaka, Okuda, Nakagawa, Mikami, and Utsumi 2001), and the 8Sα globulin from mung bean (Itoh, Garcia, Adachi, Maruyama, Tecson-Mendoza, Mikami, and Utsumi 2006). These structures illustrate that trimeric vicilins are disc shaped (Fig. 2E, F). Vicilins lack cysteines and, therefore, contain no disulfide bonds. Their detailed subunit compositions vary considerably due to the differences in proteolytic processing and glycosylation of the monomers (Shewry, Napier, and Tatham 1995).

The best-analyzed allergenic vicilin is most likely the major peanut allergen Ara h 1 that is responsible for the majority of cases of fatal anaphylaxis induced by a plant food. Ara h 1 is one of the main seed storage proteins of peanut and is recognized by serum IgE from over 90% of peanut allergic individuals (Burks, Williams, Helm, Connaughton, Cockrell, and O'Brien 1991). Various studies have confirmed the allergenic activity of the 7S globulins Jug r 2 from walnut (Teuber, Jarvis, Dandekar, Peterson, and Ansari 1999), Ana o 2 from cashew nut (Wang, Robotham, Teuber, Tawde, Sathe, and Roux 2002), Ses i 3 from sesame (Beyer, Bardina, Grishina, and Sampson 2002), Len c 1 from lentil (Lopez-Torrejon, Salcedo, Martin-Esteban, Diaz-Perales, Pascual, and Sanchez-Monge 2003), Pis s 1 from pea (Sanchez-Monge, Lopez-Torrejon, Pascual, Varela, Martin-Esteban, and Salcedo 2004), and Cor a 11 from hazelnut (Lauer, Foetisch, Kolarich, Ballmer-Weber, Conti, Altmann, Vieths, and Scheurer 2004).

Mature 11S globulins are hexameric proteins that are initially assembled and transported through the secretory system as intermediate trimers (Shewry et al. 1995). In the protein storage vacuole, the subunits of the trimers are proteolytically processed to yield an acidic 30–40 kDa polypeptide linked by a disulfide bond to a basic polypeptide of ~20 kDa. Cleavage is accompanied by the transformation of two trimers into a mature hexameric 11S globulin (Müntz 1998). The three-dimensional structures of the trimeric precursor and the hexameric mature protein of glycinin from soybean have been determined (Adachi, Kanamori, Masuda, Yagasaki, Kitamura, Mikami, and Utsumi 2003; Adachi, Takenaka, Gidamis, Mikami, and Utsumi 2001).

The minor peanut allergen Ara h 3 was identified as the N-terminal portion of a peanut glycinin subunit (Rabjohn, Helm, Stanley, West, Sampson, Burks, and Bannon 1999). The 11S fraction of soybean proteins consists almost entirely of glycinin, the predominant soy storage globulin. Native glycinin is a 350-kDa hexamer composed of different combinations of the five subunits G1 to G5. IgE epitopes of the acidic chain of the soybean glycinin subunit G1 have been found to be similar to IgE-epitopes of the peanut glycinin Ara h 3 (Beardslee, Zeece, Sarath, and Markwell 2000). Each basic chain of the five soybean glycinin subunits reacted with IgE from soybean allergic individuals to a similar extent (Helm, Cockrell, Connaughton, Sampson, Bannon, Beilinson, Livingstone, Nielsen, and Burks 2000). Additional allergenic 11S legumins are Ber e 2 from Brazil nut (Bartolome, Mendez, Armentia, Vallverdu, and Palacios 1997), Ses i 6 from sesame (Tai, Wu, Chen, and Tzen 1999), and Fag e 1 from buckwheat (Fujino, Funatsuki, Inada, Shimono, and Kikuta 2001).

## 2.5 Expansins and Expansin-Related Allergens

Expansins are ubiquitous plant cell wall glycoproteins that catalyze cell wall loosening during cell growth and other developmental processes (reviewed in Sampedro and Cosgrove 2005). They are typically 250–275 amino acids long. Phylogenetic analysis revealed the existence of four subfamilies designated α-expansins, β-expansins, expansin-like A, and expansin-like B. Sequence identity between members of different subfamilies is only 20–40%.

Expansin structures consist of two domains preceded by a poorly conserved unstructured glycosylated N-terminal stretch (Fig. 2G). The N-terminal domain folds into a six-stranded β-barrel stabilized by three disulfide bonds (Fig. 2G, right domain). It is distantly related and structurally similar to glycoside hydrolase family 45 proteins. Despite the conservation of several residues that form the active center in those enzymes, no hydrolytic activity has been detected for expansins. Instead, they exert their cell wall loosening activity by breaking non-covalent bonds between polysaccharide molecules (Cosgrove

2000). The C-terminal expansin domain adopts an immunoglobulin-like two-layered β-sandwich structure (Fig. 2G, left domain).

Several groups of allergens are related to one or both expansin domains. Group 1 grass pollen allergens are β-expansins. They are the major allergens recognized by IgE from 90% of grass pollen allergic patients (Andersson and Lidholm 2003). Interestingly, no expansins outside the grass family have been shown to be allergenic. The groups 2 and 3 of grass pollen allergens are related to β-expansins showing 35–45% sequence identities. However, they lack the N-terminal domain and are thought to have evolved from a truncated β-expansin gene. Unlike group 1 grass pollen allergens, members of the groups 2 and 3 show no cell wall loosening activity and their biologic activity is unknown. These proteins are minor allergens and show only weak cross-reactivity with group 1 allergens (Andersson et al. 2003).

Recently, kiwellin, a cell wall protein from kiwi fruits, has been sequenced and identified as an allergen (Tamburrini, Cerasuolo, Carratore, Stanziola, Zofra, Romano, Camardella, and Ciardiello 2005). The protein is similar to ripening-related proteins from grape, potato, and rice. Furthermore, it is distantly related to the N-terminal domain of α-expansins, with only 20–25% sequence identity. Another protein family with weak homology to the N-terminal expansin domain (sequence identities of 20-30%) is the barwin family, also termed pathogenesis-related proteins PR-4, a widespread family of proteins expressed upon wounding or pathogen attack in plants (Neuhaus 1999). Barwin domains are found as the C-terminal parts of the major latex allergen Hev b 6 (prohevein) and the turnip allergen Bra r 2 (Breiteneder and Radauer 2004). All these proteins are members of a superfamily of diverse proteins sharing a double-psi β-barrel as the common structural motif (Castillo, Mizuguchi, Dhanaraj, Albert, Blundell, and Murzin 1999).

## 2.6 Polcalcins

Polcalcins are cytosolic proteins with molecular masses of 9 kDa. They contain two calcium-binding EF hand motifs (Fig. 2H). The EF hand is a ubiquitous helix-loop-helix motif found in a large number of calcium-binding proteins (Grabarek 2006). Polcalcins are specifically expressed in anthers and pollen, their biological function still being unknown (Okada, Sasaki, Ohta, Onozuka, and Toriyama 2000).

Polcalcins are minor allergens causing IgE reactivity in 10–45% of pollen-allergic patients. Allergenic polcalcins were described in plants from the families Poaceae (timothy, Bermuda grass), Oleaceae (olive, lilac), Brassicaceae (oilseed rape, turnip), Betulaceae (birch, alder), and Chenopodiaceae (white goosefoot). As members from different plant families show at least 65% sequence identity, cross-reactivity among polcalcins is considerably high (Ledesma, Barderas,

Westritschnig, Quiralte, Pascual, Valenta, Villalba, and Rodriguez 2006; Tinghino, Twardosz, Barletta, Puggioni, Iacovacci, Butteroni, Afferni, Mari, Hayek, Di Felice, Focke, Westritschnig, Valenta, and Pini 2002). Structures of three polcalcins were solved. While the allergens from timothy (Phl p 7) and goosefoot (Che a 3) pollen fold into a domain-swapped dimer (Verdino, Westritschnig, Valenta, and Keller 2002), the structure of Bet v 4 from birch pollen (Fig. 2H) was shown to be monomeric (Neudecker, Nerkamp, Eisenmann, Nourse, Lauber, Schweimer, Lehmann, Schwarzinger, Ferreira, and Rösch 2004). Structural changes accompanying binding of calcium in all EF hand proteins explain the loss of IgE binding upon calcium depletion (Engel, Richter, Obermeyer, Briza, Kungl, Simon, Auer, Ebner, Rheinberger, Breitenbach, and Ferreira 1997; Hayek, Vangelista, Pastore, Sperr, Valent, Vrtala, Niederberger, Twardosz, Kraft, and Valenta 1998; Niederberger, Hayek, Vrtala, Laffer, Twardosz, Vangelista, Sperr, Valent, Rumpold, Kraft, Ehrenberger, Valenta, and Spitzauer 1999).

Apart from the polcalcins with two EF hands, several polcalcin-related pollen allergens with three (Bet v 3 from birch) and four EF hand domains (Ole e 8 from olive, Jun o 4 from juniper, and Amb a 10 from ragweed) have been described. IgE inhibition experiments showed that IgE reactivity to these allergens is caused by cross-reactivity of IgE originally directed to polcalcins (Tinghino et al. 2002).

## 3 Common Molecular Properties of Allergens

Presently, we are only beginning to understand what molecular features and characteristics may make certain proteins more allergenic than others. The level of exposure certainly plays a part. Consequently, the various sources of allergens differ between countries. Buckwheat allergy is much more common in Asia where it is widely consumed (Park, Kang, Kim, Koh, Yum, Kim, Hong, and Lee 2000). In areas where birch pollen is common, there is a higher incidence of associated apple allergy, based on the cross-reactivity of Bet v 1 with its homolog in apple, Mal d 1 (Pauli, Oster, Deviller, Heiss, Bessot, Susani, Ferreira, Kraft, and Valenta 1996). In Mediterranean countries, where birch is absent and the nsLTP from peach is on of the major allergens, atopic patients also react with the nsLTP of apples (Salcedo et al. 2004). Seeds and nuts contain highly allergenic storage proteins, which may account for 50% or more of their total protein. Most major food allergens that sensitize via the gastrointestinal tract are present in at least 1% of the total protein content of plant foods. However, some proteins that are present in all plants in very large quantities such as the enzyme ribulose-1,5-bisphosphate carboxylase-oxygenase – it accounts for 30–40% of total leaf protein – have never been reported as allergens. In contrast, nsLTPs are potent allergens but not very abundant. Thus, the amount of protein alone does not explain its allergenicity. While abundance is an important factor, it is probably secondary to protein stability.

## 3.1 Properties of Food Allergens

A compact three-dimensional structure, ligand binding, disulfide bonds, and glycosylation contribute to protein stability (Breiteneder et al. 2005c). These factors are relevant to both the resistance of proteins to denaturation by food processing and the harsh conditions of the gastrointestinal tract. Ligand binding can have the overall effect of reducing the mobility of the polypeptide backbone, increasing both thermal stability and resistance to proteolysis. Some proteins form a cavity while others possess a tunnel into which ligands fit. nsLTPs, which possess a lipid-binding pocket, show increased stability when the pocket is occupied by fatty acids or phospholipid molecules (Douliez, Jegou, Pato, Molle, Tran, and Marion 2001). nsLTPs are remarkably thermostable, retaining their allergenic activity in processed foods such as sterilized peach juice (Brenna, Pompei, Ortolani, Pravettoni, Farioli, and Pastorello 2000), beer (Asero, Mistrello, Roncarolo, Amato, and van Ree 2001), and fermented products such as wine (Garcia-Robaina, de la Torre-Morin, Sanchez-Machin, Sanchez-Monge, Barber, and Lombardero 2001). One of the structural features clearly related to stability is the presence of disulfide bonds. Both inter- and intra-chain disulfide bridges constrain the three-dimensional fold such that perturbation of the structure by heat or chemicals is limited and frequently reversible. Important plant food allergens that have high numbers of disulfide bonds include members of the prolamin superfamily (nsLTPs, 2S albumins, cereal alpha-amylase/trypsin inhibitors) as well as of the pathogenesis-related proteins (class I chitinases, thaumatin-like proteins) (Breiteneder et al. 2005c). N-glycosylation can have a significant stabilizing effect on protein structure. There is evidence that it increases the stability of the 7S globulin of peas and its resistance to chemical denaturation (Pedrosa, De Felice, Trisciuzzi, and Ferreira 2000).

Several plant food allergens are able to associate with cell membranes or other types of lipid structures found in food (Breiteneder et al. 2005c). The allergenic 2S albumin from mustard was shown to interact with phospholipid vesicles (Onaderra, Monsalve, Mancheno, Villalba, Martinez del Pozo, Gavilanes, and Rodriguez 1994). This led to the proposition that such interactions might affect the uptake and processing of the allergen in the gastrointestinal tract, indicating that the biologic activity of these proteins plays a role in sustaining their allergenic potential. Similarly, there is emerging evidence that nsLTPs are also able to interact with lipid structures (Douliez, Sy, Vovelle, and Marion 2002; Subriade, Salesse, Marion, and Pézolet 1999).

The tendency of certain proteins to aggregate might affect their ability to sensitize by generally enhancing their immunogenicity. Both 7S and 11S globulins are highly thermostable. It seems that the cupin barrel remains intact, but the unfolding of other regions of the protein results in a loss of structure leading to the formation of large aggregates as was in detail examined for soybean globulins. Peanuts are often subjected to thermal processing at low water levels such as roasting. Thus, peanut proteins become more thermostable in low-water systems while at the same time glycation reactions cross-link individual molecules and increase their allergenic activity (Maleki and Hurlburt 2004; Mondoulet, Paty, Drumare, Ah-Leung, Scheinmann, Willemot, Wal, and Bernard 2005).

## 3.2 Properties of Pollen Allergens

There are only few publications referring to the common properties of pollen allergens. A prerequisite of a pollen protein for becoming allergenic is its rapid elution from the pollen grain after contact with the mucosa. In an examination of birch and timothy pollen, it was shown that major allergens were released from pollen grains within less than 1 min (Vrtala, Grote, Duchene, van Ree, Kraft, Scheiner, and Valenta 1993). In contrast, minor allergens such as profilins were eluted at much slower rates and heat shock protein 70, a non-allergenic protein, remained inside the pollen grain after aqueous extraction.

Allergens released from pollen grains during rainfalls frequently bind to polysaccharide granules derived from the pollen grain cytoplasm (Schäppi, Suphioglu, Taylor, and Knox 1997; Swoboda, Grote, Verdino, Keller, Singh, De Weerd, Sperr, Valent, Balic, Reichelt, Suck, Fiebig, Valenta, and Spitzauer 2004; Taylor, Flagan, Miguel, Valenta, and Glovsky 2004). Due to their small size, these particles are able to reach the lower airways and are responsible for asthma episodes in sensitized individuals after rainfalls (reviewed in D'Amato, Liccardi, D'Amato, and Holgate 2005).

Many pollen allergens families contain glycosylated allergens. Among the main families of pollen allergens, β-expansins (group 1 grass pollen allergens), Ole e 1-related allergens, polygalacturonases, berberine bridge enzymes, and pectate lyases from Cupressaceae pollen carry N-glycans. Plant glycoproteins contain N-linked glycans differing from their vertebrate counterparts by the presence of α1,3-fucosyl and β1,2-xylosyl residues. Therefore, these glycans are immunogenic and elicit the production of glycan-specific IgE that cross-reacts between unrelated glycoproteins from diverse plant sources. Similar to profilin-specific, cross-reactive IgE, the clinical relevance of glycan-specific IgE is usually low (reviewed in Malandain 2005). Some pollen allergens carry O-liked glycans. The major mugwort allergen, Art v 1, contains a hydroxyproline-rich C-terminal domain with a novel type of O-glycans, which bind IgE from 50% of Art v 1-sensitized patients (Leonard, Petersen, Himly, Kaar, Wopfner, Kolarich, van Ree, Ebner, Duus, Ferreira, and Altmann 2005). In contrast, a single arabinosylated hydroxyproline residue at the N-terminus of group 1 grass pollen allergens does not take part in IgE binding (Wicklein, Lindner, Moll, Kolarich, Altmann, Becker, and Petersen 2004).

Glycosylation of proteins significantly increases their hydrophilicity. It may therefore contribute to the rapid extractability of pollen allergens during rehydration, as described above. However, glycosylation is not a requirement for pollen protein allergenicity, as members of several important families of pollen allergens such as profilins, Bet v 1-homologs, and polcalcins are not glycosylated.

In summary, most food allergens are characterized by stability to proteolytic digestion, acidic pH, and heating, while the only common feature of pollen allergens seems to be solubility and rapid extractability from the pollen grain. These different molecular requirements explain the distinct distributions of pollen and food allergens among protein families as shown in Fig. 1.

# 4 Sequences, Structures, and Cross-Reactivity

In general, the membership of an allergen to a protein family is calculated based on its primary sequence. Sequence identities, however, do not always indicate the degree of cross-reactivity. Cross-reactivities within the various allergen families run the whole gamut from high cross-reactivity to no cross-reactivity. Cross-reactivity is a function of antibody-accessible molecular surface. In protein families where sequence conservation and surface conservation are high, cross-reactivity will be broad. This can be observed for the profilins and polcalcins. As described above, plant profilins share at least 75% of their sequences. The distribution of conserved and variable residues on the molecular surfaces of profilins shows that the degree of surface conservation is lower than of complete sequence conservation but still high enough to explain the pronounced cross-reactivity among profilins (Radauer et al. 2006b). Similarly, sequence identities of at least 65% between polcalcins from different plant families explain their cross-reactivity (Ledesma et al. 2006; Tinghino et al. 2002).

The situation for the Bet v 1 family of proteins seems to be rather unique. The molecular surface that is seen by the antibody is much more conserved among the members of this family than their primary amino acid sequence (Jenkins, Griffiths-Jones, Shewry, Breiteneder, and Mills 2005). While Bet v 1 and its homologous allergen from apple, Mal d 1, share an amino acid sequence identity of 56%, their surfaces are 71% identical. The homolog Gly m 4 from soybean has 47% sequence identity, but a surface identity of 60% to Bet v 1. This high surface conservation explains the high degree of cross-reactivity that is the underlying cause for the oral allergy syndrome observed in many birch pollen allergic patients (Vieths et al. 2002).

A similar analysis was performed for the nsLTP family of the prolamin superfamily, comparing sequences and structures from the allergenic nsLTPs of peach (Pru p 3), corn (Zea m 14) and, wheat (Jenkins et al. 2005). The peach and corn LTPs have large proportions of their surface conserved resulting in the observed IgE cross-reactivity (Pastorello, Farioli, Pravettoni, Ispano, Scibola, Trambaioli, Giuffrida, Ansaloni, Godovac-Zimmermann, Conti, Fortunato, and Ortolani 2000). In contrast, although some of the surface of wheat nsLTP is conserved compared to peach nsLTP, the wheat nsLTP has a very different surface topography. It seems therefore unlikely that anti-Pru p 3 IgE will cross-react with wheat nsLTP. The three-dimensional structures of the members of the family of trypsin and alpha-amylase inhibitors are more divergent than those of the nsLTPs. The wheat (Oda, Matsunaga, Fukuyama, Miyazaki, and Morimoto 1997) and the ragi (Gourinath, Alam, Srinivasan, Betzel, and Singh 2000) inhibitors are quite closely related; they share around 63% structurally aligned residues. In contrast, the wheat and the corn (Behnke, Yee, Trong, Pedersen, Stenkamp, Kim, Reeck, and Teller 1998) inhibitors share only around 30% structurally aligned residues (Mills et al. 2004). This being the case and the fact that the other type of cereal grain allergens, the cereal prolamins, show immense variety in their structures, which might explain why only a low level of cross-reactivity among cereal grains is observed (Jones, Magnolfi, Cooke, and Sampson 1995). 2S albumins do share the conserved disulfide structure common to all members of the prolamin superfamily, but they show a high level of sequence variation that is also obvious from the available structures of 2S albumins from peanut (Lehmann et al. 2006), castor bean (Pantoja-Uceda, Bruix, Gimenez-Gallego, Rico, and Santoro 2003), oil-seed rape (Rico, Bruix, Gonzalez, Monsalve, and Rodriguez 1996), and sunflower (Pantoja-Uceda, Shewry, Bruix, Tatham, Santoro, and Rico 2004). These proteins contain hypervariable loop regions that adopt a variety of conformations and are often the sites of IgE binding. This is probably the cause for the absence of IgE cross-reactivity between 2S albumins.

## 5 Protein Family Membership and Prediction of Allergenicity

During the past two decades, the number of identified allergens grew to several hundred. In parallel, genomics and proteomics research was extended to the organisms that represent relevant allergen sources. The development of allergen databases and associated bioinformatics tools enabled the comparison of newly identified proteins or complete proteomes or genomes to these databases (Jiang, Jasmin, Ting, and Ramachandran 2005).

Prediction of allergenicity became an important matter with the development of transgenic crops. The potential risk of introducing novel allergenic proteins into the human diet prompted the World Health Organization to develop a decision tree for the assessment of allergenicity (http://www.fao.org/ag/agn/food/pdf/allergygm.pdf). The algorithm assumes potential cross-reactivity if the query protein shows sequence identity to a known allergen of more than 35% over at least 80 amino acids or a contiguous stretch of at least six identical residues. However, this method produces many false-positive results; non-allergenic proteins erroneously identified as potential allergens (Stadler and Stadler 2003), especially when relying on identities of short peptides (Silvanovich, Nemeth, Song, Herman, Tagliani, and Bannon 2006). In order to improve the performance of allergen prediction, numerous sequence or motif-based algorithms have been developed (Björklund, Soeria-Atmadja, Zorzct, Hammerling, and Gustafsson 2005; Riaz, Hor, Krishnan, Tang, and Li 2005; Saha and Raghava 2006; Stadler et al. 2003).

A review of the literature on allergenic and non-allergenic members of certain protein families reveals the limitations of all allergenicity prediction approaches. The most interesting example is the Bet v 1 family. Despite considerable cross-reactivity between family members with sequence identities below 50%, closely related isoforms from the same species can show great variability in their IgE binding capacities (Ferreira, Hirtenlehner, Jilek, Godnik-Cvar, Breiteneder, Grimm, Hoffmann-Sommergruber, Scheiner, Kraft, Breitenbach, Rheinberger, and Ebner 1996). Given the fact that non-allergenic isoforms of Bet v 1 share as much as 95% of their sequences with the major allergenic isoform, Bet v 1a, any presently available allergenicity prediction system is doomed to failure.

Some allergen families such as the polcalcins and the expansins are members of large, ubiquitous superfamilies that share common structures. However, most proteins from these superfamilies are non-allergenic. The EF hand superfamily contains several allergen families as there are polcalcins in pollen, parvalbumins as food allergens in fish and amphibians (Wild and Lehrer 2005), troponin C in cockroaches (Hindley, Wunschmann, Satinover, Woodfolk, Chew, Chapman, and Pomes 2006), and a member of the S100 family in bovine dander (Rautiainen, Rytkonen, Parkkinen, Pentikainen, Linnala-Kankkunen, Virtanen, Pelkonen, and Mantyjarvi 1995). However, most EF hand proteins, including ubiquitous proteins such as calmodulins, are non-allergenic. A similar situation is found in the double-psi beta barrel superfamily that contains allergenic proteins such as grass pollen β-expansins, kiwellin, and some members of the barwin family. However, α-expansins and dicot β-expansins seem to be non-allergenic just like proteins from the majority of the protein families that are members of this superfamily.

# 6 Concluding Remarks

During the last two decades, hundreds of allergenic proteins from all relevant sources have been identified and their sequences made available in various allergen databases. This development offered the possibility of classifying allergens into protein families. The limited distribution of allergens to few protein families confirmed the assumption that not every protein is capable of becoming an allergen. However, the elucidation of molecular features that govern the allergenicity of a protein is just in its infancy. An interdisciplinary effort involving allergology, structural biology, and bioinformatics will be required to gain new insights in this area of research.

**Acknowledgments** The authors wish to acknowledge the support of the Austrian Science Fund grant SFB F01802.

# References

Adachi, M., Takenaka, Y., Gidamis, A.B., Mikami, B. and Utsumi, S. (2001) Crystal structure of soybean proglycinin A1aB1b homotrimer. J. Mol. Biol. 305, 291–305.

Adachi, M., Kanamori, J., Masuda, T., Yagasaki, K., Kitamura, K., Mikami, B. and Utsumi, S. (2003) Crystal structure of soybean 11S globulin: glycinin A3B4 homohexamer. Proc. Natl. Acad. Sci. U. S. A. 100, 7395–7400.

Alcocer, M.J., Murtagh, G.J., Bailey, K., Dumoulin, M., Meseguer, A.S., Parker, M.J. and Archer, D.B. (2002) The disulphide mapping, folding and characterisation of recombinant Ber e 1, an allergenic protein, and SFA8, two sulphur-rich 2S plant albumins. J. Mol. Biol. 324, 165–175.

Andersson, K. and Lidholm, J. (2003) Characteristics and immunobiology of grass pollen allergens. Int. Arch. Allergy Immunol. 130, 87–107.

Asero, R., Mistrello, G., Roncarolo, D., Amato, S. and van Ree, R. (2001) A case of allergy to beer showing cross-reactivity between lipid transfer proteins. Ann. Allergy. Asthma. Immunol. 87, 65–67.

Asero, R., Mistrello, G., Roncarolo, D., Amato, S., Zanoni, D., Barocci, F. and Caldironi, G. (2003) Detection of clinical markers of sensitization to profilin in patients allergic to plant-derived foods. J. Allergy Clin. Immunol. 112, 427–432.

Bartolome, B., Mendez, J.D., Armentia, A., Vallverdu, A. and Palacios, R. (1997) Allergens from Brazil nut: immunochemical characterization. Allergol. Immunopathol. (Madr). 25, 135–144.

Beardslee, T.A., Zeece, M.G., Sarath, G. and Markwell, J.P. (2000) Soybean glycinin G1 acidic chain shares IgE epitopes with peanut allergen Ara h 3. Int. Arch. Allergy Immunol. 123, 299–307.

Behnke, C.A., Yee, V.C., Trong, I.L., Pedersen, L.C., Stenkamp, R.E., Kim, S.S., Reeck, G.R. and Teller, D.C. (1998) Structural determinants of the bifunctional corn Hageman factor inhibitor: x-ray crystal structure at 1.95 A resolution. Biochemistry (Mosc). 37, 15277–15288.

Beyer, K., Bardina, L., Grishina, G. and Sampson, H.A. (2002) Identification of sesame seed allergens by 2-dimensional proteomics and Edman sequencing: seed storage proteins as common food allergens. J. Allergy Clin. Immunol. 110, 154–159.

Björklund, A.K., Soeria-Atmadja, D., Zorzet, A., Hammerling, U. and Gustafsson, M.G. (2005) Supervised identification of allergen-representative peptides for in silico detection of potentially allergenic proteins. Bioinformatics 21, 39–50.

Breiteneder, H., Pettenburger, K., Bito, A., Valenta, R., Kraft, D., Rumpold, H., Scheiner, O. and Breitenbach, M. (1989) The gene coding for the major birch pollen allergen Betv1, is highly homologous to a pea disease resistance response gene. EMBO J. 8, 1935–1938.

Breiteneder, H., Hoffmann-Sommergruber, K., O'Riordain, G., Susani, M., Ahorn, H., Ebner, C., Kraft, D. and Scheiner, O. (1995) Molecular characterization of Api g 1, the major allergen of celery (Apium graveolens), and its immunological and structural relationships to a group of 17-kDa tree pollen allergens. Eur. J. Biochem. 233, 484–489.

Breiteneder, H. and Radauer, C. (2004) A classification of plant food allergens. J. Allergy Clin. Immunol. 113, 821–830.

Breiteneder, H. and Mills, C. (2005a) Nonspecific lipid-transfer proteins in plant foods and pollens: an important allergen class. Curr. Opin. Allergy Clin. Immunol. 5, 275–279.

Breiteneder, H. and Mills, E.N. (2005b) Molecular properties of food allergens. J. Allergy Clin. Immunol. 115, 14–23.

Breiteneder, H. and Mills, E.N. (2005c) Plant food allergens – structural and functional aspects of allergenicity. Biotechnol. Adv. 23, 395–399.

Brenna, O., Pompei, C., Ortolani, C., Pravettoni, V., Farioli, L. and Pastorello, E.A. (2000) Technological processes to decrease the allergenicity of peach juice and nectar. J. Agric. Food Chem. 48, 493–497.

Burks, A.W., Williams, L.W., Helm, R.M., Connaughton, C., Cockrell, G. and O'Brien, T. (1991) Identification of a major peanut allergen, Ara h I, in patients with atopic dermatitis and positive peanut challenges. J. Allergy Clin. Immunol. 88, 172–179.

Castillo, R.M., Mizuguchi, K., Dhanaraj, V., Albert, A., Blundell, T.L. and Murzin, A.G. (1999) A six-stranded double-psi beta barrel is shared by several protein superfamilies. Structure 7, 227–236.

Cheng, C.S., Samuel, D., Liu, Y.J., Shyu, J.C., Lai, S.M., Lin, K.F. and Lyu, P.C. (2004) Binding mechanism of nonspecific lipid transfer proteins and their role in plant defense. Biochemistry (Mosc). 43, 13628–13636.

Cosgrove, D.J. (2000) Loosening of plant cell walls by expansins. Nature 407, 321–326.

D'Amato, G., Liccardi, G., D'Amato, M. and Holgate, S. (2005) Environmental risk factors and allergic bronchial asthma. Clin. Exp. Allergy 35, 1113–1124.

Douliez, J.P., Jegou, S., Pato, C., Molle, D., Tran, V. and Marion, D. (2001) Binding of two mono-acylated lipid monomers by the barley lipid transfer protein, LTP1, as viewed by fluorescence, isothermal titration calorimetry and molecular modelling. Eur. J. Biochem. 268, 384–388.

Douliez, J.P., Sy, D., Vovelle, F. and Marion, D. (2002) Interaction of surfactants and polymer-grafted lipids with a plant lipid transfer protein, LTP1. Langmuir 18, 7309–7312.

Dunwell, J.M. (1998) Cupins: a new superfamily of functionally diverse proteins that include germins and plant storage proteins. Biotechnol. Genet. Eng. Rev. 15, 1–32.

Dunwell, J.M., Khuri, S. and Gane, P.J. (2000) Microbial relatives of the seed storage proteins of higher plants: conservation of structure and diversification of function during evolution of the cupin superfamily. Microbiol. Mol. Biol. Rev. 64, 153–179.

Dunwell, J.M., Purvis, A. and Khuri, S. (2004) Cupins: the most functionally diverse protein superfamily? Phytochemistry 65, 7–17.

Engel, E., Richter, K., Obermeyer, G., Briza, P., Kungl, A.J., Simon, B., Auer, M., Ebner, C., Rheinberger, H.J., Breitenbach, M. and Ferreira, F. (1997) Immunological and biological properties of Bet v 4, a novel birch pollen allergen with two EF-hand calcium-binding domains. J. Biol. Chem. 272, 28630–28637.

Fedorov, A.A., Ball, T., Mahoney, N.M., Valenta, R. and Almo, S.C. (1997) The molecular basis for allergen cross-reactivity: crystal structure and IgE-epitope mapping of birch pollen profilin. Structure 5, 33–45.

Ferreira, F., Hirtenlehner, K., Jilek, A., Godnik-Cvar, J., Breiteneder, H., Grimm, R., Hoffmann-Sommergruber, K., Scheiner, O., Kraft, D., Breitenbach, M., Rheinberger, H.J. and Ebner, C. (1996) Dissection of immunoglobulin E and T lymphocyte reactivity of isoforms of the major birch pollen allergen Bet v 1: potential use of hypoallergenic isoforms for immunotherapy. J. Exp. Med. 183, 599–609.

Finn, R.D., Mistry, J., Schuster-Bockler, B., Griffiths-Jones, S., Hollich, V., Lassmann, T., Moxon, S., Marshall, M., Khanna, A., Durbin, R., Eddy, S.R., Sonnhammer, E.L. and Bateman, A. (2006) Pfam: clans, web tools and services. Nucleic Acids Res. 34, D247–251.

Fujino, K., Funatsuki, H., Inada, M., Shimono, Y. and Kikuta, Y. (2001) Expression, cloning, and immunological analysis of buckwheat (Fagopyrum esculentum Moench) seed storage proteins. J. Agric. Food Chem. 49, 1825–1829.

Garcia-Robaina, J.C., de la Torre-Morin, F., Sanchez-Machin, I., Sanchez-Monge, R., Barber, D. and Lombardero, M. (2001) Anaphylaxis induced by exercise and wine. Allergy 56, 357–358.

Genov, N., Goshev, I., Nikolova, D., Georgieva, D.N., Filippi, B. and Svendsen, I. (1997) A novel thermostable inhibitor of trypsin and subtilisin from the seeds of Brassica nigra: amino acid sequence, inhibitory and spectroscopic properties and thermostability. Biochim. Biophys. Acta 1341, 157–164.

Gourinath, S., Alam, N., Srinivasan, A., Betzel, C. and Singh, T.P. (2000) Structure of the bifunctional inhibitor of trypsin and alpha-amylase from ragi seeds at 2.2 A resolution. Acta Crystallogr. D. Biol. Crystallogr. 56, 287–293.

Grabarek, Z. (2006) Structural basis for diversity of the EF-hand calcium-binding proteins. J. Mol. Biol. 359, 509–525.

Guerbette, F., Grosbois, M., Jolliot-Croquin, A., Kader, J.C. and Zachowski, A. (1999) Lipid-transfer proteins from plants: structure and binding properties. Mol. Cell. Biochem. 192, 157–161.

Hayek, B., Vangelista, L., Pastore, A., Sperr, W.R., Valent, P., Vrtala, S., Niederberger, V., Twardosz, A., Kraft, D. and Valenta, R. (1998) Molecular and immunologic character- ization of a highly cross-reactive two EF-hand calcium-binding alder pollen allergen, Aln g 4: structural basis for calcium-modulated IgE recognition. J. Immunol. 161, 7031–7039.

Helm, R.M., Cockrell, G., Connaughton, C., Sampson, H.A., Bannon, G.A., Beilinson, V., Livingstone, D., Nielsen, N.C. and Burks, A.W. (2000) A soybean G2 glycinin allergen. 1. Identification and characterization. Int. Arch. Allergy Immunol. 123, 205–212.

Hindley, J., Wunschmann, S., Satinover, S.M., Woodfolk, J.A., Chew, F.T., Chapman, M.D. and Pomes, A. (2006) Bla g 6: a troponin C allergen from Blattella germanica with IgE binding calcium dependence. J. Allergy Clin. Immunol. 117, 1389–1395.

Hoffmann-Sommergruber, K., O'Riordain, G., Ahorn, H., Ebner, C., Laimer Da Camara Machado, M., Pühringer, H., Scheiner, O. and Breiteneder, H. (1999) Molecular char- acterization of Dau c 1, the Bet v 1 homologous protein from carrot and its cross-reactivity with Bet v 1 and Api g 1. Clin. Exp. Allergy 29, 840–847.

Itoh, T., Garcia, R.N., Adachi, M., Maruyama, Y., Tecson-Mendoza, E.M., Mikami, B. and Utsumi, S. (2006) Structure of 8Salpha globulin, the major seed storage protein of mung bean. Acta Crystallogr. D. Biol. Crystallogr. 62, 824–832.

James, J.M., Sixbey, J.P., Helm, R.M., Bannon, G.A. and Burks, A.W. (1997) Wheat alpha- amylase inhibitor: a second route of allergic sensitization. J. Allergy Clin. Immunol. 99, 239–244.

Jenkins, J.A., Griffiths-Jones, S., Shewry, P.R., Breiteneder, H. and Mills, E.N. (2005) Structural relatedness of plant food allergens with specific reference to cross-reactive allergens: an in silico analysis. J. Allergy Clin. Immunol. 115, 163–170.

Jiang, S.Y., Jasmin, P.X., Ting, Y.Y. and Ramachandran, S. (2005) Genome-wide Identifica- tion and Molecular Characterization of Ole_e_I, Allerg_1 and Allerg_2 Domain-contain- ing Pollen-Allergen-like Genes in Oryza sativa. DNA Res. 12, 167–179.

Jones, S.M., Magnolfi, C.F., Cooke, S.K. and Sampson, H.A. (1995) Immunologic cross- reactivity among cereal grains and grasses in children with food hypersensitivity. J. Allergy Clin. Immunol. 96, 341–351.

Karamloo, F., Scheurer, S., Wangorsch, A., May, S., Haustein, D. and Vieths, S. (2001) Pyr c 1, the major allergen from pear (Pyrus communis), is a new member of the Bet v 1 allergen family. J. Chromatogr. B. Biomed. Sci. Appl. 756, 281–293.

Kelly, J.D. and Hefle, S.L. (2000) 2S methionine-rich protein (SSA) from sunflower seed is an IgE-binding protein. Allergy 55, 556–560.

Kleber-Janke, T., Crameri, R., Appenzeller, U., Schlaak, M. and Becker, W.M. (1999) Selective cloning of peanut allergens, including profilin and 2S albumins, by phage display technology. Int. Arch. Allergy Immunol. 119, 265–274.

Kleine-Tebbe, J., Vogel, L., Crowell, D.N., Haustein, U.F. and Vieths, S. (2002) Severe oral allergy syndrome and anaphylactic reactions caused by a Bet v 1-related PR-10 protein in soybean, SAM22. J. Allergy Clin. Immunol. 110, 797–804.

Ko, T.P., Day, J. and McPherson, A. (2000) The refined structure of canavalin from jack bean in two crystal forms at 2.1 and 2.0 A resolution. Acta Crystallogr. D. Biol. Crystallogr. 56, 411–420.

Ko, T.P., Kuznetsov, Y.G., Malkin, A.J., Day, J. and McPherson, A. (2001) X-ray diffraction and atomic force microscopy analysis of twinned crystals: rhombohedral canavalin. Acta Crystallogr. D. Biol. Crystallogr. 57, 829–839.

Kreis, M., Forde, B.G., Rahman, S., Miflin, B.J. and Shewry, P.R. (1985) Molecular evolu- tion of the seed storage proteins of barley, rye and wheat. J. Mol. Biol. 183, 499–502.

Lauer, I., Foetisch, K., Kolarich, D., Ballmer-Weber, B.K., Conti, A., Altmann, F., Vieths, S. and Scheurer, S. (2004) Hazelnut (Corylus avellana) vicilin Cor a 11: molecular characterization of a glycoprotein and its allergenic activity. Biochem. J. 383, 327–334.

Lawrence, M.C., Izard, T., Beuchat, M., Blagrove, R.J. and Colman, P.M. (1994) Structure of phaseolin at 2.2 A resolution. Implications for a common vicilin/legumin structure and the genetic engineering of seed storage proteins. J. Mol. Biol. 238, 748–776.

Ledesma, A., Barderas, R., Westritschnig, K., Quiralte, J., Pascual, C.Y., Valenta, R., Villalba, M. and Rodriguez, R. (2006) A comparative analysis of the cross-reactivity in the polcalcin family including Syr v 3, a new member from lilac pollen. Allergy 61, 477–484.

Lehmann, K., Schweimer, K., Reese, G., Randow, S., Suhr, M., Becker, W.M., Vieths, S. and Rosch, P. (2006) Structure and stability of 2S albumin-type peanut allergens: implications for the severity of peanut allergic reactions. Biochem. J. 395, 463–472.

Leonard, R., Petersen, B.O., Himly, M., Kaar, W., Wopfner, N., Kolarich, D., van Ree, R., Ebner, C., Duus, J.O., Ferreira, F. and Altmann, F. (2005) Two novel types of O-glycans on the mugwort pollen allergen Art v 1 and their role in antibody binding. J. Biol. Chem. 280, 7932–7940.

Liscombe, D.K., MacLeod, B.P., Loukanina, N., Nandi, O.I. and Facchini, P.J. (2005) Evidence for the monophyletic evolution of benzylisoquinoline alkaloid biosynthesis in angiosperms. Phytochemistry 66, 2501–2520.

Lopez-Torrejon, G., Salcedo, G., Martin-Esteban, M., Diaz-Perales, A., Pascual, C.Y. and Sanchez-Monge, R. (2003) Len c 1, a major allergen and vicilin from lentil seeds: protein isolation and cDNA cloning. J. Allergy Clin. Immunol. 112, 1208–1215.

Malandain, H. (2005) IgE-reactive carbohydrate epitopes – classification, cross-reactivity, and clinical impact. Allerg. Immunol. (Paris). 37, 122–128.

Maleki, S.J. and Hurlburt, B.K. (2004) Structural and functional alterations in major peanut allergens caused by thermal processing. J. AOAC Int. 87, 1475–1479.

Mari, A. (2001) Multiple pollen sensitization: a molecular approach to the diagnosis. Int. Arch. Allergy Immunol. 125, 57–65.

Mari, A. (2005) Importance of databases in experimental and clinical allergology. Int. Arch. Allergy Immunol. 138, 88–96.

Markovic-Housley, Z., Degano, M., Lamba, D., von Roepenack-Lahaye, E., Clemens, S., Susani, M., Ferreira, F., Scheiner, O. and Breiteneder, H. (2003) Crystal structure of a hypoallergenic isoform of the major birch pollen allergen Bet v 1 and its likely biological function as a plant steroid carrier. J. Mol. Biol. 325, 123–133.

Maruyama, N., Adachi, M., Takahashi, K., Yagasaki, K., Kohno, M., Takenaka, Y., Okuda, E., Nakagawa, S., Mikami, B. and Utsumi, S. (2001) Crystal structures of recombinant and native soybean beta-conglycinin beta homotrimers. Eur. J. Biochem. 268, 3595–3604.

Mena, M., Sanchez-Monge, R., Gomez, L., Salcedo, G. and Carbonero, P. (1992) A major barley allergen associated with baker's asthma disease is a glycosylated monomeric inhibitor of insect alpha-amylase: cDNA cloning and chromosomal location of the gene. Plant Mol. Biol. 20, 451–458.

Menendez-Arias, L., Moneo, I., Dominguez, J. and Rodriguez, R. (1988) Primary structure of the major allergen of yellow mustard (Sinapis alba L.) seed, Sin a I. Eur. J. Biochem. 177, 159–166.

Mills, E.N., Jenkins, J.A., Alcocer, M.J. and Shewry, P.R. (2004) Structural, biological, and evolutionary relationships of plant food allergens sensitizing via the gastrointestinal tract. Crit. Rev. Food Sci. Nutr. 44, 379–407.

Mittag, D., Akkerdaas, J., Ballmer-Weber, B.K., Vogel, L., Wensing, M., Becker, W.M., Koppelman, S.J., Knulst, A.C., Helbling, A., Hefle, S.L., Van Ree, R. and Vieths, S. (2004) Ara h 8, a Bet v 1-homologous allergen from peanut, is a major allergen in patients with combined birch pollen and peanut allergy. J. Allergy Clin. Immunol. 114, 1410–1417.

Mittag, D., Vieths, S., Vogel, L., Wagner-Loew, D., Starke, A., Hunziker, P., Becker, W.M. and Ballmer-Weber, B.K. (2005) Birch pollen-related food allergy to legumes:

identification and characterization of the Bet v 1 homologue in mungbean (Vigna radiata), Vig r 1. Clin. Exp. Allergy 35, 1049–1055.

Mogensen, J.E., Wimmer, R., Larsen, J.N., Spangfort, M.D. and Otzen, D.E. (2002) The major birch allergen, Bet v 1, shows affinity for a broad spectrum of physiological ligands. J. Biol. Chem. 277, 23684–23692.

Mondoulet, L., Paty, E., Drumare, M.F., Ah-Leung, S., Scheinmann, P., Willemot, R.M., Wal, J.M. and Bernard, H. (2005) Influence of thermal processing on the allergenicity of peanut proteins. J. Agric. Food Chem. 53, 4547–4553.

Müntz, K. (1998) Deposition of storage proteins. Plant Mol. Biol. 38, 77–99.

Nakase, M., Adachi, T., Urisu, A., Miyashita, T., Alvarez, A.M., Nagasaka, S., Aoki, N., Nakamura, R. and Matsuda, T. (1996) Rice (Oryza sativa L.) alpha amylase inhibitors of 14–16 kDa are potential allergens and products of a multi gene family. J. Agric. Food Chem. 44, 2624–2628.

Neudecker, P., Schweimer, K., Nerkamp, J., Scheurer, S., Vieths, S., Sticht, H. and Rösch, P. (2001) Allergic cross-reactivity made visible: solution structure of the major cherry allergen Pru av 1. J. Biol. Chem. 276, 22756–22763.

Neudecker, P., Nerkamp, J., Eisenmann, A., Nourse, A., Lauber, T., Schweimer, K., Lehmann, K., Schwarzinger, S., Ferreira, F. and Rösch, P. (2004) Solution structure, dynamics, and hydrodynamics of the calcium-bound cross-reactive birch pollen allergen Bet v 4 reveal a canonical monomeric two EF-hand assembly with a regulatory function. J. Mol. Biol. 336, 1141–1157.

Neuhaus, J. (1999) Plant Chitinases (PR-3, PR-4, PR-8, PR-11). In: S. Datta and S. Muthukrishnan (Eds.), *Pathogenesis-related Proteins in Plants*. CRC Press, Boca Raton, pp. 77–105.

Niederberger, V., Hayek, B., Vrtala, S., Laffer, S., Twardosz, A., Vangelista, L., Sperr, W.R., Valent, P., Rumpold, H., Kraft, D., Ehrenberger, K., Valenta, R. and Spitzauer, S. (1999) Calcium-dependent immunoglobulin E recognition of the apo- and calcium-bound form of a cross-reactive two EF-hand timothy grass pollen allergen, Phl p 7. FASEB J. 13, 843–856.

Oda, Y., Matsunaga, T., Fukuyama, K., Miyazaki, T. and Morimoto, T. (1997) Tertiary and quaternary structures of 0.19 alpha-amylase inhibitor from wheat kernel determined by X-ray analysis at 2.06 A resolution. Biochemistry (Mosc). 36, 13503–13511.

Okada, T., Sasaki, Y., Ohta, R., Onozuka, N. and Toriyama, K. (2000) Expression of Bra r 1 gene in transgenic tobacco and Bra r 1 promoter activity in pollen of various plant species. Plant Cell Physiol. 41, 757–766.

Onaderra, M., Monsalve, R.I., Mancheno, J.M., Villalba, M., Martinez del Pozo, A., Gavilanes, J.G. and Rodriguez, R. (1994) Food mustard allergen interaction with phospholipid vesicles. Eur. J. Biochem. 225, 609–615.

Pantoja-Uceda, D., Bruix, M., Gimenez-Gallego, G., Rico, M. and Santoro, J. (2003) Solution structure of RicC3, a 2S albumin storage protein from Ricinus communis. Biochemistry (Mosc). 42, 13839–13847.

Pantoja-Uceda, D., Shewry, P.R., Bruix, M., Tatham, A.S., Santoro, J. and Rico, M. (2004) Solution structure of a methionine-rich 2S albumin from sunflower seeds: relationship to its allergenic and emulsifying properties. Biochemistry (Mosc). 43, 6976–6986.

Park, J.W., Kang, D.B., Kim, C.W., Koh, S.H., Yum, H.Y., Kim, K.E., Hong, C.S. and Lee, K.Y. (2000) Identification and characterization of the major allergens of buckwheat. Allergy 55, 1035–1041.

Pasquato, N., Berni, R., Folli, C., Folloni, S., Cianci, M., Pantano, S., Helliwell, J.R. and Zanotti, G. (2006) Crystal structure of peach Pru p 3, the prototypic member of the family of plant non-specific lipid transfer protein pan-allergens. J. Mol. Biol. 356, 684–694.

Pastorello, E.A., Farioli, L., Pravettoni, V., Ortolani, C., Ispano, M., Monza, M., Baroglio, C., Scibola, E., Ansaloni, R., Incorvaia, C. and Conti, A. (1999) The major allergen of peach (Prunus persica) is a lipid transfer protein. J. Allergy Clin. Immunol. 103, 520–526.

Pastorello, E.A., Farioli, L., Pravettoni, V., Ispano, M., Scibola, E., Trambaioli, C., Giuffrida, M.G., Ansaloni, R., Godovac-Zimmermann, J., Conti, A., Fortunato, D. and Ortolani, C. (2000) The maize major allergen, which is responsible for food-induced allergic reactions, is a lipid transfer protein. J. Allergy Clin. Immunol. 106, 744–751.

Pastorello, E.A., Varin, E., Farioli, L., Pravettoni, V., Ortolani, C., Trambaioli, C., Fortunato, D., Giuffrida, M.G., Rivolta, F., Robino, A., Calamari, A.M., Lacava, L. and Conti, A. (2001) The major allergen of sesame seeds (Sesamum indicum) is a 2S albumin. J. Chromatogr. B. Biomed. Sci. Appl. 756, 85–93.

Pastorello, E.A. and Robino, A.M. (2004) Clinical role of lipid transfer proteins in food allergy. Mol. Nutr. Food Res. 48, 356–362.

Pauli, G., Oster, J.P., Deviller, P., Heiss, S., Bessot, J.C., Susani, M., Ferreira, F., Kraft, D. and Valenta, R. (1996) Skin testing with recombinant allergens rBet v 1 and birch profilin, rBet v 2: diagnostic value for birch pollen and associated allergies. J. Allergy Clin. Immunol. 97, 1100–1109.

Pedrosa, C., De Felice, F.G., Trisciuzzi, C. and Ferreira, S.T. (2000) Selective neoglycosylation increases the structural stability of vicilin, the 7S storage globulin from pea seeds. Arch. Biochem. Biophys. 382, 203–210.

Poltronieri, P., Cappello, M.S., Dohmae, N., Conti, A., Fortunato, D., Pastorello, E.A., Ortolani, C. and Zacheo, G. (2002) Identification and characterisation of the IgE-binding proteins 2S albumin and conglutin gamma in almond (Prunus dulcis) seeds. Int. Arch. Allergy Immunol. 128, 97–104.

Rabjohn, P., Helm, E.M., Stanley, J.S., West, C.M., Sampson, H.A., Burks, A.W. and Bannon, G.A. (1999) Molecular cloning and epitope analysis of the peanut allergen Ara h 3. J. Clin. Invest. 103, 535–542.

Radauer, C. and Hoffmann-Sommergruber, K. (2004) Profilins. In: E.N.C. Mills and P.R. Shewry (Eds.), Plant Food Allergens. Blackwell Publishing, Oxford, UK, pp. 105–124.

Radauer, C. and Breiteneder, H. (2006a) Pollen allergens are restricted to few protein families and show distinct patterns of species distribution. J. Allergy Clin. Immunol. 117, 141–147.

Radauer, C., Willerroider, M., Fuchs, H., Hoffmann-Sommergruber, K., Thalhamer, J., Ferreira, F., Scheiner, O. and Breiteneder, H. (2006b) Cross-reactive and species-specific immunoglobulin E epitopes of plant profilins: an experimental and structure-based analysis. Clin. Exp. Allergy 36, 920–929.

Rautiainen, J., Rytkonen, M., Parkkinen, S., Pentikainen, J., Linnala-Kankkunen, A., Virtanen, T., Pelkonen, J. and Mantyjarvi, R. (1995) cDNA cloning and protein analysis of a bovine dermal allergen with homology to psoriasin. J. Invest. Dermatol. 105, 660–663.

Redfern, O., Grant, A., Maibaum, M. and Orengo, C. (2005) Survey of current protein family databases and their application in comparative, structural and functional genomics. J Chromatogr B Biomed Sci 815, 97–107.

Riaz, T., Hor, H.L., Krishnan, A., Tang, F. and Li, K.B. (2005) WebAllergen: a web server for predicting allergenic proteins. Bioinformatics 21, 2570–2571.

Rico, M., Bruix, M., Gonzalez, C., Monsalve, R.I. and Rodriguez, R. (1996) 1H NMR assignment and global fold of napin BnIb, a representative 2S albumin seed protein. Biochemistry (Mosc). 35, 15672–15682.

Saha, S. and Raghava, G.P. (2006) AlgPred: prediction of allergenic proteins and mapping of IgE epitopes. Nucleic Acids Res. 34, W202–209.

Salcedo, G., Diaz-Perales, A. and Sanchez-Monge, R. (2001) The role of plant panallergens in sensitization to natural rubber latex. Curr. Opin. Allergy Clin. Immunol. 1, 177–183.

Salcedo, G., Sanchez-Monge, R., Diaz-Perales, A., Garcia-Casado, G. and Barber, D. (2004) Plant non-specific lipid transfer proteins as food and pollen allergens. Clin. Exp. Allergy 34, 1336–1341.

Sampedro, J. and Cosgrove, D.J. (2005) The expansin superfamily. Genome Biol. 6, 242.

Sanchez-Monge, R., Gomez, L., Barber, D., Lopez-Otin, C., Armentia, A. and Salcedo, G. (1992) Wheat and barley allergens associated with baker's asthma. Glycosylated subunits of the alpha-amylase-inhibitor family have enhanced IgE-binding capacity. Biochem. J. 281, 401–405.

Sanchez-Monge, R., Lombardero, M., Garcia-Selles, F.J., Barber, D. and Salcedo, G. (1999) Lipid-transfer proteins are relevant allergens in fruit allergy. J. Allergy Clin. Immunol. 103, 514–519.

Sanchez-Monge, R., Lopez-Torrejon, G., Pascual, C.Y., Varela, J., Martin-Esteban, M. and Salcedo, G. (2004) Vicilin and convicilin are potential major allergens from pea. Clin. Exp. Allergy 34, 1747–1753.

Sandiford, C.P., Tatham, A.S., Fido, R., Welch, J.A., Jones, M.G., Tee, R.D., Shewry, P.R. and Newman Taylor, A.J. (1997) Identification of the major water/salt insoluble wheat proteins involved in cereal hypersensitivity. Clin. Exp. Allergy 27, 1120–1129.

Schäppi, G.F., Suphioglu, C., Taylor, P.E. and Knox, R.B. (1997) Concentrations of the major birch tree allergen Bet v 1 in pollen and respirable fine particles in the atmosphere. J. Allergy Clin. Immunol. 100, 656–661.

Scheurer, S., Metzner, K., Haustein, D. and Vieths, S. (1997) Molecular cloning, expression and characterization of Pru a 1, the major cherry allergen. Mol. Immunol. 34, 619–629.

Shewry, P.R., Napier, J.A. and Tatham, A.S. (1995) Seed storage proteins: structures and biosynthesis. Plant Cell 7, 945–956.

Shin, D.H., Lee, J.Y., Hwang, K.Y., Kim, K.K. and Suh, S.W. (1995) High-resolution crystal structure of the non-specific lipid-transfer protein from maize seedlings. Structure 3, 189–199.

Silvanovich, A., Nemeth, M.A., Song, P., Herman, R., Tagliani, L. and Bannon, G.A. (2006) The value of short amino acid sequence matches for prediction of protein allergenicity. Toxicol. Sci. 90, 252–258.

Stadler, M.B. and Stadler, B.M. (2003) Allergenicity prediction by protein sequence. FASEB J. 17, 1141–1143.

Subriade, M., Salesse, D., Marion, D. and Pézolet, M. (1999) Interaction of non-specific wheat lipid transfer protein with phospholipid monolyers imaged by fluorescence microscopy and studied by infrared spectroscopy. Biophys. J. 69, 974–988.

Svendsen, I., Nicolova, D., Goshev, I. and Genov, N. (1994) Primary structure, spectroscopic and inhibitory properties of a two-chain trypsin inhibitor from the seeds of charlock (Sinapis arvensis L), a member of the napin protein family. Int. J. Pept. Protein Res. 43, 425–430.

Swoboda, I., Grote, M., Verdino, P., Keller, W., Singh, M.B., De Weerd, N., Sperr, W.R., Valent, P., Balic, N., Reichelt, R., Suck, R., Fiebig, H., Valenta, R. and Spitzauer, S. (2004) Molecular characterization of polygalacturonases as grass pollen-specific marker allergens: expulsion from pollen via submicronic respirable particles. J. Immunol. 172, 6490–6500.

Tai, S.S., Wu, L.S., Chen, E.C. and Tzen, J.T. (1999) Molecular cloning of 11S globulin and 2S albumin, the two major seed storage proteins in sesame. J. Agric. Food Chem. 47, 4932–4938.

Tamburrini, M., Cerasuolo, I., Carratore, V., Stanziola, A.A., Zofra, S., Romano, L., Camardella, L. and Ciardiello, M.A. (2005) Kiwellin, a novel protein from kiwi fruit. Purification, biochemical characterization and identification as an allergen. Protein J. 24, 423–429.

Tassin, S., Broekaert, W.F., Marion, D., Acland, D.P., Ptak, M., Vovelle, F. and Sodano, P. (1998) Solution structure of Ace-AMP1, a potent antimicrobial protein extracted from onion seeds. Structural analogies with plant nonspecific lipid transfer proteins. Biochemistry (Mosc). 37, 3623–3637.

Taylor, P.E., Flagan, R.C., Miguel, A.G., Valenta, R. and Glovsky, M.M. (2004) Birch pollen rupture and the release of aerosols of respirable allergens. Clin. Exp. Allergy 34, 1591–1596.

Terras, F.R., Schoofs, H.M., De Bolle, M.F., Van Leuven, F., Rees, S.B., Vanderleyden, J., Cammue, B.P. and Broekaert, W.F. (1992) Analysis of two novel classes of plant antifungal proteins from radish (Raphanus sativus L.) seeds. J. Biol. Chem. 267, 15301–15309.

Teuber, S.S., Dandekar, A.M., Peterson, W.R. and Sellers, C.L. (1998) Cloning and sequencing of a gene encoding a 2S albumin seed storage protein precursor from English walnut (Juglans regia), a major food allergen. J. Allergy Clin. Immunol. 101, 807–814.

Teuber, S.S., Jarvis, K.C., Dandekar, A.M., Peterson, W.R. and Ansari, A.A. (1999) Identification and cloning of a complementary DNA encoding a vicilin-like proprotein, Jug r 2, from english walnut kernel (Juglans regia), a major food allergen. J. Allergy Clin. Immunol. 104, 1311–1320.

Teuber, S.S., Sathe, S.K., Peterson, W.R. and Roux, K.H. (2002) Characterization of the soluble allergenic proteins of cashew nut (Anacardium occidentale L.). J. Agric. Food Chem. 50, 6543–6549.

Thorn, K.S., Christensen, H.E., Shigeta, R., Huddler, D., Shalaby, L., Lindberg, U., Chua, N.H. and Schutt, C.E. (1997) The crystal structure of a major allergen from plants. Structure 5, 19–32.

Tinghino, R., Twardosz, A., Barletta, B., Puggioni, E.M., Iacovacci, P., Butteroni, C., Afferni, C., Mari, A., Hayek, B., Di Felice, G., Focke, M., Westritschnig, K., Valenta, R. and Pini, C. (2002) Molecular, structural, and immunologic relationships between different families of recombinant calcium-binding pollen allergens. J. Allergy Clin. Immunol. 109, 314–320.

Valenta, R., Duchene, M., Pettenburger, K., Sillaber, C., Valent, P., Bettelheim, P., Breitenbach, M., Rumpold, H., Kraft, D. and Scheiner, O. (1991) Identification of profilin as a novel pollen allergen; IgE autoreactivity in sensitized individuals. Science 253, 557–560.

Vanek-Krebitz, M., Hoffmann-Sommergruber, K., Laimer da Camara Machado, M., Susani, M., Ebner, C., Kraft, D., Scheiner, O. and Breiteneder, H. (1995) Cloning and sequencing of Mal d 1, the major allergen from apple (Malus domestica), and its immunological relationship to Bet v 1, the major birch pollen allergen. Biochem. Biophys. Res. Commun. 214, 538–551.

Verdino, P., Westritschnig, K., Valenta, R. and Keller, W. (2002) The cross-reactive calcium-binding pollen allergen, Phl p 7, reveals a novel dimer assembly. EMBO J. 21, 5007–5016.

Vieths, S., Scheurer, S. and Ballmer-Weber, B. (2002) Current understanding of cross-reactivity of food allergens and pollen. Ann. N. Y. Acad. Sci. 964, 47–68.

Vioque, J., Sanchez-Vioque, R., Clemente, A., Pedroche, J., Bautista, J. and Millan, F. (1999) Purification and partial characterization of chickpea 2S albumin. J. Agric. Food Chem. 47, 1405–1409.

Vrtala, S., Grote, M., Duchene, M., van Ree, R., Kraft, D., Scheiner, O. and Valenta, R. (1993) Properties of tree and grass pollen allergens: reinvestigation of the linkage between solubility and allergenicity. Int. Arch. Allergy Immunol. 102, 160–169.

Wang, F., Robotham, J.M., Teuber, S.S., Tawde, P., Sathe, S.K. and Roux, K.H. (2002) Ana o 1, a cashew (Anacardium occidental) allergen of the vicilin seed storage protein family. J. Allergy Clin. Immunol. 110, 160–166.

Weber, R.W. (2005) Cross-reactivity of pollen allergens: recommendations for immunotherapy vaccines. Curr. Opin. Allergy Clin. Immunol. 5, 563–569.

Wensing, M., Akkerdaas, J.H., van Leeuwen, W.A., Stapel, S.O., Bruijnzeel-Koomen, C.A., Aalberse, R.C., Bast, B.J., Knulst, A.C. and van Ree, R. (2002) IgE to Bet v 1 and profilin: cross-reactivity patterns and clinical relevance. J. Allergy Clin. Immunol. 110, 435–442.

Wicklein, D., Lindner, B., Moll, H., Kolarich, D., Altmann, F., Becker, W.M. and Petersen, A. (2004) Carbohydrate moieties can induce mediator release: a detailed characterization of two major timothy grass pollen allergens. Biol. Chem. 385, 397–407.

Wild, L.G. and Lehrer, S.B. (2005) Fish and shellfish allergy. Curr. Allergy Asthma Rep. 5, 74–79.

Witke, W. (2004) The role of profilin complexes in cell motility and other cellular processes. Trends Cell Biol. 14, 461–469.

# The Live Basophil Allergen Array (LBAA): A Pilot Study

Franco H. Falcone, Jing Lin, Neil Renault, Helmut Haas, Gabi Schramm, Bernhard F. Gibbs, and Marcos J.C. Alcocer

**Abstract** This chapter describes a pilot study aimed at combining protein microarrays with live basophil cells (live basophil allergen array). The aim is to combine the power of microarrays with the specificity and sensitivity of basophil tests, adding a biological dimension to the diagnosis of allergy. Allergens or control proteins were spotted onto FAST slides and incubated with purified peripheral blood basophils for different lengths of time. In the initial set of experiments, binding of basophils to the microarray was obtained by spotting anti-human IgE antibodies onto the array, and visualized by conventional light microscopy. Further proof-of-concept experiments used purified peripheral blood basophils that had been IgE-stripped by a short lactic acid buffer incubation and resensitized with an atopic donor's serum with known IgE specificity, in combination with the relevant pollen allergens. Activation of basophils was measured by detecting CD63, as an indirect measurement of basophil degranulation. The potential pitfalls of the live basophil allergen array are discussed in the context of basophil immunobiology.

**Keywords** Allergy · Basophils · Allergen array · Proteomics

## 1 Live Basophil Allergen Array

With the increasing numbers of patients with allergy or related diseases, the accurate detection and characterization of allergens to help diagnosis have never been so necessary. In a decade marked by the miniaturization, high-throughput, and multiplex analysis, it is natural that these concepts permeate the clinical test for allergens. Here we describe initial advances toward a modified and optimized allergen microarray in which live human basophils are included, with the intention to couple the diversity and power of protein

F.H. Falcone
Division of Molecular and Cellular Science, School of Pharmacy, University of Nottingham, Nottingham, NG7 2RD, UK
e-mail: Franco.Falcone@nottingham.ac.uk

A. Falus (ed.), *Clinical Applications of Immunomics*,
DOI: 10.1007/978-0-387-79208-8_8, © Springer Science+Business Media, LLC 2009

153

arrays with the sensitivity of basophils, adding a new biological dimension to the test. Details of this method have been described in a first pilot study elsewhere (Lin, Renault, Haas, Schramm, Vieths, Vogel, Falcone and Alcocer 2007). The objective of this chapter is to discuss the feasibility of live basophil allergen arrays in the light of basophil immunobiology.

## 2  A Short History of Tests for Allergy

### 2.1  Skin Prick Tests and Specific IgE Determination

The term allergy was coined by Clemens von Pirquet in 1906 (Pirquet 1906). However, the oldest diagnostic test for allergy, on which today's skin prick test is based, predates Pirquet's definition of allergy, as it was first described by Charles Blackley, who in 1873 described pollen as the causative agent of hay fever (Blackley 1873), and seven years thereafter described skin provocation tests with pollen extracts (Blackley 1880).

More than 125 years later, the skin prick test is still widely used in allergy diagnosis. In its current format, the main advantages of its use are that it is a cheap technique that is relatively safe and easy to perform. It is less amenable for the food allergens found, for example, in fruit or vegetables, because these allergens often do not have the required stability or are not (yet) available as commercial extracts. This has led to modified protocols such as prick-to-prick, in which the fresh fruit or vegetable to be examined is pricked just before being applied to the skin.

A test for the determination of specific IgE in serum was developed in 1967 by Johansson, the co-discoverer of IgE (Wide, Bennich and Johansson 1967), and marketed by Pharmacia Diagnostics AB, as radioallergosorbent assay (RAST) in 1974. Since then, several ELISA-based tests and more recently fluorescent-based techniques have become available. As far as automation is concerned, the initial RAST test was later replaced by the improved Pharmacia CAP System, UniCAP, and ImmunoCAP test. RAST, UniCAP, and ImmunoCAP are all based on the detection of specific IgE in the serum of allergic patients.

The main drawback of IgE recognition-binding to the allergens is that it does not necessarily imply that degranulation of mast cells/basophils also occurs in vivo. Therefore, IgE-based assays may not correlate with the clinical symptoms, with the results of the skin prick tests or with more robust clinical trials employing DBPCFC (Double-blind, placebo-controlled oral food challenge) as end points. This is illustrated by the existence of cross-reactive carbohydrate determinants (CCDs) on allergens which are known to lead to false positive results in patients whose IgE response is directed only at CCDs. This can result in discrepancies. for example, between skin testing and IgE binding, when only specific IgE binding is measured (Mari, Iacovacci, Afferni, Barletta, Tinghino, Di Felice and Pini 1999). However, this cannot be generalized, as some CCDs have the ability to induce basophil activation and degranulation, and therefore have clinical relevance (Foetisch, Westphal, Lauer, Retzek, Altmann, Kolarich, Scheurer and Vieths 2003). A further drawback of tests based on specific IgE measurement is their cost and the need for specialized equipment.

## 2.2 Basophil Activation Tests

Instead of detection of IgE-specific antibodies in serum, another type of allergy tests is based on the principle of detecting basophil activation, or the mediators released during this process, by a set of allergens rather than the corresponding specific IgE levels in serum. These tests have the advantage of being closer to the biology underlying allergic manifestations, caused by basophil and mast cell activation. They are collectively referred to as basophil activation tests (BATs).

An example for BATs is the determination of leukotrienes, a de novo synthesized product of basophil activation, with the cellular activation stimulation test (CAST)-ELISA (Medrala, Malolepszy, Medrala, Liebhart, Marszalska, Dobek, Gietkiewicz, Litwa and Pietrzak 1997). These tests, however, are relatively expensive and are sometimes said to be difficult to interpret (Burastero, Paolucci, Breda, Monasterolo, Rossi and Vangelista 2006). Assays based on histamine release have been widely used as a research tool, but due to a series of limitations they have not found their entry into clinical practice (Demoly, Lebel and Arnoux 2003) and are still in need of standardization.

A more recent set of tests uses Fluorescence Activated Cell Sorting (FACS) to detect the upregulation of specific surface activation markers, such as CD63 or CD203c on basophils in whole blood (recently discussed in Kleine-Tebbe, Erdmann, Knol, MacGlashan, Poulsen and Gibbs 2006).

A number of surface activation markers with different kinetics have been described: CD13, CD107a, CD164, and CD203c (Hennersdorf, Florian, Jakob, Baumgartner, Sonneck, Nordheim, Biedermann, Valent and Bühring 2005). CD203c, in particular, an ectonucleotide pyrophosphatase/phosphodiesterase 3 specific for basophils in peripheral blood, is thought to be present only on basophils and mast cells (lung, skin), and their $CD34^+$ progenitor cell lines (Bühring, Streble and Valent 2004). Its expression is low on resting basophils and increased upon activation. CD63 (a member of the transmembrane 4 superfamily, also known as tetraspanin family) is present on the granular membrane of neutrophils, eosinophils, and basophils (Mahmudi-Azer, Downey, and Moqbel 2002), but absent from the surface of resting basophils (Knol, Mul, Jansen, Calafat, and Roos 1991). As basophils degranulate, the granular membrane fuses with the cell membrane, resulting in the rapid appearance of CD63 on the basophil surface (Knol et al. 1991).

These tests are particularly attractive in cases where severe adverse reactions can be expected, such as in the case of nut allergens (Senna, Bonadonna, Crivellaro, Schiappoli and Passalacqua 2005), or where there are frequent discrepancies between known clinical sensitivity and skin testing, as reported for insect venom (Golden, Tracy, Freeman and Hoffman 2003). However, they may yield false results depending on basophil handling: e.g., CD203c, while being a relatively specific marker for basophils, is expressed on the cell surface by priming factors and following density-gradient centrifugation without cell degranulation. Conversely, CD63, which is not exclusively expressed on activated basophils, is present on platelets that can adhere to basophils.

## 2.3  Allergen Microarray Tests: The Next Generation?

While the aforementioned tests all have their relative merits and limitations, they are usually restricted to testing a few dozens of allergens. The development of microarray technologies, first for DNA (Barinaga 1991), and later for proteins (Lueking, Horn, Eickhoff, Bussow, Lehrach, and Walter 1999), has laid the foundations for the next generation of high-throughput tests.

With the availability of a large number of well-characterized recombinant proteins, several groups have produced protein arrays of different complexity that employ purified recombinant allergens (Harwanegg, Laffer, Hiller, Mueller, Kraft, Spitzauer and Valenta 2003). The results obtained from these prototypes are very encouraging, and a number of published examples now show that the sensitivity of these tests is similar to those from RAST, ELISA, and immunoblot tests (Hiller, Laffer, Harwanegg, Huber, Schmidt, Twardosz, Barletta, Becker, Blaser, Breiteneder, Chapman, Crameri, Duchene, Ferreira, Fiebig, Hoffmann-Sommergruber, King, Kleber-Janke, Kurup, Lehrer, Lidholm, Muller, Pini, Reese, Scheiner, Scheynius, Shen, Spitzauer, Suck, Swoboda, Thomas, Tinghino, Van Hage-Hamsten, Virtanen, Kraft, Muller and Valenta 2002; Lin, Shewry, Archer, Beyer, Niggemann, Haas, Wilson, and Alcocer 2006). The main advantage of these tests therefore is that it is possible to screen several thousands of potential allergens without incurring a loss of sensitivity.

The next step was to ask whether the power of allergen microarrays could be combined with the specificity and sensitivity of BATs, adding a potential amplification step and a biological dimension to the test. Is a live basophil allergen array (LBAA) a viable option? The proof-of-concept of immobilizing intact cells on the surface of the array has been elegantly established for the classification of different leukemias on a microarray of antibodies to CD markers (Belov, de la Vega, dos Remedios, Mulligan and Christopherson 2001). A further elegant application described by Ziauddin (Ziauddin and Sabatini 2001) consisted in transfecting live mammalian cells "on the spot" on cDNA arrays treated with lipid transfection reagent. This approach, in combination with phenotypic screening of the transfected cells on the microarray, enables the study of genes with unknown function on target cells. An example illustrating the potential of this technology was recently published by Palmer (Palmer, Miller and Freeman 2006), looking at apoptosis-inducing proteins. Cell arrays have been extended to RNA silencing (Mousses, Caplen, Cornelison, Weaver, Basik, Hautaniemi, Elkahloun, Lotufo, Choudary, Dougherty, Suh and Kallioniemi 2003). These examples demonstrated that although cell arrays are still in their infancy, they are a powerful emergent technology with considerable potential applicable to different areas.

## 3 Immunobiology of IgE-Dependent Basophil Activation

Success of the LBAA critically hinges on issues surrounding the high-affinity IgE receptor FcεRI, not only in terms of IgE binding but also in terms of stability and successful signal transduction. It is therefore useful to briefly review some key issues regarding this receptor.

### 3.1 The High-Affinity IgE Receptor FcεRI

Two receptors (FcεRI and FcεRII) bind the constant region of IgE with different affinities. FcεRI can bind monomeric IgE with very high affinity ($K_a = 2.1 \times 10^9$ /M) (Pruzansky and Patterson, 1986) whereas the low-affinity receptor FcεRII (CD23) has 100 times lower affinity, thus preferentially binds immunocomplexed IgE (Ravetch and Kinet 1991). Human FcεRI is mainly found on basophils and mast cells, where it occurs as a heterotetrameric $\alpha\beta\gamma_2$ receptor. The $\alpha$ chain, which is highly glycosylated, is responsible for binding of IgE via Cε3 and Cε4 domains, which is further stabilized by the Cε2 domain (McDonnell, Calvert, Beavil, Beavil, Henry, Sutton, Gould and Cowburn 2001).

The two $\gamma$ chains covalently linked to each other via a disulfide bond each contain a large intracytoplasmic domain responsible for intracellular signaling. This is achieved via unique cytoplasmic sequence motifs called immunoreceptor tyrosine-based activation motifs (ITAMs). An ITAM motif is also found on the β-chain (Cambier 1995; Jouvin, Numerof, and Kinet 1995). A heterotrimeric $\alpha\gamma_2$ form of the receptor is found on antigen presenting cells such as monocytes, dendritic cells, and Langerhans cells (Novak, Kraft and Bieber 2001). Despite the lack of a β-chain, the heterotrimeric form of FcεRI is also functional (Novak et al. 2001).

FcεRI-dependent signaling has been extensively studied in basophil-like or mast cell lines, particularly in the rat basophilic leukemia cell line RBL-2H3. The classical mechanism of activation of basophils is through the cross-linking of FcεRI receptor-bound IgE by allergens. This leads to a rapid aggregation of the allergen/IgE/FcεRI $\alpha$ chain complex, followed by internalization in coated pits. In general, soluble antigens are known to be internalized while the allergens on protein microarrays are irreversibly bound to the solid phase (e.g., nitrocellulose) surface of the array slides. Therefore, we needed to address the issue whether binding to solid-phase allergens could have a negative effect on cellular activation. In order to assess this issue, it is useful to look at basophil activation mechanisms in more detail.

## 3.2 Lipid Rafts and Basophil Activation

The question addressed in the following section is the significance of membrane compartmentalization, i.e., lipid rafts, for FcεRI–dependent signaling, as this may a priori interfere with solid-phase allergen-induced activation. Cross-linking of FcεRI–bound IgE by specific allergen is known to result in rapid phosphorylation of ITAM motifs on the β and γ chain by tyrosine kinase Lyn (Eiseman and Bolen 1992). This phosphorylation increases affinity of Lyn for the β chain and of another tyrosine kinase, Syk, for the γ chain, resulting in Syk binding and activation. Two models have been proposed to account for the early interactions between Lyn and the high-affinity IgE receptor. The first model only considers protein–protein interactions, and in this model, Lyn is loosely associated with cytoplasmic domains of non-aggregated, monomeric FcεRI; IgE-dependent cross-linking would result in the aggregation of several receptors, and this juxtaposition facilitates cross-chain transphosphorylation of ITAMs by Lyn (Vonakis, Haleem-Smith, Benjamin and Metzger 2001).

The alternative model takes into consideration the possible functions of the plasma membrane (Holowka and Baird 2001). While tyrosine phosphorylation is still carried out by the same kinases in this model, initial phosphorylations by Lyn and subsequently Syk depend on protein–lipid interactions. Monomeric IgE bound to FcεRI on resting cells is weakly associated with a mixture of cholesterol, sphingolipids, and glycero-phospholipids, forming the so-called liquid ordered ($L_o$) phase. This phase is characterized by detergent insolubility, a gel-like consistency with retained high lateral and rotational mobility at physiological temperatures. These properties depend on the presence of choles-terol, which has an ordering effect on the acyl chains of glycerophospholipids and sphingolipids. Lyn is normally anchored to the cytoplasmic side of the membrane bilayer via myristoyl and palmitoyl chains, and associated with the same type of lipids surrounding FcεRI. Upon cross-linking of FcεRI by allergen, Lyn-containing clusters are thought to rapidly coalesce with FcεRI-containing clusters, and this proximity to result in facilitated phosphorylation and signal transduction.

How does this relate to the live basophil array? At the membrane level, activation of basophils results in the clustering of FcεRI in lipid rafts in certain areas of the cell only, while the uncrosslinked receptors are evenly distributed on the whole cell surface in weak association with lipid rafts. Upon cross-linking, the association with lipid rafts becomes stronger and coalescence with Lyn-containing lipid rafts begins, initiating the signalling cascade that ulti-mately results in degranulation. Thus FcεRI aggregation after stimulation occurs only on specific, restricted areas, which can be visualized by confocal microscopy. Pierini and co-authors have shown that triggering of RBL-2H3 cells with IgE-specific ligands that had been coated onto 6-μm Sepharose beads (i.e., particles of a size comparable to the cells themselves) failed to induce a sustained $Ca^{2+}$ response, which is a prerequisite for degranulation (Pierini, Holowka and Baird 1996). Instead, the 6-μm Sepharose beads initiated phagocytosis by the cell line. This raises the crucial question whether basophil activation via IgE-specific allergens irreversibly bound to the microarray surface, are indeed able to induce stable $Ca^{2+}$ gradients sufficient to trigger degranulation, or whether similar effects as seen with the coated beads, possibly caused by the physicochemical constraints of lipid raft association, would impede basophil activation on the microarray. Thus a strategy for the detection of basophil activation on the microarray was needed.

## 4 Basophil Binding and Monitoring of Activation

### 4.1 Basophil Binding in the LBAA

Due to the uncertainties related to the movements of lipid rafts and before the activation experiments were started, the ability of basophilic cells to bind to the immobilized proteins on the array surface was investigated.

The expected concentration of cells on the spot was calculated by a very simplistic mathematical approach. The diameter of a single protein spot on a FAST slide is 150 μm, with a pitch of 300 μm between spots. Assuming that the protein will coat the slide homogeneously, i.e., without leaving large gaps, this means that the robotically spotted protein will cover a surface of $\pi r^2 = 17,662\ \mu m^2$ for each spot. Assuming a 10-μm diameter for the basophil cell, and assuming that the basophil will flatten to a half-sphere, the contact interface per cell would be $\pi r^2 = 78,5\ \mu m^2$; therefore, at least 225 basophils would be able to bind to each spot (see Fig. 1). Basophils are known to become "sticky" when activated (homotypic adhesion) (Knol, Kuijpers, Mul and Roos 1993) and this could ensure a tight packing of cells on the spots.

To test whether basophil binding to a protein microarray is possible at all, we incubated highly purified peripheral blood basophils (96% purity at $3 \times 10^6$ cell/ slide) with FAST array slides on which different concentrations of anti-IgE had been spotted on manually. The manual spotting results in a spot size of

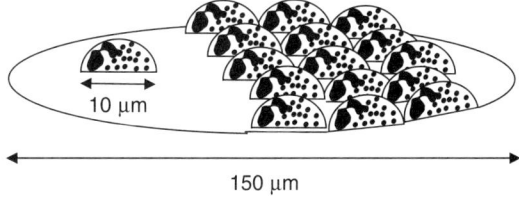

**Fig. 1** Schematic diagram illustrating basophils binding to spotted allergen on the protein array

10 μm

150 μm

approximately 650 μm, i.e., larger than the 150 μm obtained by robotics. Binding of basophils, as shown in Fig. 2, was visible after 10 min of incubation and reached a plateau after about 1 h. There was no binding to unrelated control proteins such as serum albumin at the same concentrations (not shown). While increasing incubation times appeared to increase basophil binding to the slide, prolonged incubation also seemed to decrease background binding (i.e., directly to the slide). No fixation steps were used in the process. These results clearly showed that IgE-dependent binding of basophils to solid phase–bound proteins is technically possible.

If specific allergen/extracts are spotted onto the FAST slides, the situation becomes more complex. This may be illustrated by the following consideration: peripheral blood basophils (PBBs) of healthy donors have a median

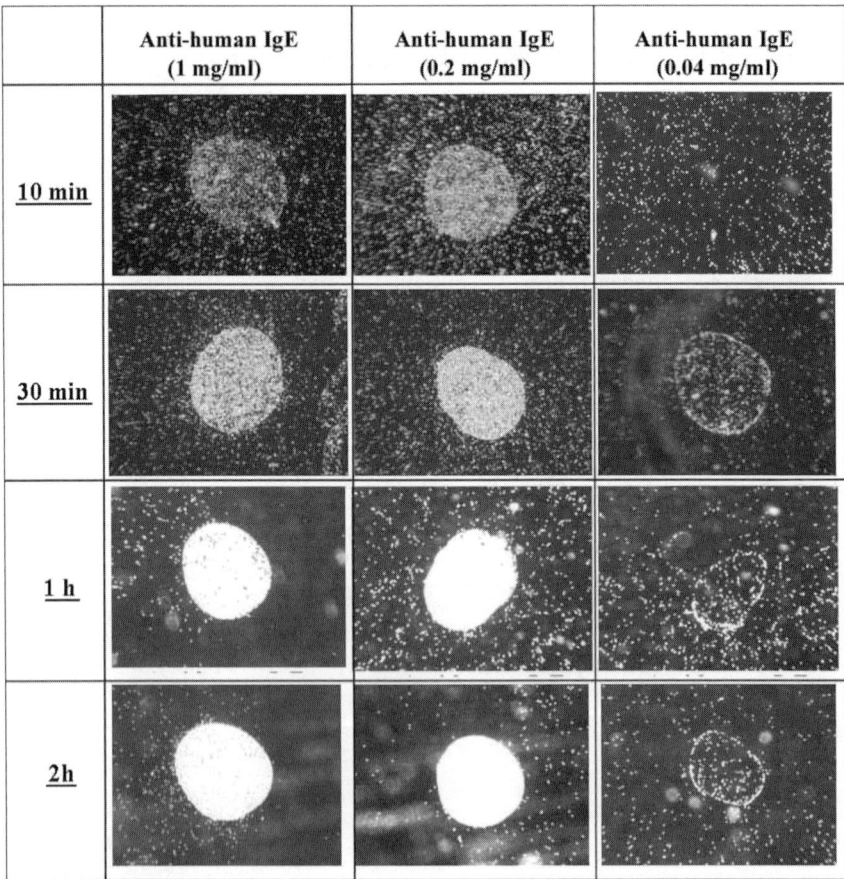

**Fig. 2** Time course of binding of purified peripheral blood basophils to different concentrations of anti-IgE manually spotted onto FAST slides. Cells were detected by light microscopy at 40× magnification

value of 100,000 IgE receptors per cell (Kleine-Tebbe et al. 2006), equally distributed on their surface ($4\pi r^2 = 314$ µm$^2$) before activation. This would mean that there would be a median of 25,000 receptors on the assumed semi-spherical contact interface between each cell and the array surface. A minimum of 200 receptors (median 2,000) need to be cross-linked in order to achieve half-maximal response, depending on the intrinsic sensitivity of the donor. However, only a subfraction of IgE receptors will carry IgE specific for the spotted allergen.

Furthermore, depending on whether recombinant or purified allergens or complex mixtures, e.g., food extracts, are spotted on the array, there will be a further reduction in viable interactions between the basophil and the solid-phase protein. Similar considerations apply to the detection of allergen-specific IgE on "conventional", cell-free allergen arrays, reducing sensitivity. This explains why the binding of basophils to allergen on the array appears weaker and "patchier" than to spotted anti-IgE (see Fig. 3).

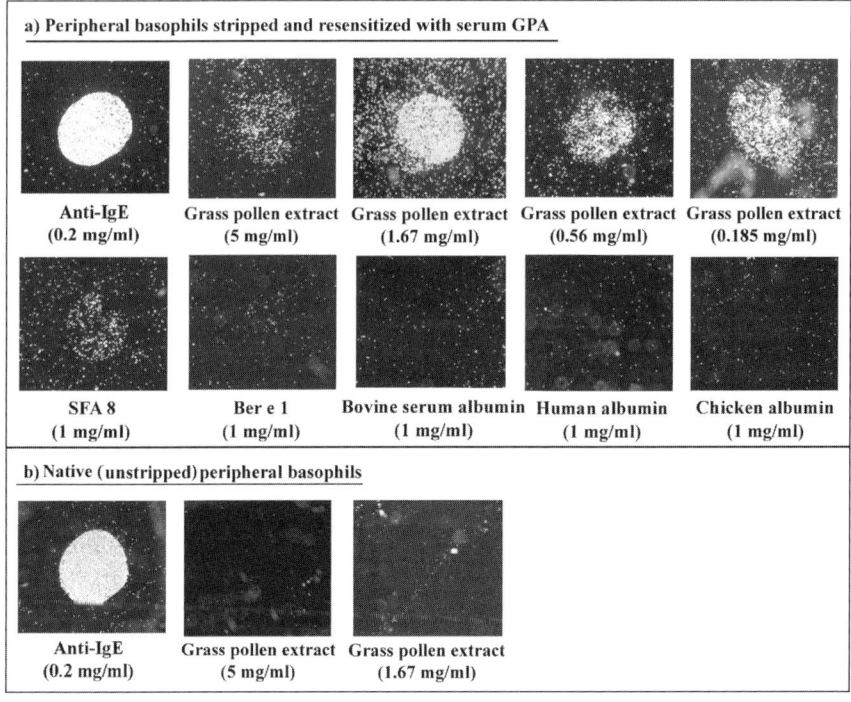

**Fig. 3** Cell binding of purified human peripheral blood basophils to anti-human IgE, allergens, or control proteins on FAST slides. Basophils of a non-atopic donor were stripped, resensitized with heterologous atopic donor's serum, and incubated with the slides (3 × 10$^6$cells/slide) in culture medium for 2 h. Images of each protein spot captured using light microscope with 40× magnification. (**a**) Stripped and resensitized basophils; (**b**) native basophils (unstripped)

From the above, it appears that when using purified PBBs, the extent of basophil activation may be variable, depending on the IgE receptor occupancy by specific IgE, intrinsic sensitivity (see next section), and the type of allergen preparation spotted.

The main advantage of using basophils, rather than attempting to directly detect specific IgE in serum samples of patients, is the added information on actual biological cross-linking, i.e., on the in vivo relevance of the respective interaction. Moreover, a higher sensitivity of the LBAA by further biological amplification of the signal is to be expected once the technique has been optimized.

## 4.2 Detection of Basophil Activation

The initial results (see Fig. 2 and Fig. 3) had shown that basophils could bind to the microarray when either anti-IgE antibodies or specific allergens were spotted onto it. Although this would suggest binding via specific IgE, alternative binding mechanisms also need to be taken into consideration. Some proteins contained in food or other allergen preparations may lead to non-IgE-dependent binding, but not necessarily activation, of basophils, e.g., in the case of lectins binding to carbohydrate structures on surface receptors. Such a binding would lead to a false-positive result if only judged by binding of basophils. The more complicated issue would be encountered where these lectins possess carbohydrate specificities that match the carbohydrate side chains of either IgE or the FcεRI α-chain. Indeed, lectins contained in certain foods, such as Concanavalin A (from the jack bean *Canavalia ensiformis*), PHA-E (from the common bean *Phaseolus vulgaris*), and LCA (agglutinin from the common lentil *Lens culinaris*) are able to bind to IgE and to induce full basophil activation and degranulation (Haas, Falcone, Schramm, Haisch, Gibbs, Klaucke, Poppelmann, Becker, Gabius and Schlaak 1999). Other lectins also found in food, such as SBA (agglutinin from soybean *Glycine max*) or HPA (agglutinin found in the edible snail *Helix pomatia*), whose specificities do not match the carbohydrate side chains of IgE, do not result in basophil activation (Haas et al. 1999), but theoretically still have the potential to bind basophils via other surface-exposed glycans.

This led to the question of how to detect basophil activation on the array. The classical products of basophil activation, either preformed in granules or de novo synthesized, such as histamine or leukotrienes, would rapidly diffuse away from the specific spot on which activation has occurred, and therefore are not useful for in situ detection of activation. This raised the question whether known surface activation markers were useful for the detection of basophil activation "on the spot".

Among the activation markers described for basophils in recent years, (Hennersdorf et al. 2005; Bühring, Seiffert, Giesert, Marxer, Kanz, Valent and

Sano 2001), the best characterized are CD63 and CD203c. CD203c is rapidly upregulated upon FcεRI cross-linking (within 10 min). It is also expressed on resting basophils, albeit at lower levels. The rapid upregulation indicates that the increased CD203c is recruited from an intracellular store rather than from de novo synthesis. CD63 is a single-chain, type 3 membrane-spanning glycoprotein member of the TM4 superfamily, expressed on the intracellular membranes of lysosomes, endosomes, and granules, as well as in Weibel Palade bodies of the vascular endothelium. It rapidly appears on degranulated neutrophils and basophils (see Fig. 4), but is usually not found on the cytoplasmic membrane of resting cells (Knol et al. 1991). Although there are controversies as to which of the markers is best for the detection of basophils, it is clear that both are a priori suitable for the detection of activation in our LBAA.

As shown in Fig. 5, purified basophils when binding to anti-IgE on the FAST slide, show $Ca^{2+}$-dependent CD63 upregulation. Thus basophil activation on the LBAA can be detected using the surface activation marker CD63.

Taken together, the data shown so far demonstrate that binding and activation of PBBs on allergen arrays is possible and can be detected. But how reliable is such a device using basophils obtained from various donors?

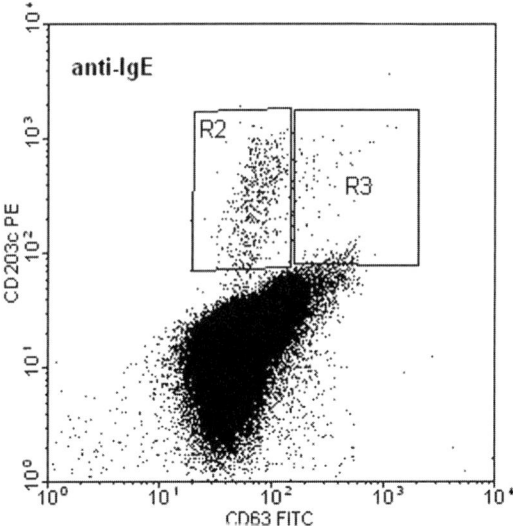

**Fig. 4** Histogram illustrating the detection of basophil activation in whole blood samples using flow cytometry. 100 μl of anticoagulated whole blood were treated with a polyclonal anti-human IgE antibody for 10 min (*right panel*), or left untreated (*left panel*), and cells stained with anti-CD63 FITC and anti CD-203c-PE antibodies. Typically, the majority of CD203c positive basophils will not be detectable in whole blood before stimulation, as the expression levels on resting basophils are low. IgE-dependent stimulation rapidly upregulates CD203c (now seen in Region R2) on basophils, and will induce degranulation and CD63 expression (Region R3) in a fraction of the activated basophils. This will depend on whether or not optimal conditions were achieved for each individual donor with a given dose of anti-IgE

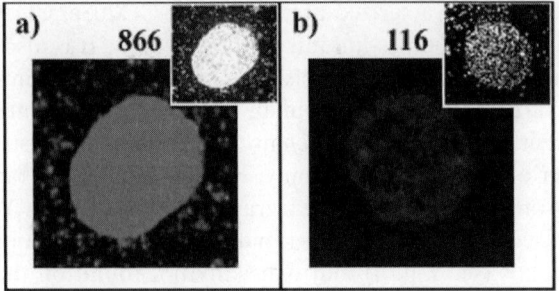

**Fig. 5** Activation of PBB bound to anti-human IgE on FAST slide in the presence (**a**) and absence (**b**) of $Ca^{2+}$. Human basophils purified from peripheral blood were incubated with the FAST slide spotted with anti-human IgE. Basophils were bound to anti-human IgE spot were scanned with GenePix 4000B scanner using 532 nm laser after staining with anti-human CD63. (**a**) Human basophils purified from peripheral blood incubated with the slides (3 × $10^6$cells/slide) in culture medium for 2 h at 37°C. (**b**) Human basophils purified from peripheral blood incubated with the slides (3 × $10^6$cells/slide) in $Ca^{2+}/Mg^{2+}$-free DPBS buffer with 0.5 % BSA and 0.2 mM EDTA for 2 h at 37°C. Insets: Images of anti-human IgE spots (0.2 mg/ml) on replicate slides were captured using light microscopes with 40× magnification. The fluorescence intensity of each spot was analyzed using the GenePix software and is shown above the fluorescent image (*See* Color Insert)

## 5 Potential Pitfalls of the LBAA

A long known feature of basophil responses to IgE-dependent stimuli is the high donor-dependent variability of the response, which depends on a series of factors. As these factors may critically impinge on the success of the live basophil array, it is useful to briefly discuss them here.

Firstly, dose-response curves to IgE-dependent stimuli, such as anti-IgE antibodies or specific allergens, are bell-shaped (see Fig. 6). Older explanations

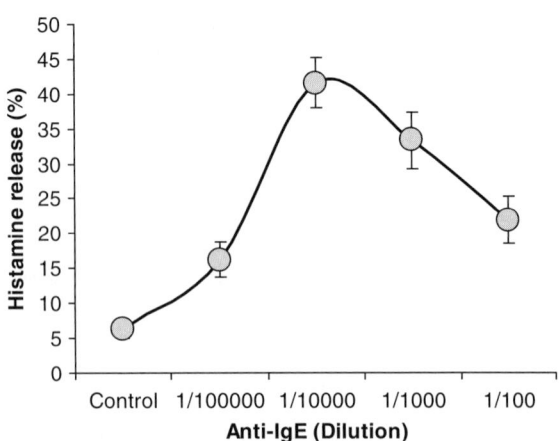

**Fig. 6** Dose-response curve of anti-IgE stimulated, purified peripheral blood basophils ($n = 14$ experiments). The curve is typically *bell-shaped*

of this phenomenon were based on the assumption that excess polyvalent antigen would lead to the formation of non-activating monovalent allergen-IgE/FcεRI interactions. A recent study, however, shows that supraoptimal IgE-dependent stimulation involves phosphorylation of src homology 2 domain-containing 5' phosphatase (SHIP), which results in lower mediator release (Gibbs, Räthling, Zillikens, Huber and Haas 2006), demonstrated in the descending arm of the stimulation curve.

These dose-response curves are highly variable between different donors, resulting in optimal stimulation conditions that can vary by several orders of magnitude between donors (Kleine-Tebbe, et al. 2006). This variability, termed **basophil sensitivity**, depends on the FcεRI receptor density, which in turn is known to be regulated by total serum IgE levels, on the composition of receptor-bound IgE, i.e., the percentage of specific vs. non-specific IgE, as well as on intrinsic basophil sensitivity, which also varies from donor to donor. As a result of these factors, the same basophils will have a different dose-response curve for different allergens. There are currently no studies addressing the issue of basophil sensitivity with solid phase–bound allergens. However, if the situation with soluble allergens is reflected in the microarrays, one may consider the possibility of basophil activation not being detectable despite binding at the lower end of the sensitivity spectrum.

The second factor to be taken into consideration, termed **basophil reactivity** (a.k.a basophil releasability), is a marked difference in the strength of activation (e.g., expressed as percentage of total histamine release) between individual donors (Fig. 7). Based on their reactivity, donors can be classed as low releasers on one end of the spectrum and high responders or releasers at the other.

The third factor that needs to be taken into account when dealing with peripheral blood basophils is the existence of a subset (10–20%) of basophil donors in the normal population, called "non-releasers", who show very little or no response at all to IgE-dependent stimulation (Nguyen, Gillis, and

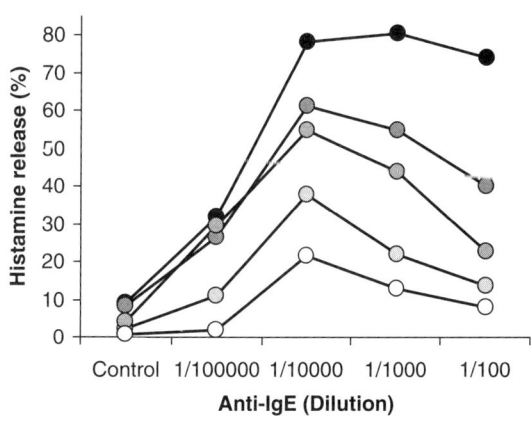

**Fig. 7** Basophil reactivities from five different donors vary individually despite similar basophil sensitivities. Basophils were stimulated with various dilutions of anti-IgE, and histamine levels (expressed as percentage of total histamine content) were determined in the supernatant

MacGlashan Jr. 1990). This has been attributed to deficient Syk expression at the protein level, while mRNA level appear to be intact (Kepley, Youssef, Andrews, Wilson, and Oliver 1994). Interestingly, Syk protein deficiency could be reversed by culturing the non-releaser basophils for four days in IL-3, restoring Syk protein expression levels (Kepley, Youssef, Andrews, Wilson, and Oliver 2000). This suggests that when studying basophils in the LBAA, the status may change depending on the time of sampling, such as in the reported case of a non-responder converting to a normal responder during the period of study (Kepley et al. 1994).

Thus basophils derived from these donors may bind to the allergen array in an IgE-dependent manner without being activated. This in turn may be taken as a false-negative result and thus compromise the sensitivity of the LBAA. Replacement of peripheral blood basophils with a suitable cell line (either of human origin or stably transfected with the human FcεRI receptor subchains) will help circumvent many of the potential pitfalls posed by varying basophil sensitivity and releasability.

# 6 Conclusion

Our proof-of-principle data show that combining live basophils with an allergen array is a feasible proposition. Purified basophils incubated with the allergen array displayed strong, specific binding to solid phase–bound anti-IgE and allergens, as well as $Ca^{2+}$-dependent activation as measured by fluorescent detection of CD63 upregulation.

However, the issues of basophil sensitivity, reactivity, and the existence of non-releasers, in combination with the technical challenges encountered with purification of basophils (low yields, cost, time), would make use of a LBAA too cumbersome for routine allergy laboratories. Despite these considerations, we think that substitution of peripheral blood basophils with a suitable basophil cell line will avoid these potential pitfalls and result in a relatively straightforward, reliable system. First pilot studies with cell lines, described elsewhere (Lin, Renault, Haas, Schramm, Vieths, Vogel, Falcone and Alcocer 2007), have been carried out with success.

Taken together, the live basophil allergen array, despite being still in its early steps of development, has the potential to add a new high-throughput tool to the technologies currently available for the screening of IgE reactivities, with the added value of a higher biological relevance than its cell-free counterpart.

**Acknowledgments** Helmut Haas and Bernhard F. Gibbs have been supported in part by a grant from the Deutsche Forschungsgemeinschaft (SFB/TRR 22). Franco Falcone is supported by the Wellcome Trust (Grant no. GR065978MA). Lin Jing and Marcos Alcocer are supported in part by a University of Nottingham's New Lecturers Fund.

# References

Barinaga, M.W. (1991) "DNA chip" speed genome initiative. Science 253, 1489.

Belov, L., de la Vega, O., dos Remedios, C.G., Mulligan, S.P. and Christopherson, R.I. (2001). Immunophenotyping of leukemias using a cluster of differentiation antibody microarray. Cancer Res. 61, 4483–4489.

Blackley C. (1873) Experimental Researches on the Cause and Nature of Catarrhus Aestivus, (London).

Blackley, C.H. (1880) Hay Fever; its Causes, Treatment, and Effective Prevention. Ballière, (London).

Bühring, H.J., Streble, A. and Valent, P. (2004) The basophil-specific ectoenzyme E-NPP3 (CD203c) as a marker for cell activation and allergy diagnosis. Int Arch Allergy Immunol 133(4), 317–329.

Bühring, H.J., Seiffert, M., Giesert, C., Marxer, A., Kanz, L., Valent, P. and Sano, K. (2001) The basophil activation marker defined by antibody 97A6 is identical to the ectonucleotide pyrophosphatase/-phosphodieste-rase 3. Blood 97, 3303–3305.

Burastero, S.E., Paolucci, C., Breda, D., Monasterolo, G., Rossi, R.E., and. Vangelista L. (2006) Unreliable Measurement of Basophil Maximum Leukotriene Release with the Bühlmann CAST 2000 Enzyme-Linked Immunosorbent Assay Kit. Clin Vaccine Immunol 13(3), 420–422.

Cambier, J.C. (1995) Antigen and Fc receptor signaling. The awesome power of the immunoreceptor tyrosine-based activation motif (ITAM). J Immunol 155, 3281–3285.

Demoly, P., Lebel, B. and Arnoux, B. (2003) Allergen-induced mediator release tests. Allergy 58, 553–558.

Eiseman, E., and Bolen. J.B. (1992) Nature 355, 78–80.

Foetisch, K., Westphal, S., Lauer, I., Retzek M., Altmann, F., Kolarich, D., Scheurer, S. and Vieths, S. (2003) Biological activity of IgE specific for cross-reactive carbohydrate determinants. J Allergy Clin Immunol 111, 889–896.

Gibbs, B.F., Räthling, A., Zillikens, D., Huber, M. and Haas, H. (2006) Initial FcεRI-mediated signal strength plays a key role in regulating basophil signaling and deactivation. J Allergy Clin Immunol 118, 1060–1067.

Golden, D.B., Tracy, J.M., Freeman, T.M. and Hoffman, D.R. (2003) Negative venom skin test results in patients with histories of systemic reaction to a sting. J Allergy Clin Immunol 112, 495–498.

Haas, H., Falcone, F.H., Schramm, G., Haisch, K., Gibbs, B.F., Klaucke, J., Poppelmann, M., Becker, W.M., Gabius, H.J. and Schlaak. M. (1999) Dietary lectins can induce in vitro release of IL-4 and IL-13 from human basophils. Eur J Immunol 29, 918–97.

Harwanegg, C., Laffer, S., Hiller, R., Mueller, M.W., Kraft, D., Spitzauer, S. and Valenta, R. (2003). Microarrayed recombinant allergens for diagnosis of allergy. Clin Exp Allergy 33, 7–13.

Hennersdorf, F., Florian, S., Jakob, A., Baumgartner, K., Sonneck, K., Nordheim, A., Biedermann, T., Valent, P. and Bühring, H.J. (2005) Identification of CD13, CD107a, and CD164 as novel basophil-activation markers and dissection of two response patterns in time kinetics of IgE-dependent upregulation. Cell Res 15, 325–335.

Hiller R., Laffer S., Harwanegg C., Huber M., Schmidt W.M., Twardosz A., Barletta B., Becker W.M., Blaser K., Breiteneder H., Chapman M., Crameri R., Duchene M., Ferreira F., Fiebig H., Hoffmann-Sommergruber K., King T.P., Kleber-Janke T., Kurup V.P., Lehrer S.B., Lidholm J., Muller U., Pini C., Reese G., Scheiner O., Scheynius A., Shen H.D., Spitzauer S., Suck R., Swoboda I., Thomas W., Tinghino R., Van Hage-Hamsten M., Virtanen T., Kraft D., Muller M.W., Valenta R. Microarrayed allergen molecules: diagnostic gatekeepers for allergy treatment. FASEB J 2002 Mar; 16(3), 414–6.

Holowka, D. and Baird, B. (2001) FcεRI as a paradigm for a lipid raft-dependent receptor in hematopoietic cells. Semin Immunol. 13, 99–105.

Jouvin, M.H., Numerof, R.P., Kinet, J.P. (1995) Signal transduction through the conserved motifs of the high affinity IgE receptor Fc epsilon RI. Semin Immunol 7, 29–35.

Kepley, C.L., Youssef, L., Andrews, R.P., Wilson, B.S. and Oliver, J.M. (1994) Syk deficiency in nonreleaser basophils. J Allergy Clin Immunol 104, 279–284.

Kepley, C.L., Youssef, L., Andrews, R.P., Wilson, B.S. and Oliver, J.M. (2000) Multiple defects in Fc epsilon RI signaling in Syk-deficient nonreleaser basophils and IL-3-induced recovery of Syk expression and secretion. J Immunol 165, 5913–5920.

Kleine-Tebbe, J., Erdmann, S., Knol, E.F., MacGlashan, D.W., Poulsen, L.K. and Gibbs, B.F. (2006) Diagnostic tests based on human basophils: potential, pitfalls and perspectives. Int Arch Allergy Immunol 141, 79–90.

Knol, E. F., Mul, F. P., Jansen, H., Calafat, J. and Roos, D. (1991) Monitoring human basophil activation via CD63 monoclonal antibody 435. J Allergy Clin Immunol 88, 328–338.

Knol, E.F., Kuijpers, T.W., Mul, F.P. and Roos, D. (1993) Stimulation of human basophils results in homotypic aggregation. A response independent of degranulation. J Immunol 151, 4926–4933.

Lin, J., Renault, N., Haas, H., Schramm, G., Vieths, S., Vogel, L., Falcone, F.H. and Alcocer, M.J. (2007) A novel tool for the detection of allergic sensitization combining protein microarrays with human basophils. Clin Exp Allergy. 37, 1854–1862.

Lin, J., Shewry, P.R., Archer, D.B., Beyer, K., Niggemann, B., Haas, H., Wilson, P., Alcocer, M.J. (2006). "The Potential Allergenicity of Two 2S Albumins from Soybean (Glycine max): A Protein Microarray Approach." Int Arch Allergy Immunol 141(2), 91–102.

Lueking, A., Horn, M., Eickhoff, H., Bussow, K., Lehrach, H. and Walter, G. (1999) Protein Microarrays for gene expression and antibody screening. Anal Biochem 270, 103–111.

Mahmudi-Azer, S., Downey, G.P. and Moqbel, R. (2002) Translocation of the tetraspanin CD63 in association with human eosinophil mediator release. Blood, 99 (11), 4039–4047.

Mari, A., Iacovacci, P., Afferni, C., Barletta, B., Tinghino, R., Di Felice, G. and Pini, C. (1999) Specific IgE to cross-reactive carbohydrate determinants strongly affect the in vitro diagnosis of allergic diseases. J Allergy Clin Immunol 103, 1005–1.

McDonnell J.M., Calvert, R., Beavil, R.L., Beavil, A.J., Henry, A.J., Sutton, B.J., Gould, H.J. and Cowburn (2001) The structure of the IgE Cε2 domain and its role in stabilizing the complex with its high-affinity receptor FcεRIα. Nat Struct Biol 8, 437–441.

Medrala, W., Malolepszy, J., Medrala, A.W., Liebhart, J., Marszalska, M., Dobek, R., Gietkiewicz, K., Litwa, M. and Pietrzak, E. (1997) CAST-ELISA test-a new diagnostic tool in pollen allergy. J Investig Allergol Clin Immunol 7, 32–35.

Mousses, S., Caplen, N.J., Cornelison, R., Weaver, D., Basik, M., Hautaniemi, S., Elkahloun, A.G., Lotufo, R.A., Choudary, A., Dougherty, E.R., Suh, E. and Kallioniemi O. (2003) RNAi microarray analysis in cultured mammalian cells. Genome Res 13(10), 2341–2347.

Nguyen, K.L., Gillis, S. and MacGlashan, D.W. Jr. (1990) A comparative study of releasing and nonreleasing human basophils: nonreleasing basophils lack an early component of the signal transduction pathway that follows IgE cross-linking. J Allergy Clin Immunol 85, 1020–1029.

Novak, N., Kraft, S. and Bieber, T. (2001) IgE receptors. Curr Opin Immunol 13, 721–726.

Palmer, E.L., Miller, A.D. and Freeman, T.C. (2006) Identification and characterisation of human apoptosis inducing proteins using cell-based transfection microarrays and expression analysis. BMC Genomics 7, 145.

Pierini, L., Holowka, D. and Baird, B. (1996). FcεRI-mediated association of 6-μm beads with RBL-2H3 mast cells results in exclusion of signaling proteins from the forming phagosome and abrogation of normal downstream signaling. J Cell Biol 134, 1427–1439.

Pirquet, C. (1906) Klinische Studien über Vakzination und vakzinale Allergie. Münchener medizinische Wochenschrift, 53, 1457–1458.

Pruzansky, J.J. and Patterson, R. (1986) Binding constants of IgE receptors on human blood basophils for IgE. Immunology, 58, 257–262.

Ravetch, J.V. and Kinet, J.P. (1991) Fc receptors. Annu Rev Immunol 9, 457–492.

Senna, G., Bonadonna, P., Crivellaro, M., Schiappoli, M. and Passalacqua, G. (2005) Anaphylaxis due to Brazil nut skin testing in a walnut-allergic subject. J Investig Allergol Clin Immunol 5, 225–227.

Vonakis, B.M., Haleem-Smith, H., Benjamin, P. and Metzger. H. (2001) Interaction between the unphosphorylated receptor with high affinity for IgE and Lyn kinase. J Biol Chem 276, 1041–1050.

Wide, L., Bennich, H. and Johansson, S.G. (1967) Diagnosis of allergy by an in-vitro test for allergen antibodies. Lancet 2 (7526), 1105–1179.

Ziauddin, J. and Sabatini, D.M. (2001) Microarrays of cells expressing defined cDNAs. Nature 411(6833), 107–110.

# Emerging Therapies for the Treatment of Autoimmune Myasthenia Gravis

**Kalliopi Kostelidou, Anastasia Sideri, Konstantinos Lazaridis, Efrosini Fostieri, and Socrates J. Tzartos**

**Abstract** Myasthenia gravis (MG) is an autoimmune disease mediated by autoantibodies that mainly target muscle nicotinic acetylcholine receptor (AChR) and cause loss of functional AChRs in the neuromuscular junction. Despite extensive knowledge from studies on MG and its major autoantigen, the aetiology of the disease remains unclear, thus rendering therapeutic options unable to target the causative agent. Latest progress on recombinant expression of the AChR subunits has allowed for some alternative, though promising, therapeutic approaches. The scope of this chapter is to provide an overview of the recent achievements in the field of emerging therapeutics for MG, including antigen-specific therapies, which are directed at the autoimmune response.

**Keywords** Autoimmunity · Myasthenia gravis · Nicotinic acetylcholine receptor · Neuromuscular junction · Autoantibodies · Specific immunotherapy

**Abbreviations** APCs – antigen-presenting cells; APL – altered peptide ligand; AChR – nicotinic acetylcholine receptor; DFP – double filtration plasmapheresis; EAMG – experimental autoimmune MG; ECDs – extracellular domains; EOMG – early-onset MG; IA – immunoadsorption; IVIg – intravenous immunoglobulin; LNCs – lymph node cells; MIR – main immunogenic region; mAb – monoclonal antibody; MG – myasthenia gravis; MHC – major histocompatibility complex; MuSK – muscle-specific kinase; NMJ – neuromuscular junction; PDEs – phosphodiesterases; PE – plasma exchange (plasmapheresis); TCR – T cell receptor.

S.J. Tzartos
Hellenic Pasteur Institute, Department of Biochemistry; University of Patras,
Department of Pharmacy, Athens, Greece
e-mail: tzartos@pasteur.gr

A. Falus (ed.), *Clinical Applications of Immunomics*,
DOI: 10.1007/978-0-387-79208-8_9, © Springer Science+Business Media, LLC 2009

# 1 Introduction

Myasthenia gravis (MG) is an antigen-specific antibody (Ab)-mediated auto-immune disease that affects the neuromuscular junction (NMJ) postsynapti-cally. It is at present one of the best understood human autoimmune diseases. In most cases, generation of auto-antibodies targeting the nicotinic acetylcholine receptor (AChR) at the NMJ causes impairment of signal transduction, failure to respond to stimulus and muscle weakness, although there are some MG patients who do not appear to have anti-AChR Abs (Lindstrom, Seybold, Lennon, Whittingham and Duane 1976; Vincent 2002). The AChR is a hetero-pentameric channel with $(\alpha_1)_2\beta_1\gamma\delta$ subunit composition in the foetal and in adult denervated muscle and $(\alpha_1)_2\beta_1\epsilon\delta$ in the adult. The five glycoprotein subunits line a central pore, which opens upon stimulation with acetylcholine to allow entry of sodium ions (Unwin, Miyazawa, Li and Fujiyoshi 2002). Loss of functional AChR molecules caused by anti-AChR Abs is clinically presented with weakness and fatigability of voluntary muscles. The animal model for MG, the Experimental Autoimmune MG (EAMG), can be induced in animals by direct administration of anti-AChR Abs or by immunisation with heterologous or homologous AChR leading to the production of Abs against the non-cognate AChR, which cross-react with the autologous receptor molecule.

A major classification of MG is made on the basis of the presence or absence of anti-AChR Abs (anti-AChR seropositive or seronegative, respectively). The for-mer group, comprising about 85% of MG patients, is characterised by the presence of anti-AChR Abs, while the latter group represents approximately 15% of patients with generalised MG symptoms who have no detectable anti-AChR Abs in their serum. In seropositive patients, the $\alpha_1$ subunit of the receptor is the major target for auto-antibodies and especially the "main immunogenic region" (MIR), a region containing overlapping epitopes including the epitope at amino acids 67–76 (Tzartos and Lindstrom 1980; Tzartos, Barkas, Cung, Mamalaki, Marraud, Orlewski, Papanastasiou, Sakarellos, Sakarellos-Daitsiotis, Tsantili and Tsikaris1998). In around 40% of anti-AChR negative patients, the target autoantigen was identified as the muscle-specific kinase (MuSK) (Hoch, McConville, Helms, Newsom-Davis, Melms and Vincent 2001), a key player in postsynaptic NMJ differentiation and AChR clustering (Evoli, Tonali, Padua, Monaco, Scuderi, Batocchi, Marino and Bartoccioni 2003; Zhou, McConville, Chaudhry, Adams, Skolasky, Vincent, and Drachman 2004). Finally, in some of the remaining anti-AChR negative patients, low-affinity Abs can be detected by an immunofluorescent assay (Leite, Cossins, Beeson, Willcox and Vincent 2006) or by a novel ultrasensitive radioimmunoassay (Trakas and Tzartos, unpublished data).

The incidence and prevalence of MG appears to vary between different epidemiological studies. For instance, the annual incidence in the UK was found to be 30/million population/year and the point prevalence rate 400/million population (MacDonald, Cockerell, Sander and Shorvon 2000), in Estonia 4/million/year and 99/million, respectively (Ööpick, Kaasik and Jakobsen

2003) and in Virginia, USA, it was 9.1/million/year and 142/million, respectively (Phillips, Torner, Anderson and Cox 1992). In a large study in Greece involving only anti-AChR seropositive patients, the incidence was 7.4/million/year (7.14 and 7.66 for women and men, respectively) and the prevalence was 70.63/million (81.58 and 59.39, respectively) (Poulas, Tsibri, Kokla, Papanastasiou, Tsouloufis, Marinou, Tsantili, Papapetropoulos and Tzartos 2001).

In terms of age of onset, there appear to be two incidence peaks, one during the third decade, termed early onset MG (EOMG), which affects mostly women, and a second during the sixth decade, which predominantly affects men. There is, however, evidence suggesting that MG in older people, $\geq 75$ years of age, is underdiagnosed (Vincent, Clover, Buckley, Grimley Evans and Rothwell 2003). MG can also manifest as transient neonatal disease due to the transfer of maternal anti-AChR Abs to the newborn. Incidence is low and about 10–15% of affected newborns from MG mothers show symptoms, which usually resolve spontaneously after 1–3 weeks (Thanvi and Lo 2004), although in few cases the babies suffer from severe joint contractures, hypoplasia of the lungs and other deformities, a condition known as arthrogryposis multiplex congenita, and die shortly after birth due to breathing difficulties (Vincent 2002; Polizzi, Huson and Vincent 2000).

## 1.1 Clinical Features

The most characteristic feature of MG is painless, fatiguable weakness, which becomes more evident with exertion (Oosterhuis 1989; Oosterhuis 1997; Vincent, Palace, and Hilton-Jones 2001). The pattern of muscle involvement varies between individuals, although in 60% of the patients the initial symptoms involve the ocular muscles in the form of ptosis and diplopia. Upper extremity muscle weakness is more pronounced than lower extremity weakness. Bulbar and facial muscle weakness cause reduced facial expression as well as speech, chewing, and swallowing difficulties. Neck flexor weakness leads to head droop and limb weakness is confined proximally, although the small muscles of the hand are also involved. Respiratory muscle failure can be life-threatening and is the most common cause of death of MG patients, which is however unusual nowadays. Facial and bulbar muscle atrophy is a common long-term consequence of the disease, while weakness can be confined to one group of muscles for many years, such as the eye muscles (ocular myasthenia), or spread to affect skeletal muscles (generalised myasthenia) (Vincent et al. 2001), with relapses and remissions occuring during the course of the disease.

Patients with anti-MuSK Abs develop bulbar and respiratory symptoms that can be life-threatening. Axial muscle weakness could develop in the acute phase of the disease, but is rarely severe, while the clinical presentation, response to treatment and thymic pathology are similar in seronegative anti-MuSK and anti-AChR positive patients (Lauriola, Ranelletti, Maggiano, Guerriero, Punzi, Marsili, Bartoccioni and Evoli 2005).

MG development is often associated with thymic abnormalities. EOMG is usually associated with thymic hyperplasia, which involves the presence of germinal centres in the thymic medulla. Thymic tumours (thymomas) are found in 10% of MG patients usually between the ages of 40–60 (Vincent 2002). Thymomas can predate MG or be diagnosed during the course of treatment. The thymus in late-onset MG is normal or atrophic. These represent more than 60% of the patients diagnosed every year due to the increasing age of the population and the availability of diagnostic tests (Vincent et al. 2001).

MG severity is classified according to the severity of the symptoms, with the Ossserman-Genkins classification being the archetypal, which is most commonly used and defines four levels of severity: (1) symptoms limited to the ocular muscles; (2) generalised myasthenia that is further divided into mild (a) and moderate symptom gravity (b); (3) generalised myasthenia with severe symptoms; and (4) myasthenic crisis with respiratory failure (Thanvi and Lo 2004). Recently, a similar classification has been proposed by the American MG Foundation, with five classes I–V, class I being ocular myasthenia and class V defined by intubation (Jaretzki, Barohn, Ernstoff, Kaminski, Keesey, Penn and Sanders 2000). Classes II–IV are each subdivided into (a) and (b) depending on whether the symptoms involve mostly the limb and axial muscles or the oropharyngeal and respiratory muscles, respectively.

## 2 Current Treatments and Treatments at the Clinical Trial Stage

The outlook for MG patients has dramatically improved in the last 30–40 years, with mortality rates near zero, with the optimal treatment (Witte, Cornblath, Parry, Lisak and Schatz 1984). The methods currently in use are classified into symptomatic, aimed at managing the consequences of the autoantibody attack, rather than the attack itself, and immunomodulatory treatments. Therapy which is directed at the immune system compared to therapy maximising the alleviation of the MG symptoms, aims at inducing and maintaining remission with the least possible cost-to-benefit ratio. Symptomatic treatments include mainly (1) the use of anticholinesterase drugs and (2) avoidance of situations and drugs that exarcebate MG, while the immunomodulatory treatments include mainly (1) thymectomy, (2) oral corticosteroids, (3) non-steroidal immunosuppressants, (4) plasmapheresis, (5) immunoadsorption and (6) intravenous immunoglobulin (IVIg).

### 2.1 Anticholinesterase Drugs

The pathogenesis of autoimmune MG, as mentioned above, consists of auto-antibody attack on the AChR, resulting in the reduction of available AChR at the NMJ. The first line of symptomatic treatment for MG includes anticholinesterase drugs, which inhibit the acetylcholinesterase activity at the NMJ and therefore increase the concentration of available acetylcholine, thus enhancing neuromuscular transmission. *Pyridostigmine* and *neostigmine*, are the most commonly used anticholinesterase drugs. They are effective early in the disease or in patients with mild MG that have adequate numbers of AChRs at the NMJ. Over time, increasing doses are required to achieve the same effect, which eventually reminishes even at maximal doses (Keesey 2004). Reported side effects include abdominal cramping, increased salivation, and diarrhea (Kothari 2004). Recently, antisense technology was recruited to selectively block the synthesis of the acetylcholinesterase protein, through use of an anti-sense, synthetic, complementary oligonucleotide to the mRNA of the specific acetylcholinesterase gene; a Phase IIa study is in progress in Israel, but results are not yet available (Sieb 2005; Boneva, Hamra-Amitay, Wirguin and Brenner 2006). Two additional symptomatic therapies for which limited data are available include *ephedrine*, which is difficult to obtain, and *3,4-diaminopyridine* (Lundh, Nillson and Rosen 1985; Sieb and Engel 1993). Both have similar effect to anticholinesterase drugs in that they increase the concentration of acetylcholine that is released from the presynaptic nerve terminal. However, they are neither as effective nor as safe as pyridostigmine (Richman and Agius 2003).

## 2.2 Other Symptomatic Treatments

Other symptomatic treatments involve avoidance of NMJ-compromised situations, such as overexertion and unnecessary fatigue, as well as drugs with direct pharmacological effect on the NMJ, such as aminoglycoside antibiotics and penicillamine (Barrons 1997; Keesey 2004). It was demonstrated that amino-glycoside antibiotics competitively restrain the release of acetylcholine from the presynaptic membrane and therefore block neuromuscular transmission (Schlesinger, Krampfl, Haeseler, Dengler and Bufler 2004; Liu and Hu 2005). D-penicillamine, an anti-rheumatoid arthritis drug, has a highly reactive sulfhydryl group and has been shown to readily bind and reduce cysteine residues close to the ligand-binding site in the $\alpha$ and $\gamma$ subunits of the AChR (Dawkins, Christiansen and Garlepp 1981; Penn, Low, Jaffe, Luo and Jacques 1998; Hill, Moss, Wordsworth, Newsom-Davis and Willcox 1999). It can, therefore, initiate or worsen the autoimmune response to AChR. Advances in artificial ventilation and the administration of improved antibiotics for respiratory infections have dramatically decreased mortality due to respiratory muscle failure, the most common cause of death of MG patients, from 40% to 5% (Grob, Arsura, Brunner and Namba 1987). Whole-body cooling in patients with MG has also been promising for reducing weakness and fatigue and improving muscle strength (Mermier, Schneider, Gurney, Weingart and Wilmerding 2006).

## 2.3 Thymectomy

The removal of the thymus gland (thymectomy) has been used for the long-term treatment of MG patients (Buckingham, Howard, Bernatz, Payne, Harrison, O'Brien and Weiland 1976). The thymus is the main site for the development of mature T-lymphocytes that are involved in the initiation and maintenance of the autoimmune response against AChR. Antibody production is mediated by AChR-specific $CD4^+$ T cells, which facilitate B cell activation (Levinson and Wheatley 1996; Schwendimann, Burton, and Minagar 2005). The effect of thymectomy is not apparent until after about a year, and its full effect may not be sensed for the first 5 years (Perlo, Poskanzer, Schwab, Viets, Osserman and Genkins 1966; Buckingham et al. 1976). Thymectomy is recommended in MG patients with EOMG or thymoma-associated MG, and its efficiency has been a matter of debate (Zieliński, Kużdżal, Szlubowski and Soja 2004). The risk of the treatment lies mainly in the operative and post-operative period, reflecting that the earlier it is performed during the course of the disease, the more effective it is for the patient (DeFilippi, Richman and Ferguson 1994). There is 25% lower remission rate in thymectomised patients with disease onset after the age of 50 years compared to younger patients (Aarli 1999).

## 2.4 Oral Corticosteroids

Immunosuppressive therapies, such as oral corticosteroids, are considered the current standard for the treatment of moderate-to-severe MG. Corticosteroids have multiple effects on macrophages as well as B and T cells (Schwendimann et al. 2005). They act by reducing both the expression of inflammatory cytokines and adhesion molecules and the trafficking of inflammatory cells (Richman and Agius 2003), but their use is escorted by complications like weight gain, hypertension, diabetes, appearance changes, anxiety/depression/insomnia (steroid psychosis), glaucoma, osteoporosis, cataracts, ulcer and gastrointestinal perforations, myopathy, opportunistic infections and avascular necrosis of large joints (Rivner 2002; Prakash, Ratnagopal, Puvanendran and Lo 2006). The most commonly used corticosteroid is *prednisone*. Administration of corticosteroids is associated with an eventual decrease in circulating titers of anti-AChR antibodies even though clinical improvement precedes the measurable decrease in pathogenic antibodies.

## 2.5 Non-Steroidal Immunosuppressants

Non-steroidal immunosuppressants, such as *azathioprine, cyclosporine A* and *cyclophosphamide* are also used for the treatment of MG patients. *Azathioprine*

# Color Insert

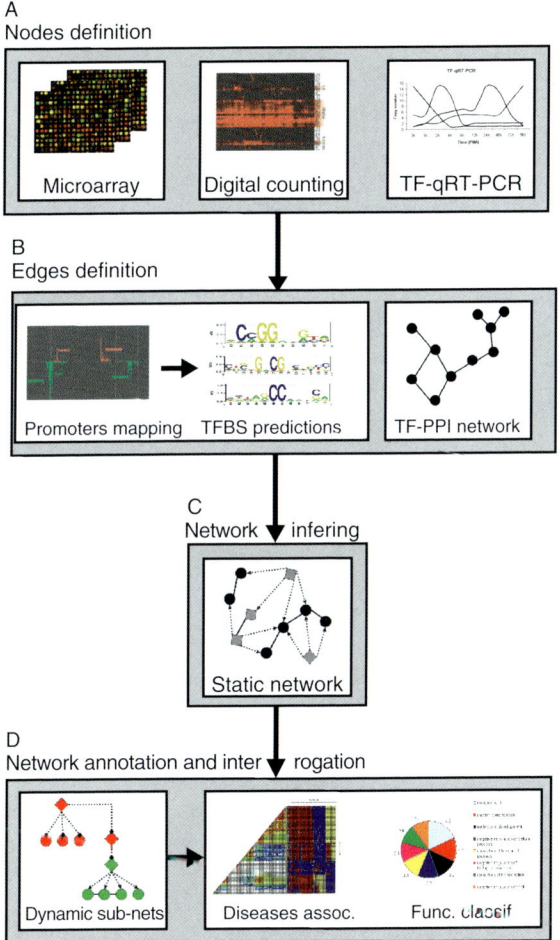

**Chapter 1, Fig. 1** Integrative approach to infer transcription regulatory networks. (**A**) The main players are defined based on their expression in the systems (nodes of the network). (**B**) Promoter mapping associated with TFBS predictions are used to infer regulatory interaction and combined with datasets for physical interactions such as protein–protein interactions (edges of the network). (**C**) Nodes and edges are combined and represented as network topology. (**D**) The network is annotated and validated based on previous knowledge or experimental validation

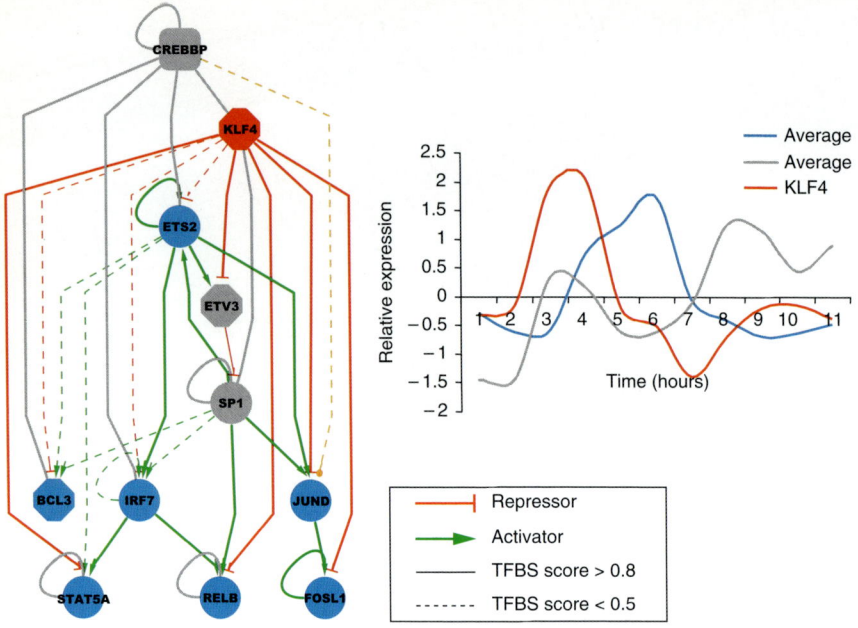

**Chapter 1, Fig. 2** Gene regulatory circuits activities in macrophage after stimulation with LPS. By integrating several datasets (expression, PPI, previous knowledge and promoter analysis), it is possible to infer dynamics and hierarchy of gene regulation; this particular case shows a beautiful feedback loop where the main players are the transcription factors KLF4, SP1 and ETV3. Green arrowed edges = gene activation, red arrowed edges = gene repression, grey edge = protein–protein interactions. The trickiness of the edges is proportional to the TFBS prediction confidence score. In the right panel is shown the expression profile of TFs and regulated genes composing the network

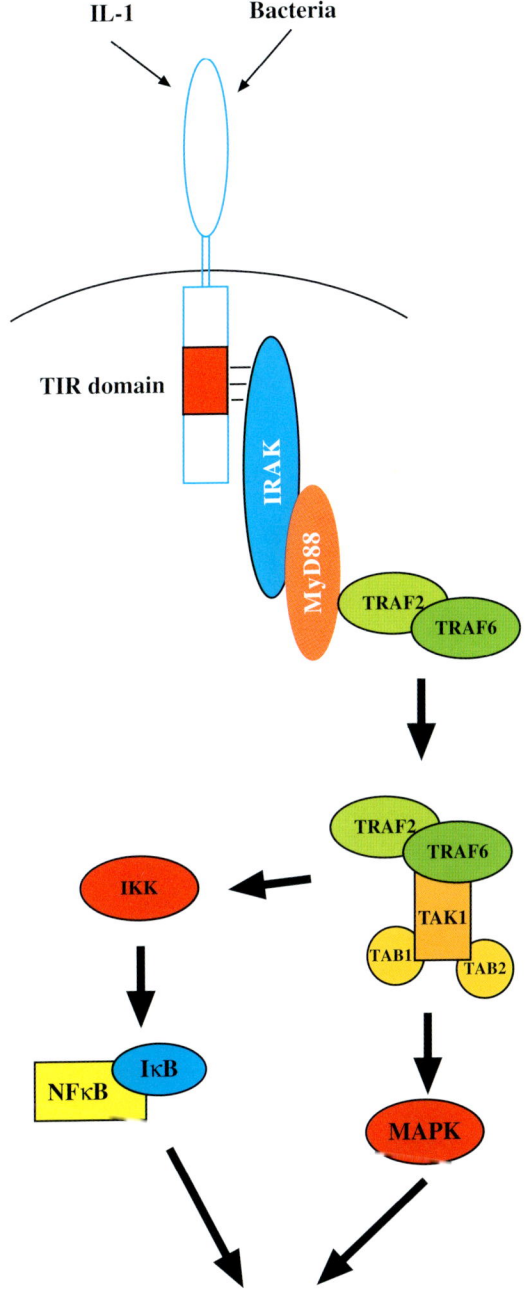

**Chapter 2, Fig. 1** Mediators of inflammatory signal transduction

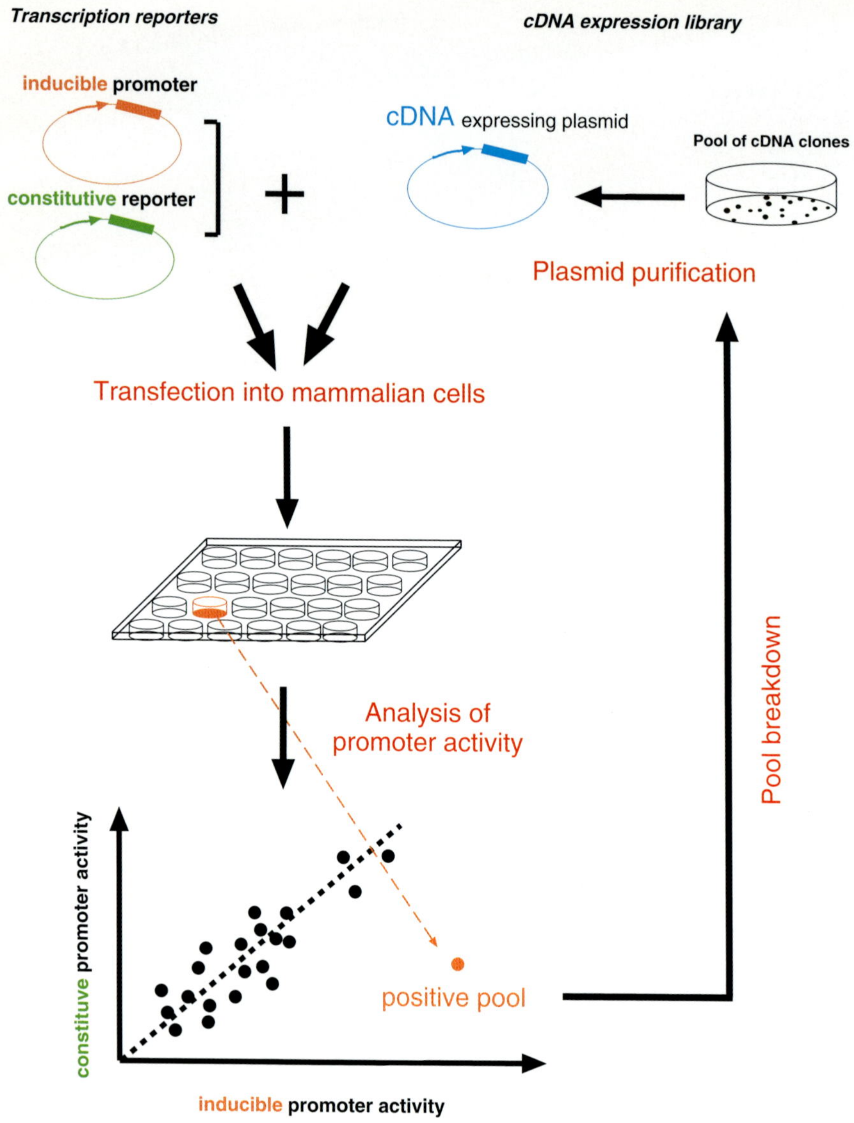

*Transcription reporters*                    *cDNA expression library*

**inducible** promoter

**constitutive** reporter

**+**

**cDNA** expressing plasmid

**Pool of cDNA clones**

**Plasmid purification**

**Transfection into mammalian cells**

**Analysis of
promoter activity**

constitutve promoter activity

inducible promoter activity

**positive pool**

**Pool breakdown**

**Chapter 2, Fig. 2** Transcription expression cloning

## Coverage of GAIA HIV B7 peptides--individually and in aggregate-- across time, countries, and clades

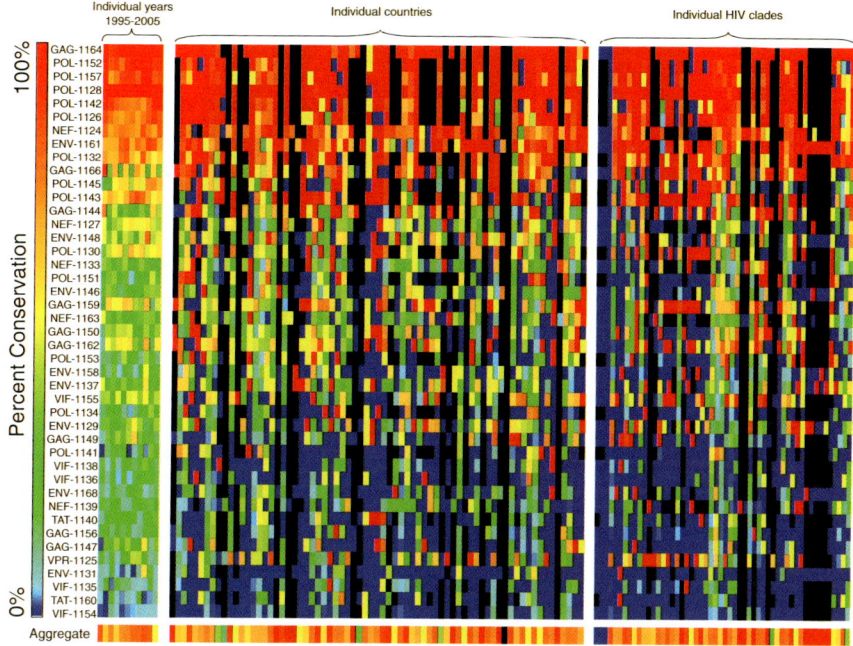

**Chapter 3, Fig. 4** GAIA HIV A2 peptides — individually and in aggregate — percentage coverage of strains by year, country, and clade. Each row of the matrix denotes a specific peptide; the peptide's protein of origin is included within the peptide ID. Each column of the matrix denotes a specific year, country, or clade, grouped as indicated. The percentage coverage of strains is represented on a color gradient, with warm tones indicating values above 50% and cool tones indicating values below 50%. The *bottom* row of the matrix shows the percentage of each protein set that contains at least one peptide from our pool. *Black boxes* indicate that no isolates of the protein are available for that year, clade, or country. The *bottom* row represents the aggregate percent coverage for the epitope set. Each cell of the matrix represents the percentage coverage per peptide, except for the bottom row cells, which represent the aggregate percentage coverage for the peptide set. Column headers are listed here for space considerations: left to right, the year columns are 1995, 1996, 1997, 1998, 1999, 2000, 2001, 2002, 2003, 2004, and 2005; aggregate coverage of strains by year ranges from 47% (2000) to 63% (2004). The countries left to right are: Angola, Argentina, Australia, Belgium, Benin, Bolivia, Brazil, Botswana, Belarus, Canada, The democratic republic of the Congo (Zaire), Congo, Cote d'Ivoire, Chile, Cameroon, China, Colombia, Cuba, Cyprus, Germany, Djibouti, Dominican Republic, Ecuador, Estonia, Spain, Ethiopia, France, Gabon, United Kingdom, Georgia, Ghana, Gambia, Equatorial Guinea, Greece, Hong Kong, Israel, India, Italy, Japan, Kenya, Republic of Korea, Mali, Myanmar, Namibia, Niger, Nigeria, Netherlands, Norway, Portugal, Russian federation, Rwanda, Sweden, Senegal, Somalia, Chad, Thailand, Trinidad and Tobago, Taiwan, United Republic of Tanzania, Ukraine, Uganda, United States, Uruguay, Uzbekistan, Venezuela, Vietnam, Yemen, South Africa, Zambia, and Zimbabwe; aggregate coverage of strains by country ranges from 33% (Gabon) to 94% (Italy). The clade columns left to right are: 0206, 0209, 1819, A, AAD, AB, AC, ACD, ACDGKU, AD, ADF, ADGU, ADHK, AE, AF, AFG, AFU, AG, AGH, AGHU, AGJ, AGU, AHJU, AU, B, BC, BF, BG, C, CD, CGU, CK, CPX, CU, D, DF, DK, DO, DU, F, G, GHJKU, GK, GKU, GU, H, J, JKU, JU, K, N, O, U. The HLA-A2 peptides together cover 51, 51, 53, and 44% of clades A, B, C, and D, respectively. On average, the chimeric clade sequences were 50% covered by our HLA-A2 restricted epitopes

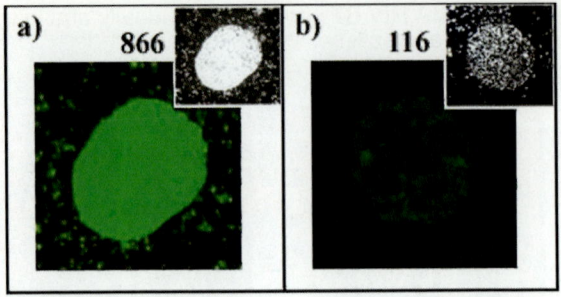

**Chapter 8, Fig. 5** Activation of PBB bound to anti-human IgE on FAST slide in the presence (**a**) and absence (**b**) of $Ca^{2+}$. Human basophils purified from peripheral blood were incubated with the FAST slide spotted with anti-human IgE. Basophils were bound to anti-human IgE spot were scanned with GenePix 4000B scanner using 532 nm laser after staining with anti-human CD63. (**a**) Human basophils purified from peripheral blood incubated with the slides ($3 \times 10^6$ cells/slide) in culture medium for 2 h at 37°C. (**b**) Human basophils purified from peripheral blood incubated with the slides ($3 \times 10^6$ cells/slide) in $Ca^{2+}/Mg^{2+}$-free DPBS buffer with 0.5 % BSA and 0.2 mM EDTA for 2 h at 37°C. Insets: Images of anti-human IgE spots (0.2 mg/ml) on replicate slides were captured using light microscopes with 40× magnification. The fluorescence intensity of each spot was analyzed using the GenePix software and is shown above the fluorescent image

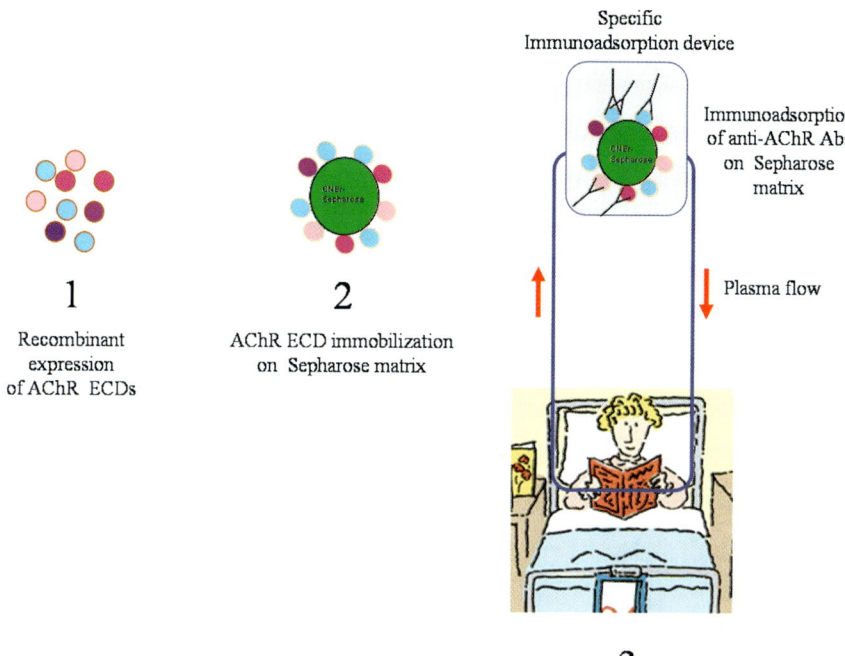

**1**

Recombinant
expression
of AChR ECDs

**2**

AChR ECD immobilization
on Sepharose matrix

Specific
Immunoadsorption device

Immunoadsorption
of anti-AChR Abs
on Sepharose
matrix

Plasma flow

**3**

**Chapter 9, Fig. 2** Schematic representation of the specific immunoadsorption technique. The recombinant expression of the AChR ECDs (**1**) is followed by their immobilisation on Sepharose beads (**2**) to construct an insoluble matrix, which will be used in an immunoadsorption device to withhold anti-AChR Abs from MG patients (**3**)

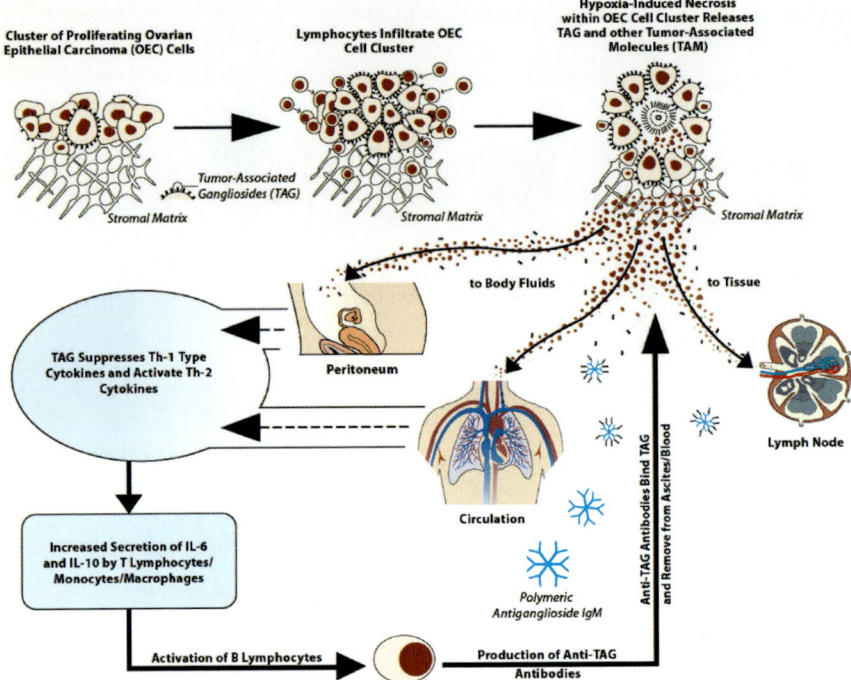

**Chapter 11, Fig. 4** Events taking place during tumor progression, which are relevant to glycoimmunomics of human cancer. Using ovarian epithelial cancer as a model, the events leading to immune response to tumor-associated gangliosides (TAG) are narrated as follows: The 1st event is proliferation of tumor cells. The 2nd event is infiltration of a variety of lymphocytes into proliferating tumor cells. The 3rd event is release or shedding of tumor-associated molecules (TAM) and TAG consequent to necrosis resulting due to lack or insufficient oxygen or hypoxia. The 4th event is spreading of TAM and TAG into the stroma or tumor microenvironment, lymph nodes or ascetic fluid, and circulation. The 5th event is TAG-mediated suppression of a variety of Th-1 type cell-mediated immune functions, and at the same time activation of Th-2 cytokines. The 6th event refers to an increased secretion of IL-6 and IL-10 by T cells, monocytes, and macrophages in the sera of patients during tumor progression. The 7th event is that these cytokines activate B lymphocytes, to produce anti-ganglioside antibody response without T-cell help, a unique phenomenon in glycoimmunomics of human cancer. The 8th event is the production of IgM, which can be polymeric and without a J chain, due to involvement of CD5 + B cells. The 9th event is the elimination of immunosuppressive gangliosides by anti-TAG antibodies (imuran) inhibits purine synthesis and therefore cell proliferation (Elion 1972; Mertens, Hertel, Reuther and Ricker 1981). It affects rapidly dividing cell populations, such as lymphocytes, and has fewer side effects than corticosteroids. The main problem associated with azathioprine is the delayed onset (up to 6 months) of clinical improvement. If long-term immunosuppression is necessary, steroids combined with azathioprine are recommended (Palace, Newsom-Davis and Lecky 1998; Skeie, Apostolski, Evoli, Gilhus, Hart, Harms, Hilton-Jones, Melms, Verschuuren and Horgel 2006).

*Cyclosporine A* has been examined as a steroid-sparing agent for steroid-dependent patients. It inhibits T cells via blocking calcineurin-mediated cytokine signalling and is comparable to azathioprine (Tindall, Phillips, Rollins, Wells and Hall 1993). It can, therefore, be used as an alternative to azathioprine in patients who are intolerant to the latter (Lavrnic, Vujic, Rakocevic-Stojanovic, Stevic, Basta, Pavlovic, Trikic and Apostolski 2005).

*Cyclophosphamide* is a strong immunosuppressive alkylating agent that acts on DNA and inhibits cell proliferation. Its effect is greater on B rather than T cells. It is, therefore, an excellent candidate for antibody-induced immune diseases (Zhu, Cupps, Whalen and Fauci 1987). It is as effective as corticosteroids in inducing remission, but its side effects are more severe and include opportunistic infections, dose-associated neoplasms and bladder toxicity (Perez, Buot, Mercado-Danguilan, Bagabaldo and Renales 1981; Niakan, Harati and Rolak 1986). High dose of cyclophosphamide (50 mg/ kg/ day intravenously for 4 days) followed by the administration of granulocyte colony stimulating factor have proven effective for the treatment of MG patients who did not respond to conventional immunotherapeutic treatments or could not tolerate their side effects (refractory MG) (Drachman, Jones and Brodsky 2003a). The side effects of high-dose cyclophosphamide include transient nausea, reversible alopecia and febrile neutropenia (Smith, Stevens and Fuller 2001). Hematopoietic stem cells are not affected because they express high levels of aldehyde dehydrogenase, which inactivates the active metabolite of cyclophosphamide, whereas lymphocytes and committed hematopoietic progenitors are rapidly killed.

Additional drugs are being considered for MG treatment, such as *methotrexate*, which has been used at high dose to treat MG patients that were not responsive to other drugs (Gerber and Steinberg 1976; Rivner 2002). The side effects of high dosage include rash, photosensitivity, haematuria and dose-dependent neurotoxicity. Alternatively, *mycophenolate mofetil*, a prodrug of mycophenolic acid, successfully used in patients with allogeneic transplants and other immune-related diseases, is currently undergoing phase III clinical trials as an adjunctive immunosuppressant for the treatment of severe, refractory and high-dose steroid-dependent MG (Scneider-Gold, Hartung and Gold 2006; Prakash et al. 2006). It inhibits the de novo purine biosynthesis pathway by blocking the enzyme inosine monophosphate dehydrogenase and is highly specific for proliferating lymphocytes, because they lack the "salvage pathway" for nucleotide synthesis (Allison and Eugui 2000; Richman and Agius 2003).

The mean onset of clinical improvement is around 10 weeks, with maximum improvement achieved after 27 weeks (Vincent and Leite 2005). *Leflunomide* (HWA 486) has also been used in an array of autoimmune disorders and has proven equivalent to or better than cyclosporine A. It is an immunosuppressant that inhibits the enzyme dihydro-orotate dehydrogenase, which is required for the de novo synthesis of pyrimidines (Burkhardt and Kalden 2005). Studies on rats immunised with AChR and treated with leflunomide from the day of disease induction demonstrated suppression of the development of EAMG (Vidic-Dankovic, Kosec, Damjanovic, Apostolski, Isakovic and Bartlett 1995). Leflunomide could, therefore, be beneficial for the treatment of MG in humans. *Tacrolimus* (FK506), a new macrolide immunosuppressant with similar action to cyclosporine A, is currently used in clinical trials at low doses as a corticosteroid-sparing medication and has proven to be as effective as cyclosporine A with fewer side effects (Tada, Shimolata, Tada, Oyake, Igarashi, Onodera, Naruse, Tanaka, Tsuji and Nishizawa 2006). The time to onset of clinical improvement was comparable to cyclosporine A (around 1 month after administration). Further studies are required to confirm its efficacy, but it appears to be well-tolerated and safe at low doses.

Biotechnological drugs that have been effective for autoimmune diseases are also being tested for MG treatment. For example, *rituximab*, a human–mouse chimeric anti-CD20 monoclonal antibody (mAb) that induces complement-dependent cytotoxicity against CD20 positive plasma cells and has been used to treat B cell malignancies and autoimmune diseases, has also been used successfully in refractory MG (Wylam, Anderson, Kuntz and Rodriguez 2003). The B-cell-directed immunosuppressant *mitoxantrone* (novantrone) (Chan, Weilbach and Toyka 2005), which also inhibits the activation of T helper cells and cytokines (Neuhaus, Kieseier and Hartung 2004), is another candidate for the treatment of MG patients (Dalakas 2006). Another drug, *etanercept*, which is inhibitory for the cytokine TNF-α that has been implicated in the pathogenesis of autoimmune diseases, has been tested successfully in some MG patients and caused exacerbations in others (Rowin, Meniggioli and Tuzun 2004).

## 2.6 Plasmapheresis

Plasmapheresis, or plasma exchange, is effective in short-term treatments and is recommended for MG exacerbations (Dalakas 2004). It is the modality where plasma from patient's blood is discarded, thus removing all circulating factors, and normal plasma or plasma constituents are fed; the technique is easily applicable and well established, especially in severe cases to induce remission and in preparation for surgery (Batocchi, Evoli, Schino and Tonali 2000). It has a proven efficacy even in patients with seronegative MG (Skeie et al. 2006), with the drawback that it unselectively removes all plasma = proteins and requires replacement with human proteins, a fact that may result in anaphylactic reactions and transfusion-transmitted infections (Dodd 1992), increasing the cost of the procedure. Plasmapheresis has a major impact on the management of MG patients because it has a fast effectiveness, is useful in myasthenic crisis, in most severe cases before thymectomy and the early postoperative period, and can decrease the occurrence of symptom deterioration during tampering or initiation of immunosuppressive drug therapy (Pinching and Peters 1976; Newsom-Davis, Pinching, Vincent and Wison 1978a; Newsom-Davis, Vincent, Wilson, Ward, Pinching and Hawkey 1978b; Kuks and Skallebaek 1998). It has been shown that daily schedule can offer greater effectiveness compared to alternate daily schedule (Yeh and Chiu 2000) and also that when combined with steroids, it constitutes a significantly superior therapeutic strategy in severe forms of MG (Morosetti, Meloni, Iani, Caramia, Galderisi, Palombo, Gallucci, Bernardi and Casciani 1998). Its adverse effects, present in approximately 20% of the patients, seem to be related to the fluid shift during the process and the need for replacement fluids, expanding the risk for blood-borne pathogen transmission (Kuks and Skallebaek 1998). Another problem is vascular access, especially when the patient requires indwelling catheters, with complications like blood clotting, thrombosis and severe infections (Kuks and Shallebaek 1998).

An alternative to plasmapheresis is *double filtration plasmapheresis (DFP)*, which is based on the differential filtration of plasma through use of two filters: the first filter separates the blood into blood cells and plasma and the second filter is the ultrafiltration filter that fractionates the plasma components according to their molecular weight, in the range of 20 or 30 to 200 kDa (Yang, Kenpe, Yamaji, Tsuda and Hashimoto 2002). The technique removes high-molecular-weight molecules semi-selectively, blocking γ-globulin fractions while letting albumin fractions pass through. The procedure removes almost 90% of the IgG from the plasma while minimising albumin loss, but requiring fluid replacement to compensate for the loss (Yeh, Chen and Chiu 2004). DFP is more selective in IgG removal and safer than PE and has become widespread in Japan (Takamori and Ide 1996). A study on 2502 treatments confirmed the safety of the treatment, also bringing forward the frequency of major complications, with hemolysis occurring in almost 20% of the patients, but with no allergic reactions or mortality associated with the process (Ych ct al. 2004). Interestingly, Lyu et al. (Lyu, Chen and Hsieh 2002) performed a retrospective review of two equivalent groups of Guillain-Barré patients in Taiwan undergoing either PE or DFP; it was shown that the PE group demonstrated better changes in disability scores and a shorter time to the onset of these changes than the DFP subjected group, allowing the authors to postulate that substances which are removed by the non-selective PE, might remain after DFP resulting in continued injury. To our knowledge, such a comparative study has not been done for MG patients and remains to be performed.

## 2.7 Immunoadsorption

A promising alternative to plasmapheresis is *immunoadsorption*, where a column bearing an adsorbent or ligand-linked carrier is used to selectively or semi-selectively withhold the anti-AChR Abs from the plasma, allowing the return of "anti-AChR Ab-cleared" plasma to the patient, thus overcoming the major disadvantage of plasma exchange, namely the indiscriminate removal of protective Abs together with the pathogenic ones. Immunoadsorption in principle is superior to PE or DFP as it does not require any replacement fluid and is currently applied only for certain diseases due to limitations of the commercially available columns. The following adsorbents are currently in use for MG therapy:

- *TR-350*. This adsorbent was developed by Yamazaki et al. (Yamazaki, Fujimori, Takahama, Inoue, Wada, Kazama, Morioka, Abe, Yamawaki and Inagaki 1982) and consists of tryptophan-linked polyvinyl alcohol gel that selectively adsorbs most large proteins, including most anti-AChR Abs through hydrophobic interaction; the technique results in amelioration of the MG symptoms with an anti-AChR Ab clearance rate of 70–90% and without concomitant decrease in albumin and thus no need for its supplementation (Shibuya, Sato, Osame, Takegami, Doi and Kawanami 1994; Grob, Simpson, Mitsumoto, Hoch, Mokhtarian, Bender, Greenberg, Koo and Nakayama 1995; Marconi, Bobbi, Pizzi, Sbrilli, Taiuti, Ronchi, Avanzi, Lombardo, Franco and Biani 1984; Yeh and Chiu 2000). The TR-350 column has also been used for the treatment of Guillain-Barré syndrome. Similarly, Yang et al. (Yang, Cheng, Yan and Yu 2004) developed a cellulose–tryptophan adsorbent and described its use for the removal of extracorporeal removal of anti-AChR antibodies from EAMG rabbit blood, with improved clinical manifestations and neuromuscular function in the animals. *PH-350*, an adsorbent produced by linking phenynalanine to the same carrier as TR-350 (polyvinyl alcohol), has been reported not to be suitable for the treatment of MG patients since the removal efficiency of anti-AChR Abs was around 35% (Takamori and Ide 1996).
- *Protein A*. The staphylococcal protein A (SPA), a major cell wall component of *Staphylococcus aureus*, is characterised by avid binding to human immunoglobulins, negligible interaction with other plasma proteins and is readily available, thus permitting its usage for extracorporeal immunoadsorption. SPA-based immunoadsorption relies on two systems, namely SPA-silica (Prosorba) and SPA-Sepharose (Immunosorba) (Matic, Bosch and Ramlow 2001). Protein A binds to the Fc portion of the IgG molecules, and its columns have been used for the treatment of MG to remove IgGs with an efficacy ranging from 50% (Somnier and Langvad 1989) up to 80% (Berta, Confalonieri, Simoncini, Bernardi, Busnach, Mantegazza, Cornelio and Antozzi 1994), without major adverse effects, being well tolerated, safe and with the great advantage of being re-usable for the same patient for a number

of treatments over a period of more than 18 months (Benny, Sutton, Oger, Bril, McAteer and Rock 1999; Matic et al. 2001), while they have been reported to be effective even in patients who do not respond to standard PE (Cornelio, Antozzi, Confalonieri, Baggi and Mantegazza 1998) or TR-350 treatment (Flachenecker, Taleghani, Gold, Grossmann, Wiebecke and Toyka 1998). The high cost of the treatment excludes its usage by a wider palette of patients (Benny et al. 1999; Cornelio et al. 1998).

– *MG-50.* This column, introduced by Takamori and co-workers (Takamori and Maruta 2001), consists of the peptide corresponding to aminoacids 183–200 of the α subunit of the *Torpedo californica* AChR immobilised on porous cellulose particles. When used on a pilot study with two MG patients whose sera were rich in anti-peptide Abs, the application resulted in an improvement in the clinical symptoms coupled to a decline not only in the anti-peptide but also in the anti-native AChR blocking antibodies, without marked reduction in the total IgG levels and thus not necessitating protein replacement. The same procedure applied to a further 22 MG patients showed an improvement on 57% and a fall in antibody titer in 69% of the patients (Takamori and Maruta 2001). A modified MG-50 column carrying the peptide corresponding to the human acetylcholine binding site on AChR instead of the *Torpedo* α183–200 peptide was introduced with similar results (Nakaji and Hayashi 2003). It is surprising, however, that a small *Torpedo* peptide can adsorb the majority of the MG patient Abs, knowing that MG sera only marginally bind to intact *Torpedo* AChR, and that the α183–200 segment is not very immunogenic.

– *Ig-Adsopak.* This column carries sheep anti-human IgG to withhold human IgG molecules. Since IgG is the dominant immunoglobulin, this immunoaffinity column should primarily adsorb IgG, without alteration of the remaining Ig molecules. However, it was shown that the column had a broader binding specificity and removed not only IgG but also IgA, IgM, and other plasma proteins, including some plasma albumin (Ptak 2004).

## *2.8 Intravenous Immunoglobulin (IVIg)*

Intravenous immunoglobulin is another short-term treatment approximately equally effective to plasmapheresis (Dalakas 2004). The supply of IgG from healthy donors to MG patients suppresses immune responses possibly by down-regulating the expression of pro-inflammatory cytokines, such as TNF-α, and upregulating the expression of anti-inflammatory cytokines (Zhu, Feferman, Maiti, Souroujon and Fuchs 2006), although their precise mechanism of action is not yet clearly understood. IVIg has less side effects than plasmapheresis, mainly fever and headache, which can be alleviated with appropriate medication (Gajdos, Chevret, Clair, Tranchant and Chastang 1998). It is recommended as an adjuvant to immunosuppressive therapies and when vascular access for plasma exchange is problematic (Dalakas 2004). IVIg has a slower onset of action compared to plasmapheresis and the effects of the treatment can last several weeks (Spring and Spies 2001). Treatment is, however, very expensive and difficulties with IgG supply have been reported (Schwendimann et al. 2005).

# 3 Emerging theRapies

The current treatments of MG are relatively effective, but have adverse side effects, due to their limited immune specificity. Several of the experimental therapeutic strategies currently under investigation aim at eliminating the specific pathogenic autoimmune response, whilst bypassing generalised immunosuppression. Below we present information on several of the emerging experimental therapies for MG, presenting the non-specific ones first, which include (1) immune system ablation, (2) gene therapy and (3) phosphodiesterase inhibitors, and leading on to the antigen-specific approaches, which include (1) suppression of MG by mucosally administered recombinant AChR fragments, (2) manipulated APCs, (3) altered peptide ligands (APLs), (4) protective anti-AChR antibody fragments, (5) anti-TCR Ab, (6) RNA aptamers and (7) anti-AChR Ab-specific plasma clearance (Fig. 1).

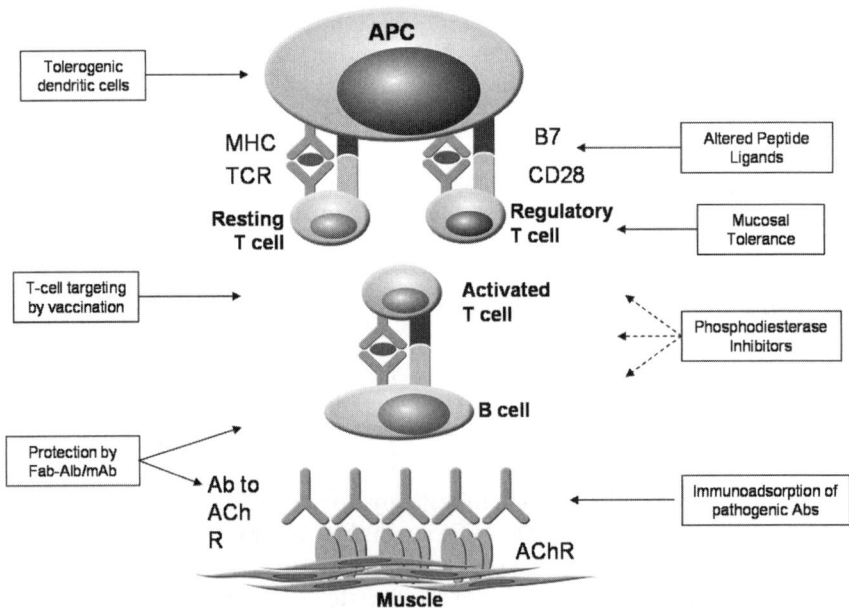

**Fig. 1** Targeting at all levels for a therapy of MG. Shown are the different ways for evading the autoimmune response, which are presented in the text. Adapted from a similar figure made by Dr S. Berrih-Aknin

## 3.1 Immune System Ablation

Immune system ablation followed by bone marrow transplantation has been effective in rats with EAMG and left the animals without residual responsiveness to AChR (Drachman et al. 2003a). However, the risks associated with bone marrow transplantation and the recurrence of autoimmune disorders in some cases have limited its application as a method of treatment.

## 3.2 Gene Therapy

Gene therapy has been used in autoimmune diseases as an alternative to systemic therapeutic treatments (Slavin, Tarner, Nakajima, Urbanek-Ruiz, McBride, Contag and Fathman 2002). A gene therapy approach for the treatment of MG targets at rapsyn, which is an AChR-anchoring protein required for maintaining high density of AChR at the postsynaptic membrane, is currently under investigation (De Baets, Stassen, Losen, Zhang and Machiels 2003). It was recently demonstrated that resistance to EAMG in aged rats correlates with increased rapsyn concentration at the NMJ (Losen, Stassen, Martínez-Martínez, Machiels, Duimel, Frederik, Veldman, Wokke, Spaans, Vincent and De Baets 2005). Differences in endogenous rapsyn expression could therefore determine disease severity and upregulation of endogenous rapsyn expression can be exploited therapeutically for the treatment of MG. Rapsyn gene therapy under a cytomegalovirus promoter by electropermeation into muscle fibres of young rats resulted in increased rapsyn expression and a parallel 40% increase in the concentration of AChR in transfected fibres (Mir, Bureau, Gehl, Rangara, Rouy, Caillaud, Delaere, Branellec, Schwartz and Scherman 1999; De Baets et al. 2003). Following passive EAMG transfer via injection of mAbs to the MIR (mAb35), no AChR loss was observed in transfected muscles, compared to 42% loss in nontransfected muscles (De Baets et al. 2003). Further experimental trials are under way, but if proven successful, rapsyn gene therapy could be used for MG patients who are on long-term immunosuppressive treatment, such as azathioprine and prednisone. Increasing rapsyn expression by modulating synaptic transcription factors could also be investigated (Rodova, Kelly, VanSaun, Daniel and Werle 2004).

Alternatively, adoptive cellular gene therapy approaches by transfection of host cells, such as T cells, dendritic cells and fibroblasts, in vitro to serve as gene delivery vehicles are also exploited (Tarner and Fathman 2001; Tarner, Slavin, McBride, Levicnik, Smith, Nolan, Contag and Fathman 2003). T cells are important mediators in the pathogenesis of MG, and the restricted heterogeneity of T-cell epitopes is promising for site-specific immunotherapies in the future (Slavin et al. 2002; Vincent 2002).

## 3.3 Phosphodiesterase Inhibitors

Phosphodiesterases (PDEs) catalyse the hydrolytic degradation of cAMP and cGMP, thus determining their intracellular levels. The major physiological importance of cAMP and cGMP as second messenger molecules mediating a broad range of cellular functions, including cellular immune responses, makes PDEs critical modulators of intracellular signalling and also potential therapeutic targets in a variety of conditions. PDEs are currently grouped into 11 families (PDE1-PDE11) differing in substrate specificity, inhibitor sensitivity, tissue distribution and regulation of activity (Lugnier 2006). PDE implication in EAMG pathogenesis has been reported only very recently (Aricha, Feferman, Souroujon and Fuchs 2006). Using DNA microarray analysis, the gene expression profiles of lymph node cells (LNC) and muscle from *Torpedo* AChR-immunised rats and control rats were compared. Overall, PDE1, PDE2, PDE3 (predominantly expressed by immune cells), PDE4 and PDE7 (highly expressed in muscle) were found to be upregulated (Aricha et al. 2006). When the general PDE inhibitor pentoxifylline (PTX) was tested for its effect on the course of EAMG, it was shown that PTX administration during the acute or chronic phase of the disease repressed EAMG progression (Aricha et al. 2006). This was evidenced by significantly lower clinical scores and weight loss, and remarkably suppressed humoral and cellular anti-AChR responses of PTX-treated rats versus controls. LNC expression of the Th1 cytokines TNF-$\alpha$, IL-12, IL-18 (but not IFN-$\gamma$), and the Th2 cytokine IL-10 were significantly reduced in PTX-treated rats, as well as muscle expression of TNF-$\alpha$, as had been shown before in independent studies (see below). Interestingly, PTX not only inhibited PDE action as expected, but also downregulated PDE expression. It also induced an upregulation of the Foxp3 transcription factor (possibly via modulation of intracellular cAMP levels in immune cells), which is essential for the function of the CD4$^+$CD25$^+$ regulatory T cells, suggesting an enhanced activity of this regulatory T cell subpopulation.

The mechanism of PTX action on EAMG treatment remains to be further clarified. Preliminary evidence points to a rather pleiotropic mechanism, at the level of both PDE expression and immunomodulation. Indeed, several studies have suggested an immunomodulatory action for PTX. This molecule can potentially interfere with Th1 differentiation pathways (Benbernou, Esnault, Potron, and Guenounou 1995), suppress the expression of pro-inflammatory cytokines (TNF-$\alpha$, IL-12 and IL-18) (Moller, Wysocka, Greenlee, Ma, Wahl, Trinchieri and Karp 1997; Samardzic, Jankovic, Stosic-Grujicic, Popadic and Trajkovic 2001; Whitehouse 2004) and intercept B and T cell proliferation in vitro (Rosenthal and Blank 1993; Kotadia, Ravindranath, Choudhry, Haque, Al-Ghoul and Sayeed 2003). Alternatively, PTX action could lie on the muscle itself, attenuating muscle fatigue possibly by stimulating the sarcoplasmic reticulum Ca$^{2+}$ pump, as suggested previously (Gonzalez-Serratos, Chang, Pereira, Castro, Aracava, Melo, Lima, Fraga, Barreiro, and Albuquerque 2001).

PTX is already commonly used in patients with peripheral vascular disorders and due to its immunomodulatory effect, it has also been tested in several autoimmune diseases including rheumatoid arthritis, multiple sclerosis, juvenile diabetes mellitus and systemic lupus erythematosus (MacDonald, Shahidi, Allen, Lustig, Mitchell and Cornwell 1994; Anaya and Espinoza 1995; Rieckmann, Weber, Gunther, Martin, Bitsch, Broocks, Kitze, Weber, Borner and Poser 1996; Galindo-Rodriguez, Bustamante, Esquivel-Nava, Salazar-Exaire, Vela-Ojeda, Vadillo-Buenfil and Avina-Zubieta 2003). The newly provided data, together with the fact that PDE inhibitors are already approved medication for other indications, make them very promising candidates for MG immunotherapy.

## 3.4 Suppression of MG by Mucosally Administered Recombinant AChR Fragments

Multiple mechanisms of tolerance are induced by mucosally administered antigen (Faria and Weiner 2005). Autoantigen admininstration can induce systemic tolerance more efficiently when administered nasally than when administered orally. Different mechanisms of tolerance are induced depending on the dosage of the administered protein with higher doses leading to anergy or deletion and lower doses inducing active cellular regulation by stimulating T cells that secrete TGF-β, IL-10 or IL-4 (Weiner 1999). Mucosal (oral or nasal) administration of *Torpedo* AChR in Lewis rats or B6 mice, prior to immunisation with AchR, resulted in a markedly decreased severity of clinical EAMG in very low doses (Wang, Qiao and Link 1993; Okumura, McIntosh and Drachman 1994; Ma, Zhang, Xiao, Link, Olsson and Link 1995; Shi, Bai, Xiao, van der Meide and Link 1998; Shi, Li, Wang, Bai, van der Meide, Link and Ljunggren 1999). Administration of AChR for the treatment of ongoing EAMG required 10 times higher dosage to cause amelioration of muscular weakness of EAMG, suggesting that, upon establishment of an abnormal immune response to AChR, very high dosages are required to arrest the progression of the pathologic process (Shi et al. 1998). The treatment protocol induced selective suppression of Th1 function, but not of Th2 or Th3 responses. Fuchs and co-workers reported the suppression of ongoing EAMG in rats through oral administration of a xenogeneic recombinant fragment of the human AChR corresponding to the 1–205 region of the α subunit extracellular domain (ECD) (Im, Barchan, Fuchs and Souroujon 1999). They later demonstrated that the spatial conformation of the administered recombinant fragments of the AChR determines their tolerogenicity, with one of the near-native structure failing to suppress ongoing EAMG (Im, Barchan, Souroujon and Fuchs 2000). To bypass possible complications from the use of the xenogeneic fragment, the same group tested the potential of either a syngeneic fragment, corresponding to the rat AChR α–ECD, or 50–60 amino acids long, partially overlapping peptides covering the entire α–ECD. Interestingly, the former preparation was observed as effective as the previously presented human fragment, while the peptide mixture was unable to replace the recombinant fragments, further advocating in favour of an indispensable conformation requirement, which the peptides cannot meet (Souroujon, Maiti, Feferman, Im, Raveh and Fuchs 2003). Link and co-workers have achieved suppression of EAMG and experimental allergic encephalomyelitis, through administration of a mixture of AChR and myelin basic protein (Wang, He, Qiao and Link 1995). Their results indicated that simultaneous tolerisation to more than one autoimmune diseases and their related autoantigens is feasible, projecting that the autoantigens do not interfere with one another when they are from different target organs. Overall, induction of mucosal tolerance has been tested with encouraging results, while its effectiveness greatly depends on the tolerogen (conformation, size, origin and concentration) and the administration route (mucosal).

## 3.5 Manipulated Antigen-Presenting Cells (APCs)

As already mentioned, MG is a T-cell-dependent autoimmune disease (Lindstrom, Shelton and Fujii 1988). Autoreactive $CD4^+$ T cells from MG patients respond to various epitopes from different AChR subunits in the context of major histocompatibility complex (MHC) class II molecules (Link, Olsson, Sun, Wang, Andersson, Ekre, Brenner, Abramsky and Olsson 1991; Protti, Manfredi, Horton, Bellone and Conti-Tronconi 1993; Wang, Okita, Howard and Conti-Fine 1998) and provide help to B cells becoming essential for anti-AChR antibody production (Rosenberg, Oshima and Atassi 1996). Hence, a strategy that could selectively eliminate the AChR-responsive T cells would probably block anti-AChR autoantibody response. However, a specific T-cell-directed therapy would have to confront the diversity of the AChR-reactive T cell repertoire, which differs from patient to patient and shows time (from onset)-dependent alterations within a single MG patient (Matsumoto, Matsuo, Sakuma, Park, Tsukada, Kohyama, Kondo, Kotorii and Shibuya 2006).

Drachman and collaborators have proposed the use of genetically engineered APCs to effectively and specifically target-and-destroy AChR-specific T cells (Drachman, Wu, Miagkov, Williams, Adams and Wu 2003b). Vaccinia virus-derived vectors were used to convert APCs (mostly T-cell-depleted splenocytes) into "guided missiles" that: (i) endogenously express the extracellular domain of *Torpedo* AChR α-subunit, process it and present the α-subunit-derived peptides in the context of MHC II, serving as the guidance system for detecting and activating AChR-responsive T cells and (ii) express Fas ligand (FasL) to selectively interact with activated AChR-specific T cells bearing Fas and kill them via apoptosis. The genetically manipulated APCs were demonstrated in vitro to express the appropriate gene products and to selectively eliminate the AChR-specific T cells via the Fas/FasL pathway, while, importantly, leaving unaffected T cells to an unrelated antigen (Wu, Wu, Miagkov, Adams and Drachman 2001).

So far, this model was tested in vivo in a preliminary experiment where APCs converted to express an influenza hemagglutinin epitope (HA) were injected in a transgenic mouse strain with a CD4$^+$ T cell population enriched in HA-reactive T cells (40–60%) (Drachman et al. 2003b). APC administration induced a significant decrease in HA-reactive T cells in the APC-treated mice, and LNCs and splenocytes derived from the APC-treated mice showed reduced proliferation in vitro after challenge with HA (Drachman et al. 2003b). These preliminary findings suggest that after appropriate optimisation, the method could prove useful in vivo for specific T cell immunosuppression.

## 3.6 Altered Peptide Ligands (APLs)

Synthetic peptides capable of blocking T cell activation can also be used as therapeutic agents for treating autoimmune diseases. Altered peptide ligands (APLs) are peptide variants derived from the original antigenic peptide with substitutions at particular residue(s) (Yachi, Ampudia, Zal and Gascoigne 2006). APLs can be grouped into different categories depending on the extent of T cell activation, which ranges from non-stimulation to full activation. Studies on T cell responses to APLs have shown that TCRs can distinguish between MHC-bound ligands that differ even by a single amino acid residue, resulting in different effector functions (Sloan-Lancaster and Allen 1996; Alberola-Ila, Takaki, Kerner and Perlmutter 1997; Faber-Elmann, Paas-Rozner, Sela and Mozes 1998; Gascoigne, Zal and Alam 2001). Such small changes can convert fully activating ligands into partially activating (partial agonists) or inhibitory ones (antagonists) (Faber-Elmann et al. 1998). MHC-binding of APLs of myasthenogenic peptides can be preserved or even improved, whereas interactions with the TCR are distorted, so that the APLs will compete with the original myasthenogenic antigen (Zisman, Katz-Levy, Dayan, Kirshner, Paas-Rozner, Karni, Abramsky, Brautbar, Fridkin, Sela and Mozes 1996).

The α subunit of the AChR is predominant for T cell epitopes; the peptides p195–212 and p259–271 in particular could stimulate the proliferation of peripheral blood lymphocytes of MG patients and were found to be immunodominant T cell epitopes in SLJ and BALB/c mice, respectively (Brocke, Brautbar, Steinman, Abramsky, Rothbard, Neumann, Fuchs and Mozes 1988; Brocke, Dayan, Rothbard, Fuchs and Mozes 1990). A single mutation of Met-207 to Ala in the p195–212 peptide and Glu-262 to Lys in the p259–271 peptide produced altered peptides that significantly inhibited the proliferative responses of p195–212 and p259–271-primed T cells, respectively (Katz-Levy, Kirshner, Sela and Mozes 1993). Furthermore, a peptide composed of the two analogues in tandem repeat (Lys-262-Ala-207) termed dual APL, inhibited the proliferation of T cells responsive to both p195–212 and p259–271 peptides, while oral administration in C57BL/6 and BALB/c mice resulted in down-regulation of EAMG clinical manifestations caused by immunisation with *Torpedo* AChR or by pathogenic T cell lines, respectively (Katz-Levy, Paas-Rozner, Kirshner, Dayan, Zisman, Fridkin, Wirguin, Sela and Mozes 1997; Katz-Levy, Dayan, Wirguin, Fridkin, Sela and Mozes 1998; Paas-Rozner, Dayan, Paas, Changeux, Wirguin, Sela, and Mozes 2000). The amelioration of the clinical symptoms correlated with a reduction in the AChR-specific antibody titers and immunomodulation of the secreted cytokines, such that the secretion of pathogenic Th-1 type cytokines (IFN-γ and IL-2) was down-regulated, whereas that of immunosuppressive Th-2/Th-3 type cytokines (TGF-β, IL-4 and IL-10) was upregulated (Faber-Elmann, Grabovsky, Dayan, Sela, Alon and Mozes 2001; Paas-Rozner et al. 2000; Paas-Rozner, Sela and Mozes 2001; Aruna, Sela, and Mozes 2006b). The secretion of immunosuppressive cytokines may not only affect the specifically primed lymphocytes, but other lymphocytes as well in the microenvironment (bystander suppression), there-fore leading to effective immunomodulation (Sela 1999). T cell adhesion and the secretion of matrix metalloproteinase MMP-9, both of which are involved in the pathogenesis of autoimmune diseases, were also inhibited (Faber-Elmann et al. 2001). In addition, the dual APL was effective in ameliorating MG responses in 21 out of 22 patients, judging by the inhibition of the proliferation of patients' peripheral blood lymphocytes in response to *Torpedo* AChR (Dayan, Sthoeger, Neiman, Abarbanel, Sela and Mozes 2004; Sela and Mozes, 2004). Administration of dual APL to SJL mice caused an upregulation of the CD4$^+$CD25$^+$ T cells, which have been shown to suppress T cell prolifera-tion in many autoimmune and inflammatory conditions (Wraith, Nicolson and Whitley 2004). Depletion of this subpopulation hindered the effect of the dual APL on T cell proliferation, indicating that these cells play a central role in the suppressive action of the dual APL (Paas-Rozner, Sela and Mozes 2003). The CD4$^+$CD25$^+$ cells specifically inhibited the proliferation of the myasthenogenic peptide-specific T cells (Aruna, Sela and Mozes 2005). The molecular mechan-ism responsible appears to involve factors such as Foxp3 and cytotoxic T lymphocyte-associated antigen 4 (CTLA-4), the expression of which is upre-gulated in the CD4$^+$CD25$^+$ cells, and Fas-FasL induced apoptosis (Dayan et al. 2004; Ben-David, Sela and Mozes 2005; Aruna, Ben-David, Sela and Mozes 2006a). The dual APL is, therefore, specific for inhibiting MG-related responses without causing adverse immune responses (Sela 1999), and is a potential candidate for the treatment of MG patients.

## 3.7 Protective anti-AChR Antibody Fragments

The use of anti-MIR antibody fragments (Fab or single-chain Fv, scFv) to protect AChR from pathogenic autoantibody action has been previously elaborated (Tzartos, Sophianos and Efthimiadis 1985; Mamalaki, Trakas and Tzartos 1993; Tsantili, Tzartos and Mamalaki 1999; Papanastasiou, Poulas, Kokla and Tzartos 2000). The non-pathogenic anti-AChR antibody fragments must be univalent, thereby not causing AChR degradation via antigenic modulation, and lacking the complement-binding site. The way of action of the protective antibody fragments consists in competing with the pathogenic anti-AChR autoantibodies for AChR binding, eventually preventing anchoring of the latter on AChR surface. However, the potential immunogenicity of rodent Abs, when applied to man in the clinic, necessitates the construction of humanised or human Abs. Humanised monoclonal Abs are constructed by transferring the complementarity-determining regions of a protective animal Ab into a human Ab scaffold and parallel substitutions in residues of the frame-work regions. Whole human mAbs, on the other hand, are usually produced either by the immunisation of humanised mice (mice with human Ig genes), followed by hybridoma technology, or by the phage-Ab-display technology.

In this way, a humanised anti-AChR scFv has been constructed from a rodent anti-MIR mAb (Papanastasiou, Mamalaki, Eliopoulos, Poulas, Liolitsas and Tzartos 1999), and fully human anti-AChR Fabs have been isolated from MG patient-derived phage-display antibody libraries (Farrar, Portolano, Willcox, Vincent, Jacobson, Newsom-Davis, Rapoport and McLachlan 1997; Graus, de Baets, van Breda Vriesman, and Burton 1997; Fostieri, Tzartos, Berrih-Aknin, Beeson and Mamalaki 2005) or after immunisation of transgenic mice expressing human immunoglobulin loci with recombinant AChR α-subunit (Protopapadakis, Kokla, Tzartos and Mamalaki 2005). Among them, two sets of human anti-AChR Fabs, one against the MIR and the other against an immunodominant region close to the MIR, effectively compete with a great majority of MG patients' sera for binding to AChR and also significantly protect AChR against loss due to antigenic modulation in cell culture (Graus et al. 1997; Fostieri et al. 2005). Interestingly, the two sets of anti-AChR Fabs seem to compete with distinct anti-AChR autoantibody subpopulations (Fostieri, Kostelidou, Poulas and Tzartos 2006). This suggests that the combination of several protective anti-AChR antibody fragments to create a panel of AChR "protectors" shielding the most prevailing immunogenic AChR epitopes is probably a prerequisite for the development of an efficient therapeutic pool.

An AChR-specific Fab immunotherapy could prove useful mostly for short-term therapy, e.g. in life-threatening crises. However, there is a major restriction in its application. The short in vivo half-life of Fabs makes them unsuitable for the treatment of patients. Attempts to prolong their half-life in rats by conjugation to polyethylene glycol have yielded some promising results (Trakas and Tzartos 2001). Alternative approaches aiming at the construction of non-complement-fixating, intact, non-pathogenic anti-AChR antibodies are currently in progress (Stassen, Machiels, Fostieri, Tzartos, Berrih-Aknin, Bosmans, Parren and De Baets 2003).

## 3.8 Anti-TCR Antibodies

Targeting at MG-specific subsets of T cells is another possible therapeutic approach. In the thymus of MG patients with the HLA-DR3 haplotype, there is a selection bias for Vβ5.1 T cell receptor (TCR) expressing CD4+ cells (Truffault, Cohen-Kaminsky, Khalil, Levasseur and Berrih-Aknin 1997). When lymphocytes from the thymus of such patients were engrafted in SCID mice, treatment of the animals with anti-Vβ5.1 antibodies prevented the manifestation of symptoms, such as reduction in muscle AChR content. The signs of pathogenicity were also averted by in vitro depletion of the engrafted thymic cells of the Vβ5.1+ T cell subpopulation prior to transfer (Aissaoui, Klingel-Schmitt, Couderc, Chateau, Romagne, Jambou, Vincent, Levasseur, Eymard, Maillot, Galanaud, Berrih-Aknin and Cohen-Kaminsky.1999). Taken together, these results point strongly towards a role for Vβ5.1 expressing cells in the production of pathogenic autoantibodies in HLA-DR3+ patients. It would thus be possible to target the Vβ5.1+ T cells in an effort to minimise the amount of anti-AChR antibodies produced. The rational behind T cell vaccination would be to use the TCR of the specific disease-associated T cell subpopulation, in order to be neutralised by the patients' own immune system. This approach has shown promising results in animal experimental autoimmune diseases such as adjuvant and collagen-induced arthritis, murine lupus and type I diabetes (reviewed in Cohen 2001). Interestingly, most HLA-DR3+ MG patients showed spontaneously elevated levels of anti-Vβ5.1 antibodies (Jambou, Zhang, Menestrier, Klingel-Schmitt, Michel, Caillat-Zucman, Aissaoui, Landemarre, Berrih-Aknin and Cohen-Kaminsky 2003). Furthermore, there was a strong correlation between anti-Vβ5.1 antibody titre and disease severity, with higher antibody levels associated with milder disease, suggestive of a protective role for these antibodies. In vitro, anti-Vβ5.1 Abs, derived from EOMG patients' sera, inhibited proliferation of Vβ5.1 expressing T cells, supporting the notion of a regulatory function. Therefore, the Vβ5.1 TCR presents an attractive candidate as a target for vaccination in HLA-DR3+ MG patients to instigate or boost an already existing anti-TCR humoral response.

A somewhat different approach to target autoreactive T cells has been the use of peptides complementary to the MIR on the AChR as vaccines for the production of anti-idiotypic antibodies (anti-Id) against TCRs that specifically recognise these epitopes (Araga, LeBoeuf and Blalock 1993; Araga, Xu, Nakashima, Villain and Blalock 2000). It was found that such anti-Id can inhibit proliferation of sensitised lymphocytes in response to AChR, have a protective effect against EAMG development in Lewis rats and minimise AChR loss in the muscle of challenged animals. This method has the advantage of not relying on the knowledge of the TCR sequence, but rather on the sequence of the target auto-antigen, in this case epitopes on the AChR molecule. A monoclonal antibody was subsequently isolated by this technique, termed CTCR8, directed against the Vβ15 containing TCR, responsible for recognising the T cell dominant epitope of the Lewis rat AChR α subunit (residues 100–116) (Xu, Villain, Galin, Araga and Blalock 2001). CTCR8 treatment resulted in a decrease in IFN-γ production by an AChR α 100–116 specific T cell line. Taking it a step further, CTCR8 was administered in Lewis rats with ongoing EAMG and found to improve the disease outcome, as it reduced disease severity, as well as mortality rates from 71% in control animals to 14% in CTCR8 treated, and even in some cases caused complete reversal of the symptoms.

## 3.9 RNA Aptamers

Aptamers are short oligonucleotides of 100 bases or less with the ability to bind to specific target molecules; their small size allows rapid, reproducible, large-scale synthesis, which when necessary can be combined to introduction of modifications to improve their pharmacokinetics and delay their degradation (reviewed in Pestourie, Tavitian and Duconge 2005). Aptamers tht recognize anti-AChR Abs can therefore be used as decoys in a way similar to the use of peptides or anti-idiotypic antibodies, with many therapeutical benefits, since they also show very high specificity and affinity for their target sites, but lack immunogenicity.

Lee and Sullenger (Lee and Sullenger 1997) identified an RNA aptamer that specifically recognised both the anti-MIR mAb198 and the autoantibodies from MG patient sera. Further in vitro selection resulted in the development of larger aptamers that displayed increased affinity for the autoantibodies and could protect human cells from antigenic modulation by mAb198 as well as patient autoantibodies albeit with lower efficiency (Hwang and Lee 2002). This protective effect was also observed in EAMG induced by mAb198 administration, as repeated injection of aptamers greatly improved the clinical state and muscle AChR content of diseased animals (Hwang, Han and Lee 2003). However, in the aforementioned studies, although aptamers were relatively potent at inhibiting the mAb198-induced antigenic modulation of AChR on human cells, they showed a limited ability to protect from the effects caused by patient sera (Hwang and Lee 2002; Hwang et al. 2003). This was probably due to the original selection of the RNAs (ability to specifically recognise the mAb198), while they cross-reacted poorly with autoantibodies from patients' sera, which are often of polyclonal origin and may recognise multiple epitopes on AChRs. Efficient use of RNA aptamers for the treatment of MG, therefore, requires that an extensive mixture of aptamers with an assortment of binding specificities for anti-AChR antibodies is used.

## 3.10 Anti-AChR Ab-Specific Plasma Clearance

Since MG is caused by circulating anti-AChR antibodies, the extracorporeal removal of plasma (and antibody within) should reduce the autoimmune attack at the NMJ. The therapeutic apheresis techniques currently employed for the treatment of MG aiming at the removal of the pathogenic antibodies present serious drawbacks as described earlier. Immunoadsorption on matrix-immobilized AChR columns has long been considered as a putative therapy. However, its application has been hindered by the limited available native human AChR. An attractive alternative is the use of "immunoadsorbent" carrying recombinant, native-like AChR or its parts to allow for specific adsorption of anti-AChR Abs in patient sera (Fig. 2).

An immunoadsorbent where the functional fragment of AChR α subunit 1–205 fused to the maltose-binding protein (MBP) was covalently bound to amylose resin (through the MBP), was demonstrated to have binding capacity for AChR Abs in anti-AChR mice and up to 96% specific removal efficiency of anti-AChR antibodies from some human MG patient sera, with a low decrease in total IgA, IgM, IgG levels and practically no decrease in serum albumin levels (Guo, Li, Xu and Yuan 2005). Nevertheless, our group has observed that *E.coli*-expressed AChR domains are selectively efficient for immunoadsorption of Abs from some, but not many MG patients, compared to *P. pastoris*-derived subunits, making the combined use of procaryotic and

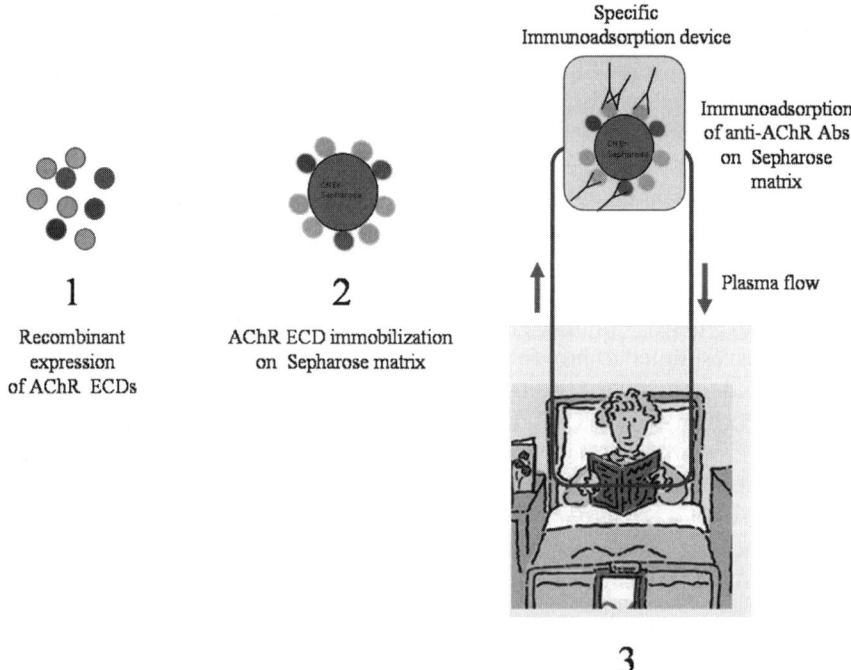

**Fig. 2** Schematic representation of the specific immunoadsorption technique. The recombinant expression of the AChR ECDs (**1**) is followed by their immobilisation on Sepharose beads (**2**) to construct an insoluble matrix, which will be used in an immunoadsorption device to withhold anti-AChR Abs from MG patients (**3**) (*See* Color Insert)

eukaryotic systems more promising for the development of efficient anti-AChR immunoadsorbents.

A specific Ab-apheresis therapeutic approach for MG is under development in our lab, through the construction of immunoadsorbent columns carrying AChR fragments with native-like conformational features. Recombinantly expressed ECDs (amino acids 1∼210) of the human AChR α, β, γ and ε subunits in the yeast *Pichia pastoris* expression system were used for the construction of immunoadsorbent columns by their immobilization on Sepharose beads; the columns were subsequently used for the selective ex vivo elimination of patients' anti-AChR Abs (Psaridi-Linardaki, Mamalaki, Remoundos and Tzartos 2002; Psaridi-Linardaki, Trakas, Mamalaki and Tzartos 2005; Kostelidou, Trakas, Zouridakis, Bitzopoulou, Sotiriadis, Gavra and Tzartos 2006) (Fig. 3). Each of the immunoadsorbents could remove different levels of autoantibodies from a set of 64 random MG positive sera, on average as follows: α-ECD 35%, β-ECD 22%, γ-ECD 21% and finally, ε-ECD 18%, while the combination of all four subunits (α, β, γ, ε) showed that they could act either in an additive fashion or could at least remove a considerably higher percentage of autoantibodies than that of any individual immunoadsorbent (Psaridi-Linardaki et al. 2005; Kostelidou et al., unpublished data). These results suggest that the combined use of all subunits for the preparation of an immunoadsorbent column could indeed result in the specific removal of the majority of autoantibodies from a

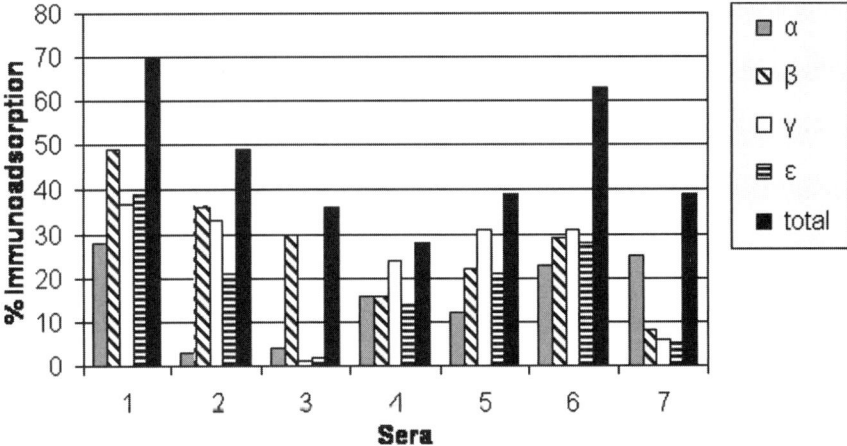

**Fig. 3** Immunoadsorption of anti-AChR autoantibodies from 7 MG sera, using CNBr-Sepharose immobilized α-, β-, γ- and ε-ECDs (*first four columns*), or a combination of all the four subunits (*last column*). Samples from MG sera (20 fmoles) were incubated with excess of Sepharose-immobilised ECDs (1 μg). The unbound anti-AChR Abs present in the supernatants were measured by RIA and the percentage of ECD-absorbed anti-AChR Abs was calculated. Each bar shows the percentage of absorbed Abs by the specified ECD. The combined use of all ECDs consistently removes more Abs than any individual ECD, although this is often lower than the sum of individual ECD columns

large portion of MG sera. Even though the elimination of the anti-AChR antibodies may result in overproduction of novel antibodies, such an activity should be milder than that in plasmapheresis where most immunoglobulins are eliminated including possible T cell inhibitory immunoglobulins.

Certain additional characteristics of the immunoadsorption experiments also deserve mentioning. Incubation of the ECDs for several hours with human plasma showed that the immobilized proteins were neither proteolyzed nor dissociated from their Sepharose matrix (Psaridi-Linardaki et al. 2005). This is very important since release of the immobilised ECD (or its fragments) into the plasma and subsequently into the patient could potentially enhance the immune response. In fact, in vivo preliminary immunoadsorption experiments in two rabbits (one previously immunised and one non-immunised) showed that the immunoadsorbed plasma returned to the rabbits neither hyperimmunised/immuunised them nor did it affect their health (Trakas and Tzartos, unpublished data). We further found that approximately 1 μg of ECD could remove at least 2 pmoles of anti-AChR antibodies, which suggests that about 5–10 mg of ECD would be sufficient for the clearance of the blood of a moderate titer patient. Finally, we found that the immunoadsorption columns could be recycled at least five times, making the approach more realistic. Overall, these results suggest that the combined use of all subunits would remove the large majority of autoantibodies from most MG sera, thus offering the option for the design and application of a novel, effective and antigen-specific therapy for MG (Psaridi-Linardaki et al., 2005). It appears that a combination of procaryotic and eukaryotic expression systems are more promising for the development of efficient anti-AChR immunoadsorbents.

## 4 Future Perspective

The triggering agent for MG remains unidentified and this renders very difficult the design of new approaches aiming at controlling/preventing/blocking the aetiological factors of the disease.

Most current therapeutic protocols were first employed on an empirical basis and have significantly reduced the morbidity and mortality rates of MG patients. Today, as knowledge accumulates on the immunopathology of MG, the role of the thymus and defects in immunoregulation, the application of prospective therapies and novel technologies are promising for the development of an antigen-specific therapy for MG in the future.

**Acknowledgments** Work in the authors' laboratories is supported by grants from the QoL program of the European Commission, the Muscular Dystrophy Association (MDA), the Association Française contre les Myopathies (AFM) and the Greek General Secretariat of Research and Technology (GSRT).

# References

Aarli, J. A. (1999) Late onset myasthenia gravis, a changing scene. *Arch Neurol* **56**, 25–27

Aissaoui, A., Klingel-Schmitt, I., Couderc, J., Chateau, D., Romagne, F., Jambou, F., Vincent, A., Levasseur, P., Eymard, B., Maillot, M. C., Galanaud, P., Berrih-Aknin, S., and Cohen-Kaminsky, S. (1999) Prevention of autoimmune attack by targeting specific T-cell receptors in a severe combined immunodeficiency mouse model of myasthenia gravis. *Ann Neurol* **46**, 559–567

Alberola-Ila, J., Takaki, S., Kerner, J. D., and Perlmutter, R. M. (1997) Differential signaling by lymphocyte antigen receptors. *Annu Rev Immunol* **15**, 125–154

Allison, A. C., and Eugui, E. M. (2000) Mycophenolate mophetil ans its mechanisms of action. *Immunopharmacology* **47**, 85–118

Anaya, J. M., and Espinoza, L. R. (1995) Phosphodiesterase inhibitor pentoxifylline: an antiinflammatory/immunomodulatory drug potentially useful in some rheumatic diseases. *J Rheumatol* **22**, 595–599

Araga, S., LeBoeuf, R. D., and Blalock, J. E. (1993) Prevention of experimental autoimmune myasthenia gravis by manipulation of the immune network with a complementary peptide for the acetylcholine receptor. *Proc Natl Acad Sci USA* **90**, 8747–8751

Araga, S., Xu, L., Nakashima, K., Villain, M., and Blalock, J. E. (2000) A peptide vaccine that prevents experimental autoimmune myasthenia gravis by specifically blocking T cell help. *Faseb J* **14**, 185–196

Aricha, R., Feferman, T., Souroujon, M. C., and Fuchs, S. (2006) Overexpression of phosphodiesterases in experimental autoimmune myasthenia gravis: suppression of disease by a phosphodiesterase inhibitor. *Faseb J* **20**, 374–376

Aruna, B. V., Sela, M., and Mozes, E. (2005) Suppression of myasthenogenic responses of a T cell line by a dual altered peptide ligand by induction of CD4+CD25+ regulatory cells. *Proc Natl Acad Sci USA* **102**, 10285–10290

Aruna, B. V., Ben-David, H., Sela, M., and Mozes, E. (2006a) A dual altered peptide ligand down-regulates myasthenogenic T cell responses and reverses experimental autoimmune myasthenia gravis via up-regulation of Fas-FasL-mediated apoptosis. *Immunology* **118**, 413–424

Aruna, B. V., Sela, M., and Mozes, E. (2006b) Down-regulation of T cell responses to AChR and reversal of EAMG manifestations in mice by a dual altered peptide ligand via induction of CD4+ CD25+ regulatory cells. *J Neuroimmunol* **177**, 63–75

Barrons, R. W. (1997) Drug-induced neuromuscular blockade and myasthenia gravis. *Pharmacotherapy* **17**, 1220–1232

Batocchi, A. P., Evoli, A., Di Schino, C., and Tonali, P. (2000) Therapeutic apheresis in myasthenia gravis. *Ther Apher* **4**, 275–279

Ben-David, H., Sela, M., and Mozes, E. (2005) Down-regulation of myasthenogenic T cell responses by a dual altered peptide ligand via CD4+CD25+-regulated events leading to apoptosis. *Proc Natl Acad Sci USA* **102**, 2028–2033

Benbernou, N., Esnault, S., Potron, G., and Guenounou, M. (1995) Regulatory effects of pentoxifylline on T-helper cell-derived cytokine production in human blood cells. *J Cardiovasc Pharmacol* **25**(2), S15–19

Benny, W. B., Sutton, D. M., Oger, J., Bril, V., McAteer, M. J., and Rock, G. (1999) Clinical evaluation of a staphylococcal protein A immunoadsorption system in the treatment of myasthenia gravis patients. *Transfusion* **39**, 682–687

Berta, E., Confalonieri, P., Simoncini, O., Bernardi, G., Busnach, G., Mantegazza, R., Cornelio, F., and Antozzi, C. (1994) Removal of antiacetylcholine receptor antibodies by protein-A immunoadsorption in myasthenia gravis. *Int J Artif Organs* **17**, 603–608

Boneva, N., Hamra-Amitay, Y., Wirguin, I., and Brenner, T. (2006) Stimulated-single fiber electromyography monitoring of anti-sense induced changes in experimental autoimmune myasthenia gravis. *Neurosci Res* **55**, 40–44

Brocke, S., Brautbar, C., Steinman, L., Abramsky, O., Rothbard, J., Neumann, D., Fuchs, S., and Mozes, E. (1988) In vitro proliferative responses and antibody titers specific to human acetylcholine receptor synthetic peptides in patients with myasthenia gravis and relation to HLA class II genes. *J Clin Invest* **82**, 1894–1900

Brocke, S., Dayan, M., Steinman, L., Rothbard, J., and Mozes, E. (1990) Inhibition of T cell proliferation specific for acetylcholine receptor epitopes related to myasthenia gravis with antibody to T cell receptor or with competitive synthetic polymers. *Int Immunol* **2**, 735–742

Buckingham, J. M., Howard, F.M. Jr, Bernatz, P.E., Payne, W.S., Harrison, E.G. Jr, O'Brien, P.C. and Weiland, L.H. (1976) The value of thymectomy in myasthenia gravis: a computer-assisted matched study. *Ann Surg* **184**, 453–458

Burkhardt, H., and Kalden, J. R. (2005) Xenobiotic immunosuppressive agents: therapeutic effects in animal models of autoimmune diseases. *Rheumatol Int* **17**, 85–90

Chan, A., Weilbach, F. X. and Toyka, K. V. (2005) Mitoxantrone induces cell death in peripheral blood leucocytes of multiple sclerosis patients. *Clinic Exp Immunol* **139**, 152–158

Cohen, I. R. (2001) T-cell vaccination for autoimmune disease: a panorama. *Vaccine* **20**, 706–710

Cornelio, F., Antozzi, C., Confalonieri, P., Baggi, F., and Mantegazza, R. (1998) Plasma treatment in diseases of the neuromuscular junction. *Ann NY Acad Sci* **841**, 803–810

Dalakas, M. C. (2004) The use of intravenous immunoglobulin in the treatment of autoimmune neuromuscular diseases: evidence-based indications and safety profile. *Pharmacol Ther* **102**, 177–193

Dalakas, M. C. (2006) B cells in the pathophysiology of autoimmune neurological disorders: A credible therapeutic target. *Pharmacol Ther* **112**(1), 57–70.

Dawkins, R. L., Christiansen, F. T., and Garlepp, M. J. (1981) Autoantibodies and HLA antigens in ocular, generalized and penicillamine-induced myasthenia gravis. *Ann NY Acad Sci* **377**, 372–384

Dayan, M., Sthoeger, Z., Neiman, A., Abarbanel, J., Sela, M., and Mozes, E. (2004) Immunomodulation by a dual altered peptide ligand of autoreactive responses to the acetylcholine receptor of peripheral blood lymphocytes of patients with myasthenia gravis. *Hum Immunol* **65**, 571–577

De Baets, M., Stassen, M., Losen, M., Zhang, X. and Machiels, B. (2003) Immunoregulation in experimental autoimmune myasthenia gravis- about T cells, antibodies and endplates. *Ann NY Acad Sci* **998**, 308–317

DeFilippi, V. J., Richman, D.P. and Ferguson, M.K. (1994) Transcervical thymectomy for myasthenia gravis. *Ann Thorac Surg* **57**, 194–197

Dodd, R. Y. (1992) The risk of transfusion-transmitted infection. *N Engl J Med* **327**, 419–421

Drachman, D. B., Jones, R.D. and Brodsky, R.A. (2003a) Treatment of refractory myasthenia: Rebooting with high dose cyclophosphamide. *Ann Neur* **53**, 29–34

Drachman, D. B., Wu, J. M., Miagkov, A., Williams, M. A., Adams, R. N., and Wu, B. (2003b) Specific immunotherapy of experimental myasthenia by genetically engineered APCs: the "guided missile" strategy. *Ann NY Acad Sci* **998**, 520–532

Elion, G. B. (1972) Significance of azathioprine metabolites. *Proc R Soc Med* **65**, 257–160

Evoli, A., Tonali, P. A., Padua, L., Monaco, M. L., Scuderi, F., Batocchi, A. P., Marino, M., and Bartoccioni, E. (2003) Clinical correlates with anti-MuSK antibodies in generalized seronegative myasthenia gravis. *Brain* **126**, 2304–2311

Faber-Elmann, A., Paas-Rozner, M., Sela, M., and Mozes, E. (1998) Altered peptide ligands act as partial agonists by inhibiting phospholipase C activity induced by myasthenogenic T cell epitopes. *Proc Natl Acad Sci USA* **95**, 14320–14325

Faber-Elmann, A., Grabovsky, V., Dayan, M., Sela, M., Alon, R., and Mozes, E. (2001) An altered peptide ligand inhibits the activities of matrix metalloproteinase-9 and phospholipase C, and inhibits T cell interactions with VCAM-1 induced in vivo by a myasthenogenic T cell epitope. *Faseb J* **15**, 187–194

Faria, A. M., and Weiner, H. L. (2005) Oral tolerance. *Immunol Rev* **206**, 232–259

Farrar, J., Portolano, S., Willcox, N., Vincent, A., Jacobson, L., Newsom-Davis, J., Rapoport, B., and McLachlan, S. M. (1997) Diverse Fab specific for acetylcholine receptor epitopes from a myasthenia gravis thymus combinatorial library. *Int Immunol* **9**, 1311–1318

Flachenecker, P., Taleghani, B. M., Gold, R., Grossmann, R., Wiebecke, D., and Toyka, K. V. (1998) Treatment of severe myasthenia gravis with protein A immunoadsorption and cyclophosphamide. *Transfus Sci* **19**, 43–46

Fostieri, E., Tzartos, S. J., Berrih-Aknin, S., Beeson, D., and Mamalaki, A. (2005) Isolation of potent human Fab fragments against a novel highly immunogenic region on human muscle acetylcholine receptor which protect the receptor from myasthenic autoantibodies. *Eur J Immunol* **35**, 632–643

Fostieri, E., Kostelidou, K., Poulas, K., and Tzartos, S. J. (2006) Recent advances in the understanding and therapy of myasthenia gravis. *Future Neurology*, **1**, 799–817.

Gajdos, P., Chevret, S., Clair, B., Tranchant, C., Chastang, C.. (1998) Plasma exchange and intravenous immunoglobulin in autoimmune myasthenia gravis. *Ann NY Acad Sci* **841**, 720–726

Galindo-Rodriguez, G., Bustamante, R., Esquivel-Nava, G., Salazar-Exaire, D., Vela-Ojeda, J., Vadillo-Buenfil, M., and Avina-Zubieta, J. A. (2003) Pentoxifylline in the treatment of refractory nephrotic syndrome secondary to lupus nephritis. *J Rheumatol* **30**, 2382–2384

Gascoigne, N. R., Zal, T., and Alam, S. M. (2001) T-cell receptor binding kinetics in T-cell development and activation. *Expert Rev Mol Med* **2001**, 1–17

Gerber, N. L., and Steinberg, A. D. (1976) Clinical use of immunosuppressive drugs: part II. *Drugs* **11**, 90–112

Gonzalez-Serratos, H., Chang, R., Pereira, E. F., Castro, N. G., Aracava, Y., Melo, P. A., Lima, P. C., Fraga, C. A., Barreiro, E. J., and Albuquerque, E. X. (2001) A novel thienylhydrazone, (2-thienylidene)3,4-methylenedioxybenzoylhydrazine, increases inotropism and decreases fatigue of skeletal muscle. *J Pharmacol Exp Ther* **299**, 558–566

Graus, Y. F., de Baets, M. H., van Breda Vriesman, P. J., and Burton, D. R. (1997) Antiacetylcholine receptor Fab fragments isolated from thymus-derived phage display libraries from myasthenia gravis patients reflect predominant specificities in serum and block the action of pathogenic serum antibodies. *Immunol Lett* **57**, 59–62

Grob, D., Arsura, E. L., Brunner, N. G., and Namba, T. (1987) The course of myasthenia gravis and therapies affecting outcome. *Ann NY Acad Sci* **505**, 472–499

Grob, D., Simpson, D., Mitsumoto, H., Hoch, B., Mokhtarian, F., Bender, A., Greenberg, M., Koo, A., and Nakayama, S. (1995) Treatment of myasthenia gravis by immunoadsorption of plasma. *Neurology* **45**, 338–344

Guo, C. Y., Li, Z. Y., Xu, M. Q., and Yuan, J. M. (2005) Preparation of an immunoadsorbent coupled with a recombinant antigen to remove anti-acetylcholine receptor antibodies in abnormal serum. *J Immunol Methods* **303**, 142–147

Hill M, Moss P, W. P., Newsom-Davis J, Willcox N. (1999)complexes., T. c. r. t. D.-p. i. d.-i. m. g. r. o. m. D. p., and 1;97(1–2):146–53., J. N. J.

Hoch, W., McConville, J., Helms, S., Newsom-Davis, J., Melms, A., and Vincent, A. (2001) Auto-antibodies to the receptor tyrosine kinase MuSK in patients with myasthenia gravis without acetylcholine receptor antibodies. *Nat Med* **7**, 365 368

Hwang, B., and Lee, S. W. (2002) Improvement of RNA aptamer activity against myasthenic autoantibodies by extended sequence selection. *Biochem Biophys Res Commun* **290**, 656–662

Hwang, B., Han, K., and Lee, S. W. (2003) Prevention of passively transferred experimental autoimmune myasthenia gravis by an in vitro selected RNA aptamer. *FEBS Lett* **548**, 85–89

Im, S. H., Barchan, D., Fuchs, S., and Souroujon, M. C. (1999) Suppression of ongoing experimental myasthenia by oral treatment with an acetylcholine receptor recombinant fragment. *J Clin Invest* **104**, 1723–1730

Im, S. H., Barchan, D., Souroujon, M. C., and Fuchs, S. (2000) Role of tolerogen conformation in induction of oral tolerance in experimental autoimmune myasthenia gravis. *J Immunol* **165**, 3599–3605

Jambou, F., Zhang, W., Menestrier, M., Klingel-Schmitt, I., Michel, O., Caillat-Zucman, S., Aissaoui, A., Landemarre, L., Berrih-Aknin, S., and Cohen-Kaminsky, S. (2003) Circulating regulatory anti-T cell receptor antibodies in patients with myasthenia gravis. *J Clin Invest* **112**, 265–274

Jaretzki, A., 3rd, Barohn, R. J., Ernstoff, R. M., Kaminski, H. J., Keesey, J. C., Penn, A. S., and Sanders, D. B. (2000) Myasthenia gravis: recommendations for clinical research standards. Task Force of the Medical Scientific Advisory Board of the Myasthenia Gravis Foundation of America. *Neurology* **55**, 16–23

Katz-Levy, Y., Kirshner, S. L., Sela, M., and Mozes, E. (1993) Inhibition of T-cell reactivity to myasthenogenic epitopes of the human acetylcholine receptor by synthetic analogs. *Proc Natl Acad Sci USA* **90**, 7000–7004

Katz-Levy, Y., Paas-Rozner, M., Kirshner, S., Dayan, M., Zisman, E., Fridkin, M., Wirguin, I., Sela, M., and Mozes, E. (1997) A peptide composed of tandem analogs of two myasthenogenic T cell epitopes interferes with specific autoimmune responses. *Proc Natl Acad Sci USA* **94**, 3200–3205

Katz-Levy, Y., Dayan, M., Wirguin, I., Fridkin, M., Sela, M., and Mozes, E. (1998) Single amino acid analogs of a myasthenogenic peptide modulate specific T cell responses and prevent the induction of experimental autoimmune myasthenia gravis. *J Neuroimmunol* **85**, 78–86

Keesey, J. C. (2004) Clinical evaluation and management of myathenia gravis. *Muscle Nerve*, 484–505

Kostelidou, K., Trakas, N., Zouridakis, M., Bitzopoulou, K., Sotiriadis, A., Gavra, I., and Tzartos, S. J. (2006) Expression and characterization of soluble forms of the extracellular domains of the beta, gamma and epsilon subunits of the human muscle acetylcholine receptor. *Febs J* **273**, 3557–3568

Kotadia, B. K., Ravindranath, T. M., Choudhry, M. A., Haque, F., Al-Ghoul, W., and Sayeed, M. M. (2003) Effects of pentoxyfylline on mesenteric lymph node T-cells in a rat model of thermal injury. *Shock* **20**, 517–520

Kothari, M.J. (2004) Myasthenia gravis. *J Am Osteopath Assoc.* **104**, 377–384.

Kuks, J. B., and Skallebaek, D. (1998) Plasmapheresis in myasthenia gravis.: A survey. *Transfus Sci* **19**, 129–136

Lauriola, L., Ranelletti, F., Maggiano, N., Guerriero, M., Punzi, C., Marsili, F., Bartoccioni, E., and Evoli, A. (2005) Thymus changes in anti-MuSK-positive and -negative myasthenia gravis. *Neurology.* **64**, 536–538.

Lavrnic, D., Vujic, A., Rakocevic-Stojanovic, V., Stevic, Z., Basta, I., Pavlovic, S., Trikic, R. and Apostolski, S. (2005) Cyclosporine in the treatment of myasthenia gravis. *Acta Neurol Scand* **111**, 247–252

Lee, S. W., and Sullenger, B. A. (1997) Isolation of a nuclease-resistant decoy RNA that can protect human acetylcholine receptors from myasthenic antibodies. *Nat Biotechnol* **15**, 41–45

Leite M.I., Cossins J., Beeson D., Willcox, N. and Vincent, A. 2006. Antibodies to AChR in seronegative myasthenia gravis. *J Neuroimmunology* **178** (supp. 1: 8th International Conference of Neuroimmunology): 123.

Levinson, A. I., and Wheatley, L. M. (1996) The thymus and the pathogenesis of myasthenia gravis. *Clinical Immunol Immunopathol* **78**, 1–5

Lindstrom, J. M., Seybold, M. E., Lennon, V. A., Whittingham, S., and Duane, D. D. (1976) Antibody to acetylcholine receptor in myasthenia gravis. Prevalence, clinical correlates, and diagnostic value. *Neurology* **26**, 1054–1059

Lindstrom, J., Shelton, D., and Fujii, Y. (1988) Myasthenia gravis. *Adv Immunol* **42**, 233–284

Link, H., Olsson, O., Sun, J., Wang, W. Z., Andersson, G., Ekre, H. P., Brenner, T., Abramsky, O., and Olsson, T. (1991) Acetylcholine receptor-reactive T and B cells in myasthenia gravis and controls. *J Clin Invest* **87**, 2191–2196

Liu, C., and Hu, F. (2005) Investigation on the mechanism of exacerbation of myasthenia gravis by aminoglycoside antibiotics in mouse model. *J Huazhong Univ Sci Technolog Med Sci* **25**, 294–296

Losen, M., Stassen, M.H.W., Martínez-Martínez, P., MAchiels, B.M., Duimel, H., Frederik, P., Veldman, H., Wokke, J.H.J., Spaans, F., Vincent, A. and De Baets, M.H. (2005) Increased expression of rapsyn in muscles prevents acetylcholine receptor loss in experimental autoimmune myasthenia gravis. *Brain* **128**, 2327–2337

Lugnier, C. (2006) Cyclic nucleotide phosphodiesterase (PDE) superfamily: a new target for the development of specific therapeutic agents. *Pharmacol Ther* **109**, 366–398

Lundh, H., Nillson, O. and Rosen, I. (1985) Improvement in neuromuscular transmission in myasthenia gravis by 3,4-diaminopyrimidine. *Eur Arch Psychiatry Neurol Sci* **234**, 374–377

Lyu, R. K., Chen, W. H., and Hsieh, S. T. (2002) Plasma exchange versus double filtration plasmapheresis in the treatment of Guillain-Barre syndrome. *Ther Apher* **6**, 163–166

Ma, C. G., Zhang, G. X., Xiao, B. G., Link, J., Olsson, T., and Link, H. (1995) Suppression of experimental autoimmune myasthenia gravis by nasal administration of acetylcholine receptor. *J Neuroimmunol* **58**, 51–60

MacDonald, M. J., Shahidi, N. T., Allen, D. B., Lustig, R. H., Mitchell, T. L., and Cornwell, S. T. (1994) Pentoxifylline in the treatment of children with new-onset type I diabetes mellitus. *Jama* **271**, 27–28

MacDonald, B. K., Cockerell, O. C., Sander, J. W., and Shorvon, S. D. (2000) The incidence and lifetime prevalence of neurological disorders in a prospective community-based study in the UK. *Brain* **123**(4), 665–676

Mamalaki, A., Trakas, N., and Tzartos, S. J. (1993) Bacterial expression of a single-chain Fv fragment which efficiently protects the acetylcholine receptor against antigenic modulation caused by myasthenic antibodies. *Eur J Immunol* **23**, 1839–1845

Marconi, G., Bobbi, S., Pizzi, A., Sbrilli, C., Taiuti, R., Ronchi, O., Avanzi, G., Lombardo, R., Franco, C., and Biani, D. (1984) Plasma exchange in myasthenia gravis. *Int J Artif Organs* **7**, 297–300

Matic, G., Bosch, T., and Ramlow, W. (2001) Background and indications for protein-A based extracorporeal immunoadsorption. *Ther Apher* **5**, 394–403

Matsumoto, Y., Matsuo, H., Sakuma, H., Park, I. K., Tsukada, Y., Kohyama, K., Kondo, T., Kotorii, S., and Shibuya, N. (2006) CDR3 spectratyping analysis of the TCR repertoire in myasthenia gravis. *J Immunol* **176**, 5100–5107

Mermier, C. M., Schneider, S.M., Gurney, A.B., Weingart, H.M. and Wilmerding, M.V. (2006) Preminary results; Effect of whole-body cooling in patients with myasthenia gravis. *Med Sci Sports Exerc*, 13–20

Mertens, H. G., Hertel, G., Reuther, P. and Ricker, K. (1981) Effect of immunosuppressive drugs (azathioprine). *Ann NY Acad Sci* **377**, 691–699

Mir, L. M., Bureau, M.F., Gehl, J., Rangara, R., Rouy, D., Caillaud, J.M., Delaere, P., Branellec, D., Schwartz, B. and Scherman, D. (1999) High-efficiency gene transfer into skeletal muscle mediated by electric pulses. *Proc Natl Acad Sci USA* **96**, 4262–4267

Moller, D. R., Wysocka, M., Greenlee, B. M., Ma, X., Wahl, L., Trinchieri, G., and Karp, C. L. (1997) Inhibition of human interleukin-12 production by pentoxifylline. *Immunology* **91**, 197–203

Morosetti, M., Meloni, C., Iani, C., Caramia, M., Galderisi, C., Palombo, G., Gallucci, M.T., Bernardi, G. and Casciani, C.U. (1998) Plasmapheresis in severe forms of myasthenia gravis. *Artif Organs* **22**, 129–134

Nakaji, S., and Hayashi, N. (2003) Adsorption column for myasthenia gravis treatment: Medisorba MG-50. *Ther Apher Dial* **7**, 78–84

Neuhaus, O., Kieseier, B. C. and Hartung, H. P. (2004) Mitoxantrone (Novantrone) in multiple sclerosis: new insights. *Expert Rev Neurother* **4**, 17–26

Newsom-Davis, J., Pinching, A.J., Vincent, A. and Wilson, S.G. (1978a) Function of circulating antibody to acetylcholine receptor in myasthenia gravis: investigation by plasma exchange. *Neurology* **28**, 266–272

Newsom-Davis, J., Vincent, A., Wilson, S.G., Ward, C.D., Pinching, A.J. and Hawkey, C. (1978b) Plasmapheresis for myasthenia gravis. *N Engl J Med* **298**, 456–457

Niakan, E., Harati, Y. and Rolak, L.A. (1986) Immunosuppressive drug therapy in myasthenia gravis. *Arch Neurol* **43**, 155–156

Okumura, S., McIntosh, K., and Drachman, D. B. (1994) Oral administration of acetylcholine receptor: effects on experimental myasthenia gravis. *Ann Neurol* **36**, 704–713

Ööpick, M., Kaasik, A. E., and Jakobsen, J. (2003) A population based epidemiological study on myasthenia gravis in Estonia. *J Neurol Neurosurg Psychiatry* **74**, 1638–1643

Oosterhuis, H. J. G. H. (1989) The natural course of myasthenia gravis: a long term follow up study. *J Neurol Neurosurg Psychiatry* **52**, 1121–1127

Oosterhuis, H. (1997) *Myasthenia gravis.*, Groningen Neurological Press.

Paas-Rozner, M., Dayan, M., Paas, Y., Changeux, J. P., Wirguin, I., Sela, M., and Mozes, E. (2000) Oral administration of a dual analog of two myasthenogenic T cell epitopes down-regulates experimental autoimmune myasthenia gravis in mice. *Proc Natl Acad Sci USA* **97**, 2168–2173

Paas-Rozner, M., Sela, M., and Mozes, E. (2001) The nature of the active suppression of responses associated with experimental autoimmune myasthenia gravis by a dual altered peptide ligand administered by different routes. *Proc Natl Acad Sci USA* **98**, 12642–12647

Paas-Rozner, M., Sela, M., and Mozes, E. (2003) A dual altered peptide ligand down-regulates myasthenogenic T cell responses by up-regulating CD25– and CTLA-4-expressing CD4+ T cells. *Proc Natl Acad Sci USA* **100**, 6676–6681

Palace, J., Newsom-Davis, J. and Lecky, B. (1998) A randomised double-blind trial of prednisolone alone or with azathioprine in myasthenia gravis. *Neurology* **50**, 1778–1783

Papanastasiou, D., Mamalaki, A., Eliopoulos, E., Poulas, K., Liolitsas, C., and Tzartos, S. J. (1999) Construction and characterization of a humanized single chain Fv antibody fragment against the main immunogenic region of the acetylcholine receptor. *J Neuroimmunol* **94**, 182–195

Papanastasiou, D., Poulas, K., Kokla, A., and Tzartos, S. J. (2000) Prevention of passively transferred experimental autoimmune myasthenia gravis by Fab fragments of monoclonal antibodies directed against the main immunogenic region of the acetylcholine receptor. *J Neuroimmunol* **104**, 124–132

Penn, A. S., Low, B. W., Jaffe, I. A., Luo, L., and Jacques, J. J. (1998) Drug-induced autoimmune myasthenia gravis. *Ann NY Acad Sci* **841**, 433–449

Perez, M. C., Buot, W.L., Mercado-Danguilan, C., Bagabaldo, Z.G. and Renales, L.D. (1981) Stable remissions in myasthenia gravis. *Neurology* **31**, 32–37

Perlo, V. P., Poskanzer, D.C., Schwab, R.S., Viets, H.R., Osserman, K.E. and Genkins, G. (1966) Myasthenia gravis: evaluation of treatment in 1,355 patients. *Neurology* **16**, 431–439

Pestourie, C., Tavitian, B., and Duconge, F. (2005) Aptamers against extracellular targets for in vivo applications. *Biochimie* **87**, 921–930

Phillips, L. H., 2nd, Torner, J. C., Anderson, M. S., and Cox, G. M. (1992) The epidemiology of myasthenia gravis in central and western Virginia. *Neurology* **42**, 1888–1893

Pinching, A. J., and Peters, D. K. (1976) Remission of myasthenia gravis following plasmapheresis. *Lancet* **2**, 1373–1376

Polizzi, A., Huson, S. M., and Vincent, A. (2000) Teratogen update: maternal myasthenia gravis as a cause of congenital arthrogryposis. *Teratology* **62**, 332–341

Poulas, K., Tsibri, E., Kokla, A., Papanastasiou, D., Tsouloufis, T., Marinou, M., Tsantili, P., Papapetropoulos, T., and Tzartos, S. J. (2001) Epidemiology of seropositive myasthenia gravis in Greece. *J Neurol Neurosurg Psychiatry* **71**, 352–356

Prakash, K. M., Ratnagopal, P., Puvanendran, K. and Lo, Y.L. (2007) Mycophenolate mofetil- as an adjunctive immunosuppressive therapy in refractory myasthenia gravis: The Singapore experience. *J Clin Neurosci* **14**, 278–281

Protopapadakis, E., Kokla, A., Tzartos, S. J., and Mamalaki, A. (2005) Isolation and characterization of human anti-acetylcholine receptor monoclonal antibodies from transgenic mice expressing human immunoglobulin loci. *Eur J Immunol* **35**, 1960–1968

Protti, M. P., Manfredi, A. A., Horton, R. M., Bellone, M., and Conti-Tronconi, B. M. (1993) Myasthenia gravis: recognition of a human autoantigen at the molecular level. *Immunol Today* **14**, 363–368

Psaridi-Linardaki, L., Mamalaki, A., Remoundos, M., and Tzartos, S. J. (2002) Expression of soluble ligand- and antibody-binding extracellular domain of human muscle acetylcholine receptor alpha subunit in yeast Pichia pastoris. Role of glycosylation in alpha-bungarotoxin binding. *J Biol Chem* **277**, 26980–26986

Psaridi-Linardaki, L., Trakas, N., Mamalaki, A., and Tzartos, S. J. (2005) Specific immunoadsorption of the autoantibodies from myasthenic patients using the extracellular domain of the human muscle acetylcholine receptor alpha-subunit. Development of an antigen-specific therapeutic strategy. *J Neuroimmunol* **159**, 183–191

Ptak, J. (2004) Changes of plasma proteins after immunoadsorption using Ig-Adsopak columns in patients with myasthenia gravis. *Transfus Apher Sci* **30**, 125–129

Richman, D. P., and Agius, M. A. (2003) Treatment of autoimmune myasthenia gravis. *Neurology* **61**, 1652–1661

Rieckmann, P., Weber, F., Gunther, A., Martin, S., Bitsch, A., Broocks, A., Kitze, B., Weber, T., Borner, T., and Poser, S. (1996) Pentoxifylline, a phosphodiesterase inhibitor, induces immune deviation in patients with multiple sclerosis. *J Neuroimmunol* **64**, 193–200

Rivner, M. H. (2002) Steroid tretament for myasthenia: steroids are overutilised. *Muscle Nerve* **25**, 115–117

Rodova, M., Kelly, K.F., VanSaun, M., Daniel, J.M. and Werle, M.J. (2004) Regulation of the rapsyn promoter by kaiso and delta-catenin. *Mol Cell Biol* **24**, 7188–7196

Rosenberg, J. S., Oshima, M., and Atassi, M. Z. (1996) B-cell activation in vitro by helper T cells specific to region alpha 146–162 of Torpedo californica nicotinic acetylcholine receptor. *J Immunol* **157**, 3192–3199

Rosenthal, L. A., and Blank, K. J. (1993) Pentoxifylline- and caffeine-induced modulation of major histocompatibility complex class I expression on murine tumor cell lines. *Immunopharmacology* **25**, 145–161

Rowin, J., Meniggioli, M.N. and Tuzun, E. (2004) Etanercept treatment in corticosteroid-dependent myasthenia gravis. *Neurology* **63**, 2390–2392

Samardzic, T., Jankovic, V., Stosic-Grujicic, S., Popadic, D., and Trajkovic, V. (2001) Pentoxifylline inhibits the synthesis and IFN-gamma-inducing activity of IL-18. *Clin Exp Immunol* **124**, 274–281

Schlesinger, F., Krampfl, K., Haeseler, G., Dengler, R., and Bufler, J. (2004) Competitive and open channel block of recombinant nAChR channels by different antibiotics. *Neuromuscl Disord* **14**, 307–312

Schwendimann, R. N., Burton, E. and Minagar, A. (2005) Management of myasthenia gravis. *Am J Ther* **12**, 262–268

Sneider-Gold, C., Hartung, H.P, and Gold, R. (2006) Mycophenolate mofetil and tacrolimus: New therapeutic options in neuroimmunological diseases. *Muscle Nerve* **34**, 284–291

Sela, M. (1999) The concept of specific immune treatment against autoimmune diseases. *Int Rev Immunol* **18**, 201–216

Sela, M., and Mozes, E. (2004) Therapeutic vaccines in autoimmunity. *Proc Natl Acad Sci U S A* **101**(2), 14586–14592

Shi, F. D., Bai, X. F., Xiao, B. G., van der Meide, P. H., and Link, H. (1998) Nasal administration of multiple antigens suppresses experimental autoimmune myasthenia gravis, encephalomyelitis and neuritis. *J Neurol Sci* **155**, 1–12

Shi, F. D., Li, H., Wang, H., Bai, X., van der Meide, P. H., Link, H., and Ljunggren, H. G. (1999) Mechanisms of nasal tolerance induction in experimental autoimmune myasthenia gravis: identification of regulatory cells. *J Immunol* **162**, 5757–5763

Shibuya, N., Sato, T., Osame, M., Takegami, T., Doi, S., and Kawanami, S. (1994) Immunoadsorption therapy for myasthenia gravis. *J Neurol Neurosurg Psychiatry* **57**, 578–581

Sieb, J. P., and Engel, A. G. (1993) Ephedrine: effects on neuromuscular transmission. *Brain Res* **623**, 167–171

Sieb, J. P. (2005) Myasthenia gravis: emerging new therapy options. *Curr Opin Pharmacol* **5**, 303–307

Skeie, G. O., Apostolski, S., Evoli, A., Gilhus, N.E., Hart, I.K., Harms, L., Hilton-Jones, D., Melms, A., Verschuuren, J. and Horgel, H.W. (2006) Guidelines for the treatment of autoimmune neuromuscular transmission disorders. *Eur J Neurol* **13**, 1468–1331

Slavin, A. J., Tarner, I.H., Nakajima, A., Urbanek-Ruiz, I., McBride, J., Contag, C.H. and Fathman, C.G. (2002) Adoptive cellular gene therapy of autoimmune disease. *Autoimmun Rev* **1**, 213–219

Sloan-Lancaster, J., and Allen, P. M. (1996) Altered peptide ligand-induced partial T cell activation: molecular mechanisms and role in T cell biology. *Annu Rev Immunol* **14**, 1–27

Smith, G. D. P., Stevens, D.L. and Fuller, G.N. (2001) Myasthenia gravis, corticosteroids and osteoporosis prophylaxis. *J Neurol* **248**, 151

Somnier, F. E., and Langvad, E. (1989) Plasma exchange with selective immunoadsorption of anti-acetylcholine receptor antibodies. *J Neuroimmunol* **22**, 123–127

Souroujon, M. C., Maiti, P. K., Feferman, T., Im, S. H., Raveh, L., and Fuchs, S. (2003) Suppression of myasthenia gravis by antigen-specific mucosal tolerance and modulation of cytokines and costimulatory factors. *Ann NY Acad Sci* **998**, 533–536

Spring, P. J., and Spies, J. M. (2001) Myasthenia gravis: options and timing of immunomodulatory treatment. *BioDrugs* **15**, 173–183

Stassen, M. H., Machiels, B. M., Fostieri, E., Tzartos, S. J., Berrih-Aknin, S., Bosmans, E., Parren, P. W., and De Baets, M. H. (2003) Characterization of a fully human IgG1 reconstructed from an anti-AChR Fab. *Ann NY Acad Sci* **998**, 399–400

Tada, M., Shimohata, T., Tada, M., Oyake, M., Igarashi, S., Onodera, O., Naruse, S., Tanaka, K., Tsuji, S. and Nishizawa, M. (2006) Long-term therapeutic efficacy and safety of low-dose tacrolimus (FK506) for myasthenia gravis. *J Neurol Sci* **247**, 17–20

Takamori, M., and Ide, Y. (1996) Specific removal of anti-acetylcholine receptor antibodies in patients with myasthenia gravis. *Transfus Sci* **17**, 445–453

Takamori, M., and Maruta, T. (2001) Immunoadsorption in myasthenia gravis based on specific ligands mimicking the immunogenic sites of the acetylcholine receptor. *Ther Apher* **5**, 340–350

Tarner, I. H., and Fathman, C. G. (2001) Gene therapy in autoimmune disease. *Curr Opin Immunol* **13**, 676–682

Tarner, I. H., Slavin, A.J., McBride, J., Levicnik, A., Smith, R., Nolan, G.P., Contag, C.H. and Fathman, C.G. (2003) Treatment of autoimmune disease by adoptive cellular gene therapy. *Ann NY Acad Sci* **998**, 512–519

Thanvi, B. R., and Lo, T. C. (2004) Update on myasthenia gravis. *Postgrad Med J* **80**, 690–700

Tindall, R. S., Phillips, J.T., Rollins, J.A., Wells, L. and Hall, K. (1993) A clinical therapeutic trial of cyclosporine in myasthenia gravis. *Ann NY Acad Sci* **681**, 539–551

Trakas, N., and Tzartos, S. J. (2001) Conjugation of acetylcholine receptor-protecting Fab fragments with polyethylene glycol results in a prolonged half-life in the circulation and reduced immunogenicity. *J Neuroimmunol* **120**, 42–49

Truffault, F., Cohen-Kaminsky, S., Khalil, I., Levasseur, P., and Berrih-Aknin, S. (1997) Altered intrathymic T-cell repertoire in human myasthenia gravis. *Ann Neurol* **41**, 731–741

Tsantili, P., Tzartos, S. J., and Mamalaki, A. (1999) High affinity single-chain Fv antibody fragments protecting the human nicotinic acetylcholine receptor. *J Neuroimmunol* **94**, 15–27

Tzartos, S. J., and Lindstrom, J. M. (1980) Monoclonal antibodies used to probe acetylcholine receptor structure: localization of the main immunogenic region and detection of similarities between subunits. *Proc Natl Acad Sci USA* **77**, 755–759

Tzartos, S. J., Sophianos, D., and Efthimiadis, A. (1985) Role of the main immunogenic region of acetylcholine receptor in myasthenia gravis. An Fab monoclonal antibody protects against antigenic modulation by human sera. *J Immunol* **134**, 2343–2349

Tzartos, S. J., Barkas, T., Cung, M. T., Mamalaki, A., Marraud, M., Orlewski, P., Papanastasiou, D., Sakarellos, C., Sakarellos-Daitsiotis, M., Tsantili, P., and Tsikaris, V. (1998) Anatomy of the antigenic structure of a large membrane autoantigen, the muscle-type nicotinic acetylcholine receptor. *Immunol Rev* **163**, 89–120

Unwin, N., Miyazawa, A., Li, J., and Fujiyoshi, Y. (2002) Activation of the nicotinic acetylcholine receptor involves a switch in conformation of the alpha subunits. *J Mol Biol* **319**, 1165–1176

Vidic-Dankovic, B., Kosec, D., Damjanovic, M., Apostolski, S., Isakovic, K. and Bartlett, R.R. (1995) Leflunomide prevents the development of experimentally induced myasthenia gravis. *Int J Immunopharmac* **17**, 273–281

Vincent, A., Palace, J. and Hilton-Jones, D. (2001) Myasthenia gravis. *The Lancet* **357**, 2122–2128

Vincent, A. (2002) Unravelling the pathogenesis of myasthenia gravis.

Vincent, A., Clover, L., Buckley, C., Grimley Evans, J., and Rothwell, P. M. (2003) Evidence of underdiagnosis of myasthenia gravis in older people. *J Neurol Neurosurg Psychiatry* **74**, 1105–1108

Vincent, A., and Leite, M. I. (2005) Neuromuscular junction autoimmune disease: muscle specific kinase antibodies and treatments for myasthenia gravis. *Curr Opin Neurol* **18**, 519–525

Wang, Z. Y., Qiao, J., and Link, H. (1993) Suppression of experimental autoimmune myasthenia gravis by oral administration of acetylcholine receptor. *J Neuroimmunol* **44**, 209–214

Wang, Z. Y., He, B., Qiao, J., and Link, H. (1995) Suppression of experimental autoimmune myasthenia gravis and experimental allergic encephalomyelitis by oral administration of acetylcholine receptor and myelin basic protein: double tolerance. *J Neuroimmunol* **63**, 79–86

Wang, Z. Y., Okita, D. K., Howard, J., Jr., and Conti-Fine, B. M. (1998) T-cell recognition of muscle acetylcholine receptor subunits in generalized and ocular myasthenia gravis. *Neurology* **50**, 1045–1054

Weiner, H. L. (1999) Induction of oral tolerance to the acetylcholine receptor for treatment of myasthenia gravis. *J Clin Invest* **104**, 1667–1668

Whitehouse, M. W. (2004) Anti-TNF-alpha therapy for chronic inflammation: reconsidering pentoxifylline as an alternative to therapeutic protein drugs. *Inflammopharmacology* **12**, 223–227

Witte, A. S., Cornblath, D. R., Parry, G. J., Lisak, R. P., and Schatz, N. J. (1984) Azathioprine in the treatment of myasthenia gravis. *Ann Neurol* **15**, 602–605

Wraith, D. C., Nicolson, K. S., and Whitley, N. T. (2004) Regulatory CD4+ T cells and the control of autoimmune disease. *Curr Opin Immunol* **16**, 695–701

Wu, J. M., Wu, B., Miagkov, A., Adams, R. N., and Drachman, D. B. (2001) Specific immunotherapy of experimental myasthenia gravis in vitro: the "guided missile" strategy. *Cell Immunol* **208**, 137–147

Wylam, M. E., Anderson, P.M., Kuntz, N.L. and Rodriguez, V. (2003) Successful treatment of refractory myasthenia gravis using rituximab: a pediatric case report. *J Pediatr* **143**, 674–677

Xu, L., Villain, M., Galin, F. S., Araga, S., and Blalock, J. E. (2001) Prevention and reversal of experimental autoimmune myasthenia gravis by a monoclonal antibody against acetylcholine receptor-specific T cells. *Cell Immunol* **208**, 107–114

Yachi, P. P., Ampudia, J., Zal, T., and Gascoigne, N. R. (2006) Altered peptide ligands induce delayed CD8-T cell receptor interaction – a role for CD8 in distinguishing antigen quality. *Immunity* **25**, 203–211

Yamazaki, Z., Fujimori, Y., Takahama, T., Inoue, N., Wada, T., Kazama, M., Morioka, M., Abe, T., Yamawaki, N. and Inagaki, K. (1982) Efficiency and biocompatibility of a new immunosorbent. *Trans Am Soc Artif Intern Organs* **28**, 318–323

Yang, K. S., Kenpe, K., Yamaji, K., Tsuda, H. and Hashimoto, H. (2002) Plasma adsorption in critical care. *Ther Apher* **6**, 184–188

Yang, L., Cheng, Y., Yan, W. R., and Yu, Y. T. (2004) Extracorporeal whole blood immunoadsorption of autoimmune myasthenia gravis by cellulose tryptophan adsorbent. *Artif Cells Blood Substit Immobil Biotechnol* **32**, 519–528

Yeh, J. H., and Chiu, H. C. (2000) Comparison between double-filtration plasmapheresis and immunoadsorption plasmapheresis in the treatment of patients with myasthenia gravis. *J Neurol* **247**, 510–513

Yeh, J. H., Chen, W.H. and Chiu, H.C. (2004) Complications of double-filtration plasmapheresis. *Transfusion* **44**(11), 1621–1625

Zhou, L., McConville, J., Chaudhry, V., Adams, R. N., Skolasky, R. L., Vincent, A., and Drachman, D. B. (2004) Clinical comparison of muscle-specific tyrosine kinase (MuSK) antibody-positive and -negative myasthenic patients. *Muscle Nerve* **30**, 55–60

Zhu, L. P., Cupps, T.R., Whalen, G. and Fauci, A.S. (1987) Selective effects on cyclophosphamide therapy on activation, proliferation and differentiation of human B cells. *J Clin Inv* **79**, 1082–1090

Zhu, K. Y., Feferman, T., Maiti, P.K., Souroujon, M.C. and Fuchs, S. (2006) Intravenous immunoglobulin suppresses experimental myasthenia gravis: immunological mechanisms. *J Neuroimmunol* **176**, 187–197

Zieliński, M., Kużdżal, J., Szlubowski, A., and Soja, J. (2004) Comparison of late results of basic transsternal and extended transsternal thymectomies in the treatment of myasthenia gravis. *Ann Thorac Surg* **78**, 253–258

Zisman, E., Katz-Levy, Y., Dayan, M., Kirshner, S. L., Paas-Rozner, M., Karni, A., Abramsky, O., Brautbar, C., Fridkin, M., Sela, M., and Mozes, E. (1996) Peptide analogs to pathogenic epitopes of the human acetylcholine receptor alpha subunit as potential modulators of myasthenia gravis. *Proc Natl Acad Sci USA* **93**, 4492–4497

# New Diagnostic and Therapeutic Options for the Treatment of Multiple Sclerosis

Paolo Riccio, Heinrich Haas, Grazia Maria Liuzzi, and Rocco Rossano

**Abstract** Multiple sclerosis (MS) is a multifactorial disease and its treatment can be based on different strategies. We are presently approaching the MS problem at three different levels: (1) Developing new carrier systems for the targeted delivery of therapeutical and diagnostic agents to the sites of inflammation at the blood brain barrier (BBB); (2) Modulating expression and activity of matrix metalloproteinases (MMPs) involved in MS; (3) Choosing an appropriate diet able to influence the progression of the disease.

(1) A promising basis for new therapeutic and diagnostic approaches in the context of MS is provided by the use of nanoparticulate lipidic carrier agents, specifically developed in our laboratories, for tissue-selective delivery. Such drug carriers can improve solubility characteristics, modulate serum interactions, and affect pharmacokinetics in a controlled way following standard iv drug administration. With the technology for selectively targeting inflamed tissue at the BBB by electrically charged carriers as presented here, a new diagnostic approach for the determination of BBB lesions might become available. Treatment of MS could be improved by local accumulating of established drugs, thus largely reducing systemic exposure of patients to them.

(2) MMPs are important mediators of tissue damage in inflammatory demyelinating diseases and are implicated in the pathogenesis of both MS and HIV- associated dementia (HAD). In particular, gelatinase B (MMP-9) has a crucial role in blood brain barrier (BBB) opening, leukocyte recruitment, myelin disruption, and TNF-$\alpha$ release in the course of MS. We have suggested that CSF increase of MMP-9 levels in the course of MS is due to its intrathecal synthesis. This observation – and the decrease of TIMP-1, the natural inhibitor of MMP-9 – indicates a peculiar involvement of MMP-9 in the pathogenetic mechanism of MS and the importance of therapeutic strategies able to block MMP activity and/or expression. Since both interferon-$\beta$ (IFN-$\beta$) and the antiretroviral drugs AZT and

P. Riccio
University of Basilicata, Department of Biology D.B.A.F., 85100 Potenza, Italy
e-mail: paolo.riccio@unibas.it

A. Falus (ed.), *Clinical Applications of Immunomics*,
DOI: 10.1007/978-0-387-79208-8_10, © Springer Science+Business Media, LLC 2009

Indinavir, the standard treatments for MS patients or HIV-infected patients, respectively, are able to inhibit MMP-9 expression in glial cell cultures, their use could be very effective in MS treatment.

(3) With regard to diet, recent reports indicate a possible relationship between milk and MS (Riccio 2004). The milk components involved are some minor milk proteins of the fat globules membrane (MFGM). In particular, the most abundant MFGM protein, butyrophilin (BTN), can induce the Experimental Autoimmune Encephalomyelitis (EAE) and is very similar to myelin oligodendrocyte glycoprotein (MOG), a minor myelin protein and one of the putative candidate autoantigens in MS. With this in mind, we suggest healthy and functional foods that are suited for the mild treatment of MS patients and can bear to their wellness, since they are low in the MFGM proteins but also contain healthy dietary supplements.

**Keywords** Multiple sclerosis · Cationic liposomes · Matrix metalloproteinases · Nutrition · Antioxidants · MFGM proteins · Blood brain barrier

# 1 Introduction

Multiple sclerosis (MS) is a chronic multi-focal autoimmune disease of the central nervous system (CNS), characterized by inflammatory changes, blood brain barrier (BBB) breakdown, perivascular inflammation, axonal and oligodendrocyte loss, and demyelination (McQualter and Bernard 2007; McFarland and Martin 2007).

The hallmark of MS is the breakdown of the myelin sheath. Myelin is the multi-lamellar, lipid-rich membranous process produced by oligodendrocytes and wrapped around nerve axons to provide segmental insulation and thereby fast and efficient conduction of nerve impulses jumping from one to another myelin-free Ranvier's node.

In the course of MS, which is the most common human disabling neurological disease in young people, the myelin sheath is broken down and many scars are produced in the brain as well as in the spinal cord.

The causes of MS remain to be ascertained, but it is clear that it is a complex and multifactorial disease: genetic predisposition, abnormal immune response and environmental (infectious and nutritional) factors have all been taken into consideration as possible causative agents, but none of these factors alone can explain the origin of this disease. Pathogenesis, progression, and therapeutic options of MS are quite heterogeneous (Lucchinetti, Bruck, Parisi, Scheithauer, Rodriguez, and Lassmann 2000; Lühder and Gold 2007).

Befitting with its complexity, MS shows clinical heterogeneity and is classified into two different clinical subtypes: relapsing-remitting MS, which is the most common, and progressive MS. This latter in turn shows different clinical courses such as primary progressive, secondary progressive, and progressive-relapsing.

Since MS is a multifactorial disease, its treatment can be based on different strategies. With this in mind, we are approaching the MS problem at different levels. In this chapter we describe three new options that appear to be very promising for the treatment and diagnosis of MS: (1) Use of cationic liposomes for specific delivery of specific drugs to the inflamed BBB sites; (2) Assessment of metalloproteinase-9 (MMP-9, or gelatinase B) levels in biological fluids for diagnostic purposes and modulation of MMP-9 expression and activity by therapeutic agents; (3) Choice of appropriate nutrients able to influence the progression of the disease and to provide nutritional complementary treatment of MS in addition to classic pharmacological therapy.

## 2 Targeting Blood-Brain Barrier Inflamed Sites by Cationic Colloidal Carriers

The present paragraph relates to MS treatment by local intervention at the blood brain barrier (BBB). BBB is the complex cerebral vascular endothelium that forms the interface between the CNS and systemic blood circulation. A fundamental function of the intact BBB is to restrict permeability to hydrophilic solutes. It is responsible for the maintenance of the homeostasis of the CNS environment and CNS protection for optimal functional activity.

In neuroinflammatory diseases such as multiple sclerosis (MS), BBB changes dramatically. The functionality as a barrier between blood circulation and CNS is broken down, as becomes manifest by an increase of permeability and massive cellular infiltration (Correale and Villa 2007)

Lymphocyte recruitment across BBB lesions is considered a critical event in disease pathogenesis (Floris, Blezer, Schreibelt, Dopp, van der Pol, Schadee-Eestermans, Nicolay, Dijkstra and de Vries 2004). Magnetic resonance imaging (MRI) and pathological studies have shown that lesions are distributed around venules and that new focal lesions associated with frank BBB leakage are preceded by subtle, progressive alterations in tissue integrity (Werring, Brassat, Droogan, Clark, Symms, Barker, MacManus, Thompson and Miller 2000).

Although in recent years the inflammatory cascade in the MS plaque and the mechanisms modulating neuroinflammatory reactions have been the matter of intense investigation, the precise sequence and consequences of the occurrences for disease progression still have to be clarified (Harris, Frank, Patronas, McFarlin and McFarland 1991; Bruck and Stadelmann 2005).

Irrespective, if BBB is considered simply a barrier to be crossed or an active mediator in the neuroinflammatory process, it is clear that the events at the BBB play a key role in the disease progression. Therefore, early and sensitive targeting of subtle changes occurring at the BBB is of high importance for diagnostic purposes, and it can be a basis for new and improved treatment options (Rousseau, Denizot, Le Jeune and Jallet 1999).

So far, BBB has been regarded predominantly in the context of MS diagnosis, in the sense that changes of permeability were observed. The increase of BBB permeability for small solutes is considered one of the earliest and most important indicators for BBB changes related to MS. In clinical practice, MRI is "the gold standard" for the paraclinical evaluation of the disease status in MS patients (Rebeles, Fink, Anzai and Maravilla 2006; Charil and Filippi 2007).

Usually, BBB lesions are detected by the infiltration of water-soluble contrast agents that have been injected into the blood stream. In various studies in animal models, elevated BBB was found also for colloidal particles, as liposomes or nanoparticles, with fluorescent, radioactive, magnetic, or other labels (Rousseau et al. 1999; Calvo, Gouritin, Villarroya, Eclancher, Giannavola, Klein, Andreux and Couvreur 2002; Manninger, Muldoon, Nesbit, Murillo, Jacobs and Neuwelt 2005).

The permeability of BBB lesions to drug carrier particles could be used as well to deliver drugs to the sites of leaky BBB (Merodio, Irache, Eclancher, Mirshahi and Villarroya 2000; Schmidt, Metselaar, Wauben, Toyka, Storm and Gold 2003), in analogy to enhanced permeation and retention at tumor vasculature (Torchilin 2000).

Schmidt Metselaar, Wauben, Toyka, Storm and Gold (2003) encapsulated the glucocorticosteroid Prednisolone in liposomes and used them in EAE rats. EAE is the experimental autoimmune encephalomyelitis, an animal model of MS. They found a higher concentration of the glucocorticosteroid in the brain and spinal cord of EAE rats compared to healthy rats. However, because accumulation is obtained by passive diffusion of particles across leaky vasculature, particle parameters like size and molecular setup must be selected for prolonged circulation time and optimized target accumulation.

A new option to specifically target altered sites of BBB in the course of neuroinflammatory diseases is by cationic colloidal carriers, i.e., positively charged liposomes, or other types of positively charged colloidal nanoparticles (Haas and Riccio 2007).

In our studies on EAE rats with a clear temporal profile of acute inflammation that represents particularly the early stages of MS, accumulation of cationic liposomes at affected BBB could be demonstrated (Haas and Riccio 2007; Cavaletti G, Cassetti A, Canta A, Galbiati S, Gilardini A,Oggioni N, Rodriguez-Menendez V,Fasano A, Liuzzi GM, Fattler U, Ries S, Nieland J, Riccio P and Haas H, unpublished results). After iv injection into the tail vein, cationic rhodamine- and gold-labeled liposomes were found in spinal cord endoneural vessels with a binding efficacy directly corresponding to the temporal profile of inflammation. With anionic liposomes, no such binding was observed, and in healthy rats neither cationic nor anionic liposomes were bound at BBB.

Cationic liposomes are known to accumulate at angiogenic (proliferating) tumor vasculature and endothelial cells in situations of chronic inflammation (Thurston, McLean, Rizen, Baluk, Haskell, Murphy, Hanahan and McDonald 1998; Campbell, Fukumura, Brown, Mazzola, Izumi, Jain, Torchilin and Munn 2002), and their efficacy has been underlined by several preclinical studies for tumor therapy and diagnosis (Krasnici, Werner, Eichhorn, Schmitt-Sody, Pahernik, Sauer, Schulze, Teifel, Michaelis, Naujoks and Dellian 2003; Schmitt-Sody, Strieth, Krasnici, Sauer, Schulze, Teifel, Michaelis, Naujoks and Dellian 2003; Kalra and Campbell 2006; Eichhorn, Luedemann, Strieth, Papyan, Ruhstorfer, Haas, Michaelis, Sauer, Teifel, Enders, Brix, Jauch, Bruns and Dellian 2007). MediGene AG is developing a technology platform (Endo-TAG$^{TM}$) of cationic colloidal carriers, in particular liposomes, for neovascular targeting, intended for applications in tumor therapy and inflammatory diseases (Michaelis and Haas 2006). EndoTAG$^{TM}$-1, a product comprising the anti-cancer drug paclitaxel and a mixture of cationic and zwitterionic lipids (Michaelis and Haas 2006), is currently undergoing clinical evaluation (phase II) in different cancer indications.

Although it is obvious that electrostatic interactions play an important role for preferential accumulation, the detailed molecular foundation for cationic neovascular targeting still has to be clarified. For tumor vasculature, the high proliferation rate and the presence of negatively charged fenestrae (Thurston et al. 1998) have been identified as a foundation for binding. An increase of negative charge density could be induced also by overexpression of anionic phospholipids (in particular, phosphatidylserine) at the luminal surface (Ran and Thorpe 2002; Ran, Downes and Thorpe 2002), activation of protein phosphorylation by cytokines such as TNF-$\alpha$ (Nwariaku, Chang, Zhu, Liu, Duffy, Halaihel, Terada and Turnage 2002), or by increase of the eicosanoids levels. Other events in the membrane organization, which may contribute to cationic liposome accumulation, are membrane fluidity changes, rearrangement of the extracellular matrix, and lipid raft formation.

It has been recently outlined that several key components in the pathophysiology of the CNS damage are also associated with angiogenesis (Kirk, Frank and Karlik 2004), and it has been hypothesized that angiogenesis plays a significant role in MS progression. However, angiogenesis is clearly only a limited part of the pathological changes due to the inflammatory reaction and it is very unlikely that it is evident in the earliest stages of the disease.

The observed binding of cationic liposomes in EAE rats indicates that already at a very early stage of acute neuroinflammation (i.e., before angiogenesis and/or chronic inflammation), the endothelial membrane organization changes and becomes more susceptible for the binding of cationic particles. In the experiments, the permeability of BBB was not elevated yet, or it was too low to allow sufficient passive intrusion of anionic liposomes. Therefore, cationic colloidal particles promise to be a sensitive new tool for the early detection of inflammatory BBB changes and drug delivery to these sites.

For MS diagnosis, cationic targeting offers the determination of inflammatory events with a mechanism that does not require elevated BBB permeability. The correlation between the profile of cationic targeting efficacy and BBB permeability still has to be clarified to detail; however, irrespective of the temporal and functional sequence, additional independent information on the morphological changes will be provided by such measurements.

For therapy, the local concentration of the active agents at the sites of affected BBB can be improved with respect to systemic administration, and selective action of the active agent at these sites will be obtained. This will enable to apply lower doses of drugs compared to systemic administration, and thus to reduce side effects or antibody production.

For example, drugs can be released, which are capable of limiting the entry of activated T cells into the CNS by blocking the activity of metalloproteinases (MMPs) such as MMP-9 and of the activity of reactive oxygen radicals. Other therapeutics that can be delivered via this new targeting approach include drugs that are currently used for the treatment of MS like interferon-β, corticosteroids, or cytostatics.

Summarizing, targeting affected BBB can provide new diagnostic and therapeutic pathways in MS treatment. Cationic colloidal carriers are an option to specifically target BBB lesions with a binding mechanism that does not require elevated permeability. Fine tuning molecular setup and charge of the carrier particles may enable to further improve the targeting characteristics for a given application.

Finally, further investigation of BBB targeting by charged carriers in comparison to other pathological changes will enable to get better insight into the course of inflammatory response and thus to optimize current treatment protocols.

## 3  The Role of MMPs in MS Pathogenesis and the Modulation of Their Expression and Activity for MS Treatment

Matrix degrading metalloproteinases (MMPs) are a family of zinc-dependent, neutral endopeptidases that are involved in the degradation and remodeling of the extracellular matrix (ECM) in a variety of physiological and pathological processes (Visse and Nagase 2003; Agrawal, Lau and Yong 2008).

More than 24 human MMPs have been identified. They share common structured domains, but differ with respect to their cellular sources, inducibility, and efficiency of substrate utilization (Cauwe, Van den Steen and Opdenakker 2007).

Structurally, MMPs are multi-domain enzymes. They all possess three common domains: the catalytic domain, the N-terminal pre-domain, and the pro-domain. The catalytic domain contains a zinc ion in the active site, essential for proteolytic activity. The pre-domain is a signal peptide, which serves for secretion. The pro-domain contains a single cysteine residue within a conserved inhibitory sequence, which is coordinated with the catalytic zinc ion, and thus keeps the enzyme in a latent pro-form. Some additional domains can be found attached to the common structure in several MMPs.

An hemopexin-like carboxy terminal domain is present in all MMPs, with the exception of matrilysins. Gelatinases contain a fibronectin-like domain inserted between the active site and the $Zn^{2+}$-binding domain, which serves for binding to the substrate gelatin (Allan, Docherty, Barker, Huskisson, Reynolds and Murphy 1995).

The expression and the activity of MMPs are tightly regulated at several levels (Nagase 1997). MMPs are mostly generally secreted as inactive pro-enzymes by a variety of cell types, including leukocytes, tumor cells, and neural cells, and are usually activated in the extracellular or in the pericellular space in response to a variety of inducers such as cytokines, growth factors, chemical agents, and physiological substances such as metal ions, reactive oxygen species, or hormones (Van den Steen, Dubois, Nelissen, Rudd, Dwek and Opdenakker 2002).

Activation of MMPs is achieved by disruption of the cysteine–zinc interaction, the so-called cysteine switch mechanism, and enzymatic removal of the propeptide. Inhibition of MMPs is obtained by specific tissue inhibitors of MMPs named TIMPs, which bind to MMPs forming a 1:1 non-covalent complex (Gomez, Alonso, Yoshiji and Thorgeirsson, 1997). The balance between MMP and TIMP regulates enzyme activity in physiological conditions. An excessive or inappropriate expression of MMPs and/or a decrease in TIMP production may contribute to different pathological conditions including multiple sclerosis (Rosenberg 2002).

Different studies indicate the involvement of MMPs and in particular of MMP-9 (gelatinase B), in MS pathogenesis. Elevated MMP-1, MMP-2, MMP-3, MMP-7, and MMP-9 expression has been found in CNS lesions and in blood monocytes from MS patients (Steinman and Gijbels 1994; Cuzner, Gveric, Strand, Loughlin, Paemen, Opdenakker and Newcombe 1996; Kouwenhoven, Ozenci, Gomes, Yarilin, Giedraitis, Press and Link 2001; Lindberg, De Groot, Montagne, Freitag, van der Valk, Kappos and Leppert 2001).

Several reports have demonstrated the upregulation of MMP-9 in CSF and serum of patients with active relapsing-remitting (RR) MS (Gijbels, Masure, Carton and Opdenakker 1992; Leppert, Ford, Stabler, Grygar, Lienert, Huber, Miller, Hauser and Kappos 1998; Liuzzi, Latronico, Fasano, Carlone and Riccio 2004a). In the acute phases of MS the increase in serum levels of MMP-9 was found to correlate with the appearance of new lesions, as shown by gadolinium enhancements on MRI, suggesting that MMP-9 might contribute to BBB opening, allowing inflammatory cells and plasma proteins to enter the MS lesions.

We have demonstrated the intrathecal production of MMP-9 in patients with relapsing-remitting MS by indexing MMP-9 in CSF and serum to albumin (Liuzzi, Trojano, Fanelli, Avolio, Fasano, Livrea and Riccio 2002). This result indicates the involvement of MMP-9 at different levels in MS pathogenesis. Indeed, the production of MMP-9 within the CNS indicates a role for this enzyme not only in lymphocyte trafficking across the BBB but also in myelin breakdown.

All together these findings suggest that MMPs can be considered as an important target for an innovative therapeutic approach in addition to the pharmacological therapy.

The treatment of MS has experienced an important change in the past decade with the introduction of several drugs such as β-interferons, glatiramer acetate (Copaxone), azathioprine, and mitoxantrone, which are able to modify the evolution of this disease. The current MS therapies primarily target the peripheral immune response, although the results of several studies indicate that the efficacy of MS drugs could be in part the result of the beneficial effect on other non-specific targets such as MMPs.

In this respect, most of the studies have been performed with interferon-β (IFN-β), the most commonly used immunomodulator in MS, of which there are three commercial preparations. In vitro studies have suggested that the clinical benefit of IFN-β treatment in MS patients may be in part a result of the ability of this drug to significantly reduce T-lymphocyte infiltration into the CNS by interfering with the production of MMP-9 (Leppert, Waubant, Bürk, Oksenberg and Hauser 1996; Stüve, Dooley, Uhm, Antel, Francis, Williams and Yong 1996).

It has been demonstrated that the treatment of MS patients with IFN-beta reduces MMP concentrations and restores the integrity of the barrier (Yong, Chabot, Stüve and Williams 1998). In this respect, we studied the effect of IFN-β–1b treatment on serum levels of MMP-9 in patients with RR-MS and demonstrated that serum levels of MMP-9, but not of MMP-2 (gelatinase A), were decreased during a 24-months follow-up and paralleled the clinical recovery of patients (Trojano, Avolio, Liuzzi, Ruggieri, Defazio, Liguori, Santacroce, Paolicelli, Giuliani, Riccio and Livrea 1999). In the same study we also noted that patients with neutralizing antibodies to IFN-β tended to have higher MMP-9 levels.

Proteolysis by MMPs is balanced by endogenous tissue inhibitors of MMPs (TIMPs). Serum and CSF levels of TIMP-1 decrease in RR-MS patients (Liuzzi et al., 2002), while IFN-β treatment increases serum levels of TIMP-1 in RR-MS patients (Gilli, Bertolotto, Sala, Hoffmann, Capobianco, Malucchi, Glass, Kappos, Lindberg and Leppert 2004). Since the imbalance between MMP and TIMP is a crucial feature in multiple sclerosis pathogenesis, we might suppose that one of the mechanisms by which IFN-β exerts its beneficial effect in multiple sclerosis is the reduction of MMP-9 expression and the increase of its endogenous tissue inhibitor, TIMP-1.

In the hypotheses that activated glial cells could be the candidates for the production of MMP-9 within the CNS and that the modulation of MMP activity and expression in glial cells may represent a therapeutic tool for the treatment of MS, we used an in vitro system to investigate whether IFN-β is able to modulate the expression of MMP-2 and MMP-9 in astrocytes and microglia. This study allowed us to demonstrate that IFN-β is able to inhibit MMP-2 and MMP-9 expression in LPS-activated astrocytes and microglia (Liuzzi et al. 2004a). The inhibitory effect of IFN-β was also demonstrated in a preliminary study performed on rat astrocytes, on the activity and expression of calpains, calcium-activated proteases that have been recently implicated in MS pathogenesis due to their ability to degrade myelin proteins and their presence, at increased levels, in activated glial/inflammatory cells of MS plaques (Latronico, Piscopo, Fasano, Riccio and Liuzzi 2004).

We have recently demonstrated that also the antiretroviral drugs zidovudine (AZT) and indinavir (IDV), used for the treatment of HIV-infected patients, are able to inhibit the expression of MMPs in astrocytes and microglia (Liuzzi et al. 2004b), as well as in mononuclear cells from HIV-positive subjects (Latronico, Liuzzi, Riccio, Lichtner, Mengoni, D'Agostino, Vullo and Mastroianni 2007). These drugs have a good penetration into the CNS, a prerequisite for the possible management of neurological diseases. This property may be particularly relevant in diseases such as MS, in which the intrathecal synthesis of MMP-9 has been demonstrated. On this basis, the result of this latter study open up the perspective of using antiretroviral drugs, more likely protease inhibitors, as potential candidates for the experimental treatment of diseases such as MS, in which the inhibition of MMP-9 could have clinical benefits.

# 4  The Relevance of Nutrition in the Course of MS

## 4.1  Background

Non-genetic, environmental factors may play an important role in MS and may influence risk, rate, and severity of the disease (Coo and Aronson 2004).

Various viral and bacterial infections have been suggested as environmental triggers of MS, but no one of them could yet be defined so far as the reliable cause of MS. Only very recently, some reports are indicating the Epstein-Barr virus as the leading trigger of MS (Lünemann, Kamradt, Martin and Münz 2007; Serafini, Rosicarelli, Franciotta, Magliozzi, Reynolds, Cinque, Andreoni, Trivedi, Salvetti, Faggioni and Aloisi 2007).

Another important environmental factor to take into account is diet. It is indeed commonly accepted, but not yet demonstrated, that nutrition is one of the possible environmental factors involved in the pathogenesis of MS, but the role of nutrition in MS treatment or prevention is at present largely disregarded (Payne 2001; Coo and Aronson 2004; Schwarz and Leweling 2005).

Food can influence the course of MS either by exacerbating the autoimmune inflammatory processes involved in MS pathogenesis or by counteracting them. Accordingly, the MS patients should avoid the intake of food containing molecules that are potentially toxic for myelin, and should prefer healthy food containing those molecules that can improve their well-being by lowering the extent of inflammation.

To understand how food molecules influence the autoimmune inflammatory processes in MS, we should briefly recall the protagonists present in the scenario of the attack on myelin and its degradation.

Autoreactive T cells, directed toward the myelin sheath, pass through the BBB and together with macrophages and microglial cells degrade the myelin sheath (McFarland and Martin 2007). Another pattern of both myelin and oligodendrocyte damage is mediated by antibodies or complement activation

(Archelos and Hartung 2000). In addition to demyelination, axonal damage occurs very early in the course of the disease (Trapp 2004).

Among the molecules involved in the inflammatory processes leading to blood brain barrier (BBB) and myelin breakdown, there are the reactive oxygen species (ROS), also called free radicals; the matrix metalloproteinases (MMPs), in particular MMP-9, also called gelatinase B; the eicosanoids, such as prosta-glandins, and tromboxanes; as well as cytokines, such as TNF-$\alpha$ and IFN-$\gamma$. The idea is that the levels of the molecules linked to inflammation could also be controlled by careful dietary choices.

The candidate autoantigens in MS are all the major myelin proteins: proteo-lipid protein (PLP) and its isoform DM20, myelin basic protein (MBP), cyclic nucleotide phosphodiesterase (CNPase) and, above all, a minor myelin protein, myelin oligodendrocyte glycoprotein (MOG). All these proteins show an auto-immune response in MS patients and induce the Experimental Allergic Ence-phalomyelitis (EAE), the animal model of MS (Gold, Linington and Lassmann 2006).

## 4.2 Diet and MS

### 4.2.1 The Role of Animal Fat

When considering the possible relationships between food and MS, the main dietary factors that have been considered most frequently since the beginning for their deleterious influence on the disease are saturated fatty acids, which are mainly of animal origin. Saturated animal fat is found in whole milk, butter, cheese, meat, salami, and in other food of animal origin.

As early as 1950, Swank suggested that the consumption of saturated animal fat is directly related to the frequency of MS. In 2003a, Swank and Goodwin reported that restriction of saturated fat induces remission of the disease and produces beneficial effects in MS patients. These effects have been ascribed (Swank and Goodwin 2003b) to the fact that saturated fat forms very large and rigid aggregates, which may be not capable of entering the smallest capillaries. The damage due to the obstructions of the capillaries may contribute to MS and may be a factor favoring aging and other diseases such as diabetes and heart disease. In addition, saturated fats decrease membrane fluidity, lead to the synthesis of cholesterol, and favor the formation of TNF-$\alpha$.

### 4.2.2 Milk and MS

Environmental (either microbial or dietary) factors may be associated with autoimmunity and myelin breakdown by molecular mimicry, i.e., the amino acid homology between the autoantigen and microbial or dietary peptides (Wekerle and Hohlfeld 2003). According to this hypothesis, molecular mimicry may disrupt immunological self-tolerance to CNS myelin antigens in genetically susceptible individuals. The best example of potential molecular mimicry between myelin autoantigens and dietary proteins is given by cow's milk.

The hypothesis of a link between milk consumption and multiple sclerosis has been taken into consideration since the mid of 1970s (Butcher, 1976, 1986). Later epidemiological studies gave some support to this hypothesis (Malosse, Perron, Sasco and Seigneurin, 1992; Malosse and Perron, 1993; Lauer, 1997). The milk molecules that could represent the factor linked to the prevalence of multiple sclerosis in some populations and be responsible for toxicity toward myelin are the proteins of the milk fat globule membrane (MFGM) (Riccio 2004).

The MFGM proteins account for only 1–2% of the total protein fraction and have no nutritional value. Their role is rather structural since they are required for the formation of the globule membrane and the protection of milk lipids from oxidation. Another important role of the MFGM proteins is informational. With the MFGM proteins, the breastfeeding mother delivers passive immunity to the newborn baby and supplies the necessary instructions for the development of intestinal mucosa, the correct activity of the metabolic machinery, and the development of the immune and nervous systems.

The case of cow's milk in human nutrition is different. Cow's milk has become one of the most common nutrients in the western world, since it is considered to be a complete and safe food, but its diffusion seems to be correlated with MS. The question arises, as to why?

Milk MFGM Proteins and Multiple Sclerosis

Composition of the protein coat around the fat globule is rather complex: about 40 different proteins, with a molecular mass from 15,000 to 240,000 Da, can be found in the membrane (Mather 2000). There are a total of only eight most representative MFGM proteins. Six of them are glycoproteins: mucin 1 (MUC-1), xanthine oxidoreductase (XOD), mucin 15 (MUC-15), CD 36 (PAS IV), butyrophilin (BTN), and PAS VI/VII (lactadherin). The two unglycosilated proteins are adipophilin (ADPH) and the fatty acid binding protein (FABP).

The protein which is most frequently suspected of the association with MS is butyrophilin (BTN), the most representative MFGM protein. BTN is a 56,560-Da transmembrane glycoprotein with a pI of 4.96, and represents some 40% of the total MFGM protein fraction. BTN is very similar to MOG. Indeed, BTN belongs to the Ig superfamily and shows more than 50% amino acid homology with a corresponding domain of MOG. In particular, MOG and BTN share a core region (aa 74–86) representing cross-reactive T cell epitopes in which 8 of the 12 residues are identical (Stefferl, Schubart, Storch, Amini, Mather, Lassmann and Linington 2000).

MOG can induce EAE in many experimental animals (Johns and Bernard, 1999; Stefferl et al., 2000). The findings that BTN not only inhibits the MOG-induced EAE but also, when injected alone into animals, induces inflammatory responses in the CNS and stimulates in vitro MOG-specific T cell responses are of outstanding interest (Stefferl et al. 2000; Mana, Goodyear, Bernard, Tomioka, Freire-Garabal and Linares 2004).

Antibody cross-reactivity between MOG and BTN in MS has recently been observed (Kennel De March, De Bouwerie, Kolopp-Sarda, Faure, Bene and Bernard 2003; Guggenmos, Schubart, Ogg, Andersson, Olsson, Mather and Linington 2004). Thus, both a cross-reactive T or B cell response could operate as an autoaggressive one.

The possible deleterious impact of MFGM proteins on human health is confirmed by the relationship between MFGM proteins and two other diseases: Autism (Vojdani, Campbell, Anyanwu, Kashanian, Bock and Vojdani 2002) and coronary heart disease (CHD) (Moss and Freed 2003). Antibodies against BTN and other MFGM proteins have been detected in these diseases.

## Concluding Remarks on Milk and MS

In conclusion, the MFGM proteins have little nutritional value but might have deleterious effects on health. Furthermore, they are mainly associated to milk saturated fat, another possibly deleterious dietary component. On these grounds, we have introduced the concept that the consumption by MS patients of the MFGM proteins should be discouraged (Riccio 2004). As assessed in our laboratory, both MFGM proteins and fat are absent in skimmed milk (unpublished) and this could be a convenient source of nutrients for the MS patient instead of whole milk. On the contrary, in a report by Spitsberg (2005), the intake of MFGM proteins and milk fat was recommended for their potential nutraceutical value. However, the reports on the relationship between human health and milk do not substantiate such an assertion at present.

## 4.3 Gut, Food Absorption, and MS

The question arises as to how food can induce autoimmunity. In animal experiments, as in the study with BTN to induce EAE, the antigen is injected and adjuvants of immunity are added to produce an immune response.

The case of nutrition is different. Every day, with our food we introduce orally a very large number of foreign molecules into our bodies, and dietary components represent potential etiological factors in the pathogenesis of allergic and autoimmune diseases. However, the food molecules that encounter our immune system at the mucosal level are usually tolerated (Strobel 2002). The peptides arrive at the mucosal surface of the gut, cross it, and are presented to T cells in the periphery. Usually, this leads to so-called oral tolerance, which is the suppression of inflammatory responses to food antigens. There are, however, some exceptions.

Oral tolerance depends indeed on several factors that can be related to genetics, dysfunction or immaturity of the immune system, age, environmental factors including microbial infections and population of the intestinal micro-flora population, and as well as the amount quantity and frequency of food antigen uptake. It has been calculated that about 5–6% of young children suffer from food allergies, with half of them suffering from milk allergies. In particular, an association between early exposure to cows' milk and the development of cows' milk protein intolerance has been reported (Strobel 2002). In this respect, the time at which the milk antigen comes in contact with the immune system is important. In fact, if this is not yet completely developed, the milk antigen may prime the immune system rather than induce tolerance. If feeding is interrupted and taken up again later in life, the food antigen could be recognized by the immune system as a non-self-peptide and promote disease susceptibility.

Some food-linked diseases might be ascribed simply to the fact that in some genetically unsuited individuals some food proteins are not properly broken down, and the derived peptides can pass through the gut and activate the immune system, causing allergies or other autoimmune diseases. Reichelt and Jensen (2004) reported increased levels of IgA and IgG antibodies in serum against gluten and gliadin, and possibly also for casein, in MS patients. This finding may indicate a increased gut permeability to certain proteins in MS patients. If confirmed, the increased gut permeability in MS patients could represent an important mechanism of delivery of peptides and proteins with potential molecular cross-reactivity with self-proteins.

### 4.3.1  Dietary Remarks on Gut and Malabsorption

To ameliorate gut function, in addition to exercise, the nutritionist should recommend the consumption of vegetables and yogurts containing probiotic microorganisms and dietary fibers. A good choice among fibers could be inulin, a soluble fiber (Gibson, Beatty, Wang and Cummings 1995; Kruse, Kleessen and Blaut 1999). If the hypothesis of the increased gut permeability in MS patients is correct, the consumption of potatoes should also be reduced since potatoes contain glycoalcaloids such as solanine and chaconine that can disrupt the membrane integrity and lead to inflammation (Patel, Schutte, Sporns, Doyle, Jewel and Fedorak 2002).

## 4.4  Oxidative Stress and Dietary Antioxidants

The rationale for the use of natural antioxidants in MS is based on the finding that oxidative stress, due to the generation of reactive oxygen species (ROS), is one of the most important components involved in inflammation and neuronal damage (Gilgun-Sherki, Melamed and Offen 2004; Steiner, Haughey, Li, Venkatesan, Anderson, Reid, Malpica, Pocernich, Butterfield and Nath 2006). On these grounds, the application of appropriate doses of antioxidants could be very useful to restore the right balance between ROS generation and their destruction.

Moreover, oxidative stress is related to higher expression of MMPs, and preliminary data indicate that antioxidants such as omega-3 polyunsaturated fatty acids (n-3 PUFA) inhibit the expression of MMPs in primary cultures of rat microglial cells (Liuzzi, Latronico, Rossano, Viggiani, Fasano and Riccio 2007).

In this context, it should be taken into account that besides well-known antioxidants such as vitamin E. vitamin C, alpha-lipoic acid (ALA), glutathione, and others, there are a number of lesser known natural antioxidants, present in fruits, vegetables and sea food, that could be very useful in lowering oxidative stress and MMP expression and activity. They are polyphenols, flavonoids, anthocyanins, tocopherols, quinones, as well as other vitamins, and oligo-elements, namely: quercetin, lycopene, resveratrol, hydroxytyrosol, n-3 PUFA, and selenium. They often have a complementary activity as antioxidants and radical scavengers.

Quercetin belongs to the group of flavonoids and has anti-inflammatory, immuno-modulating, and antiviral properties. Reduces capillary fragility and is found in onion, apple, citrus, and wine. It inhibits myelin phagocytosis by blocking the ROS released from the macrophages (Hendriks, de Vries, van der Pol, van den Berg, van Tol and Dijkstra 2003) and inhibits the expression of inflammatory cytokines (Min, Choi, Bark, Son, Park, Lee, Park, Park, Shin and Kim 2007). Furthermore, quercetin inhibits angiogenesis (Sagar, Yance and Wong 2006), reduces the neutrophil-dependent inflammation (Kanashiro, Souza, Kabeya, Azzolini and Lucisano-Valim 2007), can have a neuroprotective effect (Sharma, Mishra, Ghosh, Tewari, Basu, Seth and Sen 2007), and apparently is not toxic (Harwooda, Danielewska-Nikiela, Borzellecab, Flammc, Williamsd and Linese 2007).

Catechins are polyphenols found in green tea. They have anti-inflammatory and anti-carcinogenic activity (Friedman 2007), inhibit the activity of MMP-2, MMP-9, and MMP-12 (Demeule, Brossard, Pagé, Gingras and Béliveau 2000), and inhibit intestinal absorption of lipids (Koo and Noh 2007).

Lycopene is a carotenoid that is found in tomato, watermelon, and pink grapefruit. It is two times better than beta carotene and 100 times better than vitamin E as an antioxidant. It protects against radioactivity, cancer, and from oxidized LDL cholesterol (Bhuvaneswari and Nagini 2005; Rao and Rao 2007).

Resveratrol is found in red wine, chocolate, peanuts, and black grapes. It could be useful in order to complete the range of redox potentials needed in our opinion for the protection against oxidative stress. It has a synergic effect in combination with quercetin, vitamin E, vitamin C, and selenium in reducing the inflammatory processes related to the formation of thromboxanes, leukotrienes, and prostaglandins from arachidonic acid (Candelario-Jalil, de Oliveira, Gräf, Bhatia, Hüll, Muñoz and Fiebich 2007; Das, S. and Das, D.K. 2007). Resveratrol has a neuroprotective effect (Bureau, Longpré and

Martinoli 2007) and ameliorates EAE (Singh, Hegde, Hofseth, Nagarkatti and Nagarkatti 2007).

Hydroxytyrosol is the main antioxidant found in olive oil and a very efficient scavenger of free radicals (Rietjens, Bast and Haenen 2007).

n-3 PUFA from fish oil have been studied recently in our laboratories (Liuzzi et al. 2007). They inhibit the expression of MMP-9 in rat microglia.

Other useful compounds with antioxidant activity are: alpha-lipoic acid (ALA), melatonin, vitamin E, selenium, vitamin C, and glutathione.

ALA has an immunomodulating effects, inhibits T cell migration, and is effective in the treatment of EAE (Morini, Roccatagliata, Dell'Eva, Pedemonte, Furlan, Minghelli, Giunti, Pfeffer, Marchese, Noonan, Mancardi, Albini and Uccelli 2004).

Vitamin D3, together with sunshine, is the most important vitamin for the treatment of autoimmune diseases and MS (Munger, Zhang, O'Reilly, Hernan, Olek, Willett and Ascherio 2004; Arnson, Amital and Shoenfeld 2007; Pedersen, Nashold. Spach and Hayes 2007; Smolders, Damoiseaux, Menheere and Hupperts 2008). Vitamin B12 is useful in EAE.

### 4.4.1  Unsaturated Fatty Acids from Vegetables

Unsaturated vegetable oils are the alternative to saturated animal fat. Olive oil, in particular, should be preferred to seed oils for its optimal ratio between saturated and unsaturated fat. Olive oil contains the antioxidant hydroxytyrosol and the omega-9 (n-9) monounsaturated oleic acid. Vegetable oils contain also the essential fatty acids (FA), linoleic acid (n-6) and linolenic acid (n-3). In inflammatory diseases such MS, n-3 FA should prevail over n-6 FA. Indeed, the latter leads to the production of arachidonic acid, which is the precursor of eicosanoids such as prostaglandins, leukotrienes, and thromboxane, which are the molecules involved in inflammatory processes. Synthesis of these eicosanoids is favored not only by insulin and inhibited by aspirin, but also by the n-3 long chain polyunsaturated fatty acids (PUFA), i.e., eicosapentaenoic acid (EPA) and docosahexaenoic acid (DHA).

### 4.4.2  PUFA from Sea Food and MS

DHA is present at high concentration in the brain, but its levels decrease dramatically in MS patients. Both EPA and DHA are found in fish oil and show remarkable anti-inflammatory, anti-thrombotic, and immunomodulating activities, which are comparable with statins. They also exert a number of neuroprotective effects and have a therapeutic value in several neurological diseases (Farooqui, Ong, Horrocks, Chenc and Farooqui 2007). They inhibit the expression of MMP-9 in LPS-activated rat microglial cells (Latronico et al. 2007), in a similar way as to IFN-β does (Liuzzi et al. 2004a). Like IFN- β, which is at present one of the best therapies for MS, EPA, and DHA, also inhibits the formation of IFN-γ, which, as MMP-9, is involved in myelin breakdown. Seeds oils, from sunflower, corn, soy bean, and sesame, contain more n-6 FA than n-3 FA and their assumption should be limited in order to limit the extent of eicosanoid production. Coconut oil has a high content of saturated FA and is not indicated for MS patients. It is useful to know that the intake of PUFA should be integrated with antioxidants. Hydrogenated FAs, which are found in margarine and other treated fat, are trans FA and not cis FA, like the natural ones, and interfere with PUFA metabolism

### 4.4.3 The Role of Insulin in MS

Eating too much increases the production of free radicals and inflammation. It also increases insulin levels and insulin activates the enzyme 5-delta-desaturase (inhibited by EPA), which forms arachidonic acid (AA) from diomo-gamma-linoleic acid (DGLA). AA forms the eicosanoids involved in inflammation.

## 5 Conclusions

In conclusion, the diet of an MS patient should follow the indications of a nutritionist based on the following dietary recommendations. Balanced assumption consumption of carbohydrates (40%), proteins (30%), and fat (30%), at each meal, five times a day for a total of about 1600 Kcal.

Consumption of red meat, eggs, and animal fat (also including butter, fat cheese, and salami) should be limited, as should sugar and sweets (better chocolate), and sweetened drinks.

Consumption of reduced portions of pasta, rice and potatoes, chicken and turkey meats, ham, and pizza are recommended.

Vegetables, legumes, fruit, fish, mollusks, crustaceans, mushrooms, and soy should be preferred. Consumption of skimmed milk and yogurt instead of whole milk is recommended. Other dietary recommendations are: fibers, water, green tea, coffee, vitamins, oligo-elements, and fish oil (10–30 g/day)

Physical exercise and sunshine are beneficial.

**Acknowledgments** Funding by the Italian Foundation for Multiple Sclerosis (FISM) to P.R. 2004/R/16 and to G.M.L. 2005/R/13. The present studies were carried out within the framework and the activities of the MARIE Network of the European Science Foundation on Myelin Structure and Its Role in Autoimmunity, 2004–2006.

## References

Agrawal, S.M., Lau, L., and Yong, V.W. (2008) MMPs in the central nervous system: Where the good guys go bad. Semin- Cell Dev Biol 19, 42–51
Allan, J.A., Docherty, A.J., Barker, P.J., Huskisson, N.S., Reynolds, J.J., and Murphy, G. (1995) Binding of gelatinases A and B to type-I collagen and other matrix components. Biochem J 309, 299–306.

Archelos, J.J., and Hartung, H.-P. (2000). Pathogenetic roles of autoantibodies in neurological diseases. Trends Neurosci 23, 317–327.

Arnson, Y., Amital, H., and Shoenfeld, Y. (2007). Vitamin D and autoimmunity: new aetiological and therapeutic considerations. Ann Rheum Dis 66,1137–1142.

Bhuvaneswari, V., and Nagini, S. (2005) Lycopene: a review of its potential as an anticancer agent. Curr Med Chem Anticancer Agents 5, 627–635.

Bruck, W., and Stadelmann, C., (2005) The spectrum of multiple sclerosis: new lessons from pathology. Curr Opin Neurol 18, 221–4.

Bureau, G., Longpré F., and Martinoli, M.G. (2007) Resveratrol and quercetin, two natural polyphenols, reduce apoptotic neuronal cell death induced by neuroinflammation. J Neurosci Res 86, 403–410

Butcher, J. (1976) The distribution of multiple sclerosis in relation to the dairy industry and milk consumption. N Z Med J 83, 427–430.

Butcher, P.J. (1986) Milk consumption and multiple sclerosis: an etiological hypothesis. Med Hypotheses 19, 169.

Calvo, P., Gouritin, B., Villarroya, H., Eclancher, F., Giannavola, C., Klein, C., Andreux, J.P., and Couvreur, P. (2002)Quantification and localization of PEGylated polycyanoacrylate nanoparticles in brain and spinal cord during experimental allergic encephalomyelitis in the rat. Eur J Neurosci 15(8),1317–26

Campbell, R.B., Fukumura, D., Brown, E.B., Mazzola, L.M., Izumi, Y., Jain, R.K., Torchilin, V.P., and Munn, L.L. (2002). "Cationic charge determines the distribution of liposomes between the vascular and extravascular compartments of tumors." Cancer Research 62, 6831–6836.

Candelario-Jalil, E., de Oliveira, A.C., Gräf, S., Bhatia, H.S., Hüll, M., Muñoz, E., and Fiebich, B.L. (2007) Resveratrol potently reduces prostaglandin E2 production and free radical formation in lipopolysaccharide-activated primary rat microglia. J Neuroinflammation 4, 25.

Cauwe, B., Van den Steen, P.E., and Opdenakker, G. (2007) The biochemical, biological, and pathological kaleidoscope of cell surface substrates processed by matrix metalloproteinases. Crit Rev Biochem Mol Biol 42, 113–85.

Charil, A., and Filippi, M. (2007) Inflammatory demyelination and neurodegeneration in early multiple sclerosis. J Neurol Sci 259, 7–15.

Coo, H., and Aronson, K.J. (2004) A Systematic Review of Several Potential Non-Genetic Risk Factors for Multiple Sclerosis. Neuroepidemiology 23, 1–12.

Correale, J., and Villa, A. (2007) The blood-brain-barrier in multiple sclerosis: functional roles and therapeutic targeting. Autoimmunity 40, 148–60.

Cuzner, M.L., Gveric, D., Strand, C., Loughlin, A.J., Paemen, L., Opdenakker, G., and Newcombe, J. (1996) The expression of tissue-type plasminogen activator, matrix metalloproteases and endogenous inhibitors in the central nervous system in multiple sclerosis: comparison of stages in lesion evolution. J Neuropathol Exp Neurol 55, 1194–204.

Das, S., and Das, D.K. (2007) Anti-inflammatory responses of resveratrol. Inflamm Allergy Drug Targets 61, 68–73.

Demeule, M., Brossard, M., Pagé, M., Gingras, D., and Béliveau, R. (2000) Matrix metalloproteinase inhibition by green tea catechins. Biochim Biophys Acta 1478, 51–60.

Eichhorn, M.E., Luedemann, S., Strieth, S., Papyan, A., Ruhstorfer, H., Haas, H., Michaelis, U., Sauer, B., Teifel, M., Enders, G., Brix, G., Jauch, K.W., Bruns, C.J., and Dellian, M. (2007). "Cationic Lipid Complexed Camptothecin (EndoTAG((R))-2) Improves Antitumoral Efficacy by Tumor Vascular Targeting." Cancer Biol Ther 6(6).

Farooqui, A.A., Ong, W.Y., Horrocks, L.A., Chenc, P., Farooqui T. (2007) Comparison of biochemical effects of statins and fish oil in brain: The battle of the titans. Brain Res Rev 56, 443–471.

Floris, S., Blezer, E.L., Schreibelt, G., Dopp, E., van der Pol, S.M., Schadee-Eestermans, I.L., Nicolay, K., Dijkstra, C.D., and de Vries, H.E. (2004). Blood-brain barrier permeability and monocyte infiltration in experimental allergic encephalomyelitis: a quantitative MRI study. Brain 127, 616–27.

Friedman, M. (2007) Overview of antibacterial, antitoxin, antiviral, and antifungal activities of tea flavonoids and teas. Mol Nutr Food Res 51, 116–34.

Gibson, G.R., Beatty, E.R., Wang, X., and Cummings, J.H. (1995) Selective stimulation of bifidobacteria in the human colon by oligofructose and inulin. Gastroenterology 108, 975–982

Gilgun-Sherki, Y., Melamed, E., and Offen, D. (2004) The role of oxidative stress in the pathogenesis of multiple sclerosis: the need for effective antioxidant therapy. J Neurol 251, 261–268

Gijbels, K., Masure, S., Carton, H., and Opdenakker, G. (1992) Gelatinase in the cerebrospinal fluid of patients with multiple sclerosis and other inflammatory neurological disorders. J Neuroimmunol 41, 29–34.

Gilli, F., Bertolotto, A., Sala, A., Hoffmann, F., Capobianco, M., Malucchi, S., Glass, T., Kappos, L., Lindberg, R.L., and Leppert, D. (2004) Neutralizing antibodies against IFN-beta in multiple sclerosis: antagonization of IFN-beta mediated suppression of MMPs. Brain 127, 259–268.

Gold, R., Linington, C., and Lassmann, H. (2006). Understanding pathogenesis and therapy of multiple sclerosis via animal models: 70 years of merits and culprits in experimental autoimmune encephalomyelitis research. Brain 129, 1953–1971.

Gomez, D.E., Alonso, D.F., Yoshiji, H., and Thorgeirsson, U.P. (1997) Tissue inhibitors of metalloproteinases: structure, regulation and biological functions. Eur J Cell Biol 74, 111–122

Guggenmos, J., Schubart, A.S., Ogg, S., Andersson, M., Olsson, T., Mather, I.H., and Linington C. (2004) Antibody cross-reactivity between myelin oligodendrocyte glycoprotein and the milk protein butyrophilin in multiple sclerosis. J Immunol 1;172, 661–8.

Haas, H., and Riccio, P. (2007). Cationic colloidal carriers for delivery of active agents to the blood-brain barrier in the course of neuroinflammatory diseases. Patent filed 31 Oct. 2007 with the application number PCT/EP2007/009460,

Harris, J.O., Frank, J.A., Patronas, N., McFarlin, D.E., and McFarland, H.F. (1991) Serial gadolinium-enhanced magnetic resonance imaging scans in patients with early, relapsing-remitting multiple sclerosis: implications for clinical trials and natural history. Ann Neurol 29, 548–55.

Harwooda, M., Danielewska-Nikiela, B., Borzellecab, J.F., Flammc, G.W., Williamsd, G.M., and Linese, T.C. (2007) A critical review of the data related to the safety of quercetin and lack of evidence of in vivo toxicity, including lack of genotoxic/carcinogenic properties. Food Chem Toxicol 45, 2179–2205.

Hendriks, J.J., de Vries, H.E., van der Pol, S.M., van den Berg, T.K., van Tol, E.A., and Dijkstra, C.D. (2003) Flavonoids inhibit myelin phagocytosis by macrophages; a structure-activity relationship study. Biochem Pharmacol 65, 877–885.

Johns, T.G., and Bernard, C.C.A. (1999) The structure and function of myelin oligodendrocyte glycoprotein. J Neurochem 72, 1–9.

Kalra, A.V., and Campbell, R.B. (2006) Development of 5-FU and doxorubicin-loaded cationic liposomes against human pancreatic cancer: Implications for tumor vascular targeting. Pharm Res 23, 2809–2817.

Kanashiro, A., Souza, J.G., Kabeya, L.M., Azzolini, A.E., and Lucisano-Valim, Y.M. (2007) Elastase release by stimulated neutrophils inhibited by flavonoids: importance of the catechol group. Z. Naturforsch. [C]. 62, 357–61.

Kennel De March, A., De Bouwerie, M., Kolopp-Sarda, M.N., Faure, G.C., Bene, M.C., Bernard, C.C. (2003) Anti-myelin oligodendrocyte glycoprotein B-cell responses in multiple sclerosis. J Neuroimmunol 135, 117–25.

Kirk, S., Frank, J.A., and Karlik, S. (2004) Angiogenesis in multiple sclerosis: is it good, bad or an epiphenomenon? J Neurol Sci 217, 125–30.

Koo, S.I.,, and Noh, S.K. (2007) Green tea as inhibitor of the intestinal absorption of lipids: potential mechanism for its lipid-lowering effect. J Nutr Biochem 18, 179–83.

Kouwenhoven, M., Ozenci, V., Gomes, A., Yarilin, D., Giedraitis, V., Press, R., and Link, H. (2001) Multiple sclerosis: elevated expression of matrix metalloproteinases in blood monocytes. J Autoimmun 16, 463–70.

Krasnici, S., Werner, A., Eichhorn, M. E., Schmitt-Sody, M., Pahernik, S. A., Sauer, B., Schulze, B., Teifel, M., Michaelis, U., Naujoks, K., and Dellian, M. (2003) Effect of the surface charge of liposomes on their uptake by angiogenic tumor vessels. Int J Cancer 105, 561–567.

Kruse, H.P., Kleessen, B., and Blaut, M. (1999) Effects of inulin on faecal bifidobacteria in human subjects. Br J Nutr 82, 375–382.

Latronico, T., Piscopo, V., Fasano, A,., Riccio, P., and Liuzzi, G.M. (2004) Inhibition of calpain activity and expression by interferon-beta in rat astrocytes: implication for MS pathogenesis and treatment. J Neuroimmunol (Suppl) 154, 122.

Latronico, T., Liuzzi, G.M., Riccio, P., Lichtner, M., Mengoni, F., D'Agostino, C., Vullo V., and Mastroianni, C.M. (2007) Antiretroviral therapy inhibits matrix metalloproteinase-9 from blood mononuclear cells of HIV-infected patients. AIDS 21, 677–84.

Lauer, K. (1997). Diet and multiple sclerosis. Neurology 49, S55.

Leppert, D., Waubant, E., Bürk, M.R., Oksenberg, J.R., and Hauser, S.L. (1996) Interferon beta-1b inhibits gelatinase secretion and in vitro migration of human T cells: a possible mechanism for treatment efficacy in multiple sclerosis. Ann Neurol 40, 846–52.

Leppert, D., Ford, J., Stabler, G., Grygar, C., Lienert, C., Huber, S., Miller, K.M., Hauser, S.L., and Kappos, L. (1998) Matrix metalloproteinase-9 (gelatinase B) is selectively elevated in CSF during relapses and stable phases of multiple sclerosis. Brain 121, 2327–2334.

Lindberg, R.L., De Groot, C.J., Montagne, L., Freitag, P., van der Valk, P., Kappos, L., and Leppert, D. (2001) The expression profile of matrix metalloproteinases (MMPs) and their inhibitors (TIMPs) in lesions and normal appearing white matter of multiple sclerosis. Brain 124, 1743–53.

Liuzzi, G.M., Trojano, M., Fanelli, M., Avolio, C., Fasano, A., Livrea, P., and Riccio, P. (2002) Intrathecal synthesis of matrix metalloproteinase-9 in patients with multiple sclerosis: implication for pathogenesis. Mult Scler 8, 222–8.

Liuzzi, G.M., Latronico, T., Fasano. A., Carlone, G., and Riccio, P. (2004a) Interferon-beta inhibits the expression of metalloproteinases in rat glial cell cultures: implications for multiple sclerosis pathogenesis and treatment. Mult Scler 10, 290–7.

Liuzzi, G.M., Mastroianni, C.M., Latronico, T., Mengoni, F., Fasano, A., Lichtner, M., Vullo, V., and Riccio, P. (2004b) Anti-HIV drugs decrease the expression of matrix metalloproteinases in astrocytes and microglia. Brain 127, 398–407.

Liuzzi, G.M., Latronico, T., Rossano, R., Fasano, A., and Riccio, P (2007) Inhibitory Effect of Polyunsaturated Fatty Acids on MMP-9 Release from Microglial Cells – Implications for Complementary Multiple Sclerosis Treatment. Neurochem Res 32, 2184–2193.

Lucchinetti, C., Bruck, W., Parisi, J., Scheithauer, B., Rodriguez, M., and Lassmann, H. (2000). Heterogeneity of multiple sclerosis lesions: implications for the pathogenesis of demyelination. Ann Neurol 47, 707–17.

Lühder, F., and Gold, R., 2007. Many roads lead to Rome: heterogeneity among encephalitogenic T cell clones. J Neuroimmunol 192, 1–2.

Lünemann, J.D., Kamradt, T., Martin, R., and Münz, C. (2007) Epstein-Barr Virus: Environmental Trigger of Multiple Sclerosis? J Virol 81, 6777–6784.

Malosse, D., Perron, H., Sasco, A., and Seigneurin, J.M. (1992) Correlation between milk and dairy product consumption and multiple sclerosis prevalence: a worldwide study. Neuroepidemiol 11, 304–312.

Malosse, D., and Perron, H. 1993. Correlation analysis between bovine populations, other farm animals, house pets, and multiple sclerosis prevalence. Neuroepidemiol 12, 15.

Mana, P., Goodyear, M., Bernard, C., Tomioka, R., Freire-Garabal, M., and Linares, D. (2004) Tolerance induction by molecular mimicry: prevention and suppression of experimental autoimmune encephalomyelitis with the milk protein butyrophilin. Int Immunol 16, 489–99.

Manninger, S.P., Muldoon, L.L., Nesbit, G., Murillo, T., Jacobs, P.M., and Neuwelt, E.A. (2005) An exploratory study of ferumoxtran-10 nanoparticles as a blood-brain barrier imaging agent targeting phagocytic cells in CNS inflammatory lesions. AJNR Am J Neuroradiol 26, 2290–300.

Mather, I.H. (2000). A review and proposed nomenclature for major proteins of the milk-fat globule membrane. J Dairy Sci 83, 203–247.

McFarland, H.F., and Martin, R. (2007) Multiple sclerosis: a complicated picture of autoimmunity. Nat Immunol. 8, 913–9.

McQualter, J.L., and Bernard, C.C. (2007) Multiple sclerosis: a battle between destruction and repair. J Neurochem 100, 295–306.

Merodio, M., Irache, J.M., Eclancher, F., Mirshahi, M., and Villarroya, H. (2000) Distribution of albumin nanoparticles in animals induced with the experimental allergic encephalomyelitis. J Drug Target 8, 289–303.

Michaelis, U., and Haas, H. (2006). Targeting of Cationic Liposomes to Endothelial Tissue. Liposome Technology, Volume III: Interactions of Liposomes with the Biological Milieu, Third Edition. Gregoriadis, G.

Min, Y.D., Choi, C.H., Bark, H., Son, H.Y., Park, H.H., Lee, S., Park, J.W., Park, E.K., Shin, H.I., and Kim, S.H. (2007) Quercetin inhibits expression of inflammatory cytokines through attenuation of NF-kappaB and p38 MAPK in HMC-1 human mast cell line. Inflamm Res 56, 210–215.

Morini, M., Roccatagliata, L., Dell'Eva, R., Pedemonte, E., Furlan, R., Minghelli, S., Giunti, D., Pfeffer, U., Marchese, M., Noonan, D., Mancardi, G., Albini, A., and Uccelli, A. (2004) Alpha-lipoic acid is effective in prevention and treatment of experimental autoimmune encephalomyelitis. J Neuroimmunol 148,146–53.

Moss, M., and Freed, D. (2003) The cow and the coronary: epidemiology, biochemistry and immunology. Int J Cardiology 87, 203–216.

Munger, K.L., Zhang, S.M., O'Reilly, E., Hernan, M.A., Olek, M.J., Willett, W.C., and Ascherio, A. (2004) Vitamin D intake and incidence of multiple sclerosis. Neurology 62, 60–5.

Nagase, H. (1997) Activation mechanisms of matrix metalloproteinases. Biol Chem 378, 151–160.

Nwariaku, F.E., Chang, J., Zhu, X., Liu, Z., Duffy, S.L., Halaihel, N.H., Terada, L., and Turnage, R.H. (2002). "The role of p38 map kinase in tumor necrosis factor-induced redistribution of vascular endothelial cadherin and increased endothelial permeability." Shock 18(1), 82–5.

Patel, B., Schutte, R., Sporns, P., Doyle, J., Jewel, L., and Fedorak, R.N. (2002) Potatoglycoalkaloids adversely affect intestinal permeability and aggravate inflammatory bowel disease. Inflamm Bowel Dis 8, 340–346.

Payne, A. (2001) Nutrition and diet in the clinical management of multiple sclerosis. J Hum Nutr Dietet 14, 349–357.

Pedersen, L.B., Nashold. F.E., Spach, K.M., and Hayes, C.E. (2007) 1,25-dihydroxyvitamin D3 reverses experimental autoimmune encephalomyelitis by inhibiting chemokine synthesis and monocyte trafficking. J Neurosci Res 85, 2480–2490.

Ran, S., Downes, A., and Thorpe, P.E. (2002) Increased exposure of anionic phospholipids on the surface of tumor blood vessels. Cancer Research 62, 6132–6140.

Ran, S., and Thorpe, P.E. (2002) Phosphatidylserine is a marker of tumor vasculature and a potential target for cancer imaging and therapy. Int J Radiation Oncol Biology Physics 54, 1479–1484.

Rao, A.V., and Rao, L.G. (2007) Carotenoids and human health. Pharmacol Res 55, 207–216.

Rausch, M., Hiestand, P., Baumann, D., Cannet, C., and Rudin, M. (2003). MRI-based monitoring of inflammation and tissue damage in acute and chronic relapsing EAE. Magn Reson Med 50, 309–14.

Rebeles, F., Fink, J., Anzai, Y., Maravilla, K.R. (2006) Blood-brain barrier imaging and therapeutic potentials. Top Magn Reson Imaging 17, 107–16.

Reichelt, K.L., Jensen, D. (2004) IgA antibodies against gliadin and gluten in multiple sclerosis. Acta Neurol Scand 110, 239–41.

Riccio, P. (2004) The proteins of the milk fat globule membrane in the balance. Trends Food Sci Technol 15, 458–461.

Rietjens, S.J., Bast A., and Haenen, G.R.M.M. (2007) New Insights into Controversies on the Antioxidant Potential of the Olive Oil Antioxidant Hydroxytyrosol. J Agric Food Chem 55, 7609–7614.

Rosenberg, G.A. (2002) Matrix metalloproteinases and neuroinflammation in multiple sclerosis. Neuroscientist 8, 86–95.

Rousseau, V., Denizot, B., Le Jeune, J.J. and Jallet, P. (1999). Early detection of liposome brain localization in rat experimental allergic encephalomyelitis. Exp Brain Res 125, 255–64.

Sagar, S.M., Yance, D., and Wong, R.K. (2006) Natural health products that inhibit angiogenesis: a potential source for investigational new agents to treat cancer-Part 1. Curr Oncol 13, 14–26.

Sharma, V., Mishra, M., Ghosh, S., Tewari, R., Basu, A., Seth, P., and Sen, E. (2007) Modulation of interleukin-1beta mediated inflammatory response in human astrocytes by flavonoids: implications in neuroprotection. Brain Res Bull 73, 55–63.

Schmidt, J., Metselaar, J.M., Wauben, M.H., Toyka, K.V., Storm, G., and Gold, R. (2003). Drug targeting by long-circulating liposomal glucocorticosteroids increases therapeutic efficacy in a model of multiple sclerosis. Brain 126, 1895–904.

Schmitt-Sody, M., Strieth, S., Krasnici, S., Sauer, B., Schulze, B., Teifel, M., Michaelis, U., Naujoks, K., and Dellian, M. (2003). Neovascular targeting therapy: paclitaxel encapsulated in cationic liposomes improves antitumoral efficacy. Clin Cancer Res 9, 2335–41.

Schwarz, S., and Leweling, H. (2005) Multiple sclerosis and nutrition. Mult Scler 11, 24–32.

Serafini B, Rosicarelli B, Franciotta D, Magliozzi R, Reynolds R, Cinque P, Andreoni L, Trivedi P, Salvetti M, Faggioni A, Aloisi F. (2007) Dysregulated Epstein-Barr virus infection in the multiple sclerosis brain. J Exp Med 204, 2899–912.

Singh, N.P., Hegde, V.L., Hofseth, L.J., Nagarkatti, M., and Nagarkatti P. (2007) Resveratrol (trans-3,5,4'-trihydroxystilbene) ameliorates experimental allergic encephalomyelitis, primarily via induction of apoptosis in T cells involving activation of aryl hydrocarbon receptor and estrogen receptor. Mol Pharmacol 72, 1508–21.

Smolders, J., Damoiseaux, J., Menheere, P., and Hupperts, R. (2008) Vitamin D as an immune modulator in multiple sclerosis, a review. J Neuroimmunol. 2008 Jan 3 [Epub ahead of print].

Spitsberg, V.L. (2005) Invited Review: Bovine Milk Fat Globule Membrane as a Potential Nutraceutical. J. Dairy Sci 88, 2289–2294.

Stefferl, A., Schubart, A., Storch, M., Amini, A., Mather, I.H., Lassmann, H., and Linington, C. (2000) Butyrophilin, a milk protein, modulates the encephalitogenic T cell response to myelin oligodendrocyte glycoprotein in experimental autoimmune encephalomyelitis. J Immunol 165, 2859–2865.

Steiner, J., Haughey, N., Li, W., Venkatesan, A., Anderson, C., Reid, R., Malpica, T., Pocernich, C., Butterfield, D.A., and Nath, A. (2006). Oxidative stress and therapeutic approaches in HIV dementia. Antioxid Redox Signal. 8, 2089–2100. Review.

Steinman L, and Gijbels, K (1994) Gelatinase B producing cells in multiple sclerosis lesions. J Cell Biochem 143.

Strobel, S. (2002) Oral tolerance, systemic immunoregulation, and autoimmunity. Ann NY Acad Sci 958, 47–58.

Stüve, O., Dooley, N.P., Uhm, J.H., Antel, J.P., Francis, G.S., Williams, G., and Yong, V.W. (1996) Interferon beta-1b decreases the migration of T lymphocytes in vitro: effects on matrix metalloproteinase-9. Ann Neurol 40, 853–863.

Swank, RL. (1950) Multiple sclerosis: A correlation of its incidence with dietary fat. Am J Med Sci 220, 421–430.

Swank, R.L., Goodwin, J.W. (2003a) Review of MS patient survival on a Swank low saturated fat diet. Nutrition 19, 161–165.

Swank, R.L., and Goodwin, J.W. (2003b) How Saturated Fats May Be a Causative Factor in Multiple Sclerosis and Other Diseases. Nutrition 19, 478.

Thurston, G., McLean, J.W., Rizen, M., Baluk, P., Haskell, A., Murphy, T.J., Hanahan, D., and McDonald, D.M. (1998). Cationic liposomes target angiogenic endothelial cells in tumors and chronic inflammation in mice. J Clin Invest 101,1401–13.

Torchilin, V.P. (2000). Drug targeting. Eur J Pharm Sci 11(2), S81–91.

Trapp, B.D. (2004) Pathogenesis of multiple sclerosis: the eyes only see what the mind is prepared to comprehend. Ann Neurol 55, 455–457.

Trojano, M., Avolio, C., Liuzzi, G.M., Ruggieri, M., Defazio, G., Liguori, M., Santacroce, M.P., Paolicelli, D., Giuliani, F., Riccio, P., and Livrea, P. (1999) Changes of serum sICAM-1 and MMP-9 induced by rIFNbeta-1b treatment in relapsing-remitting MS. Neurology 53, 1402–8.

Van den Steen, P.E., Dubois, B., Nelissen, I., Rudd, P.M., Dwek, R.A., and Opdenakker, G. (2002) Biochemistry and molecular biology of gelatinase B or matrix metalloproteinase-9 (MMP-9). Crit Rev Biochem Mol Biol 37, 375–536.

Visse, R., and Nagase, H. (2003) Matrix metalloproteinases and tissue inhibitors of metallo-proteinases: structure, function, and biochemistry. Circ Res 92, 827–839.

Vojdani A., Campbell A.W., Anyanwu E., Kashanian A., Bock K., Vojdani E. (2002) Antibodies to neuron-specific antigens in children with autism: possible cross-reaction with encephalitogenic proteins from milk, Chlamydia pneumoniae and Streptococcus group A' J Neuroimmunol 129, 168–177.

Wekerle, H., and Hohlfeld, R. (2003) Molecular mimicry in multiple sclerosis. N Engl J Med 349,185–6.

Werring, D.J., Brassat, D., Droogan, A.G., Clark, C.A, Symms, M.R., Barker, G.J., MacManus, D.G., Thompson, A.J., and Miller, D.H. (2000) The pathogenesis of lesions and normal-appearing white matter changes in multiple sclerosis – A serial diffusion MRI study. Brain 123, 1667–1676.

Yong, V.W., Chabot, S., Stüve, O., and Williams, G. (1998) Interferon beta in the treatment of multiple sclerosis: mechanisms of action. Neurology 51, 682–9.

# Glycoimmunomics of Human Cancer: Relevance to Monitoring Biomarkers of Early Detection and Therapeutic Response

Mepur H. Ravindranath

**Abstract** There are several classes of glycoantigens. This chapter will be restricted to one class of glycoantigens called sialylated glycosylceramides or gangliosides. They are sparsely distributed on the surface of normal cells, whereas densely distributed on the surface of tumor cells. The gangliosides are shed or released from tumor cells into tumor microenvironment or circulation. While they suppress cell-mediated immune functions and Th-1-type cytokines, they activate or induce the secretion of Th-2 cytokines (interleukin-6 and interleukin-10) by T-lymphocytes, monocytes, and macrophages. Both IL-6 and IL-10 activate B lymphocytes to induce the production of IgM antibodies against glycoantigens. Special class of B lymphocytes called CD5+ B cells may be involved in the production of IgM antibodies against glycoantigens. These B cells produce persistent IgM and do not undergo class-switching. The IgM may be polymeric without J chain. The primary role of these IgM could be to remove the shed and released glycoantigens from tumor cells and thereby reverse the immune suppression induced by the shed tumor glycoantigens. In gender-based cancers, ganglioside GD1a is a major ganglioside. GD1a induces IL-6 and IL-10 production and enhances IgM production by human peripheral blood mononuclear cells. These observations explain why the patients with organ-confined prostate cancer have significantly elevated titers of IgM against GD1a. No such increase in anti-GD1a IgM is seen in patients with advanced cancer or in normal and healthy volunteers. IgM antibodies against other cancer-associated gangliosides do not show such augmentation. Anti-GD1a IgM could be the first glycoimmunomic signal for an early event of tumorigenesis. A better understanding of the glycoimmunomics of tumor-associated ganglio-sides may enable developing early biomarkers for human cancer, therapeutic response marker to assess the response to treatment, and also to effectively formulate glycoimmunotherapy of human cancer.

M.H. Ravindranath
Pacific Clinical Research, Santa Monica, CA 90404, USA
e-mail: mepurravi@yahoo.com

A. Falus (ed.), *Clinical Applications of Immunomics*,
DOI: 10.1007/978-0-387-79208-8_11, © Springer Science+Business Media, LLC 2009

**Keywords** Glycoimmunomics · Glycoantigens · Glycosylceramides · Gangliosides · Tumor necrosis · Immunogenicity · Immunosuppression · T-cell independent-antigens · IgM production · Natural antibodies · Autoantibodies · Siglec · Prostate cancer · Glycol-biomarkers · Anti-GD1a IgM · Anti-GT1b IgM · Early biomarkers · Therapeutic response biomarkers

# 1 Introduction

Cancer cells express a variety of cell surface molecules, which include proteins, glycoproteins, proteoglycans, glycolipoproteins, lipoproteins, lipids, glycolipids, and long-chain sugar residues. They are referred to as tumor-associated antigens. They may not be expressed by the progenitor cells that gave rise to cancer cells, but may be minimally expressed by other normal human cells. For example, neoplastic transformation of melanocytes results in melanoma. Human melanoma-associated antigen, MAGE, is not expressed by melanocytes, whereas MAGE-gene family proteins are expressed in the mitotic spermatogonia (germ cells), primary spermatocytes, and placenta, suggesting that the neoplastic transformation may trigger genes silent in the differentiated normal progenitor cells (Bodey, Kaiser and Siegel 2004). Similarly, in the case of human hepatocellular carcinoma, tumor progenitor cells express glypican-3 (GPC-3) only in the embryonic stage. GPC3 expression tends to reappear with malignant transformation (Nakatsura, Komori, Kubo, Yoshitake, Senju, Katagiri, Furukawa, Ogawa, Nakamura and Nishimura 2004).

Tumor-associated glycoantigens befit the term "Glycocalyx" because they constitute clusters of sugar chain residues on the cell surface. These glycoresidues are not only restricted to glycoproteins and proteoglycans but also include glycosyl-ceramides exposed on the outer layer of bilayered lipid membrane. These sugar residues are considered as glycoantigens when recognized by cellular or humoral immune system. The functional roles of cell surface glycoantigens are poorly understood although it is inferred that they interact with free and cell-bound macromolecules of the tumor microenviroment. The interaction of the glycoantigens with the immune system is unique and differs from that of proteins and peptides, and therefore categorized as the Glycoimmunomics.

# 2 Diversity of Glycoantigens

There are several classes of glycoantigens. They include

**(1) Proteoglycans**: They contain at least one core protein, with tandem repeats of polypeptides but with more (up to tens or hundreds) carbohydrate chains called glycosaminoglycans. The major heparin sulfate proteoglycan (HSPG) families at the cell surface are the four transmembrane syndecans (syndecan-1 through -4) and the five glypican family members (glypican-1 through -5) that are linked to the cell membrane via a phosphatidyl-inositol anchor (Blackhall, Merry, Davies and Jayson 2001). The heparan sulfate proteoglycans (HSPGs) play diverse roles in tumor biology and immune responses (Selvan, Ihrcke, and Platt 1996) by mediating adhesion, migration, and cellular responses to mitogenic and angiogenic growth factors. They have been proposed as critical regulators of tumor invasion and metastasis (Sanderson 2001). For example, in normal ovary, all of these HSPGs with the exception of syndecan-1 are expressed; on the other hand, in ovarian adenocarcinoma only syndecan-1 is expressed (Davies, Blackhall, Shanks, David, McGown, Swindell, Slade, Martin-Hirsch, Gallagher and Jayson 2004)

**(2) Glycoproteins**: They are proteins with carbohydrate attached to them. The attachment is a covalent linkage to:

- the hydroxyl (-OH) group of serine or threonine - called **"O-linked"**;
- the amino ($-NH_2$) group of asparagine - called **"N-linked"**.

The carbohydrate consists of short, usually branched chains of

- simple sugars (e.g., glucose, galactose),
- amino sugars (e.g., *N*-acetylglucosamine), and
- acidic sugars (e.g., sialic acid).

A typical example of tumor-associated glycoproteins with abundant sugar chains are referred to as MUC-series (Hanisch and Müller 2000)

**(3) Glycolipids**: In glycolipids, one or more monosaccharides are bound by a glycosidic linkage to a hydrophobic moiety such as acylglycerol, a sphingoid, a ceramide (*N*-acylsphingoid), or a prenyl phosphate. They may include glycoglycerolipid (lipids containing one or more glycerol residues), glycosphingolipid (lipids containing at least one sugar residue and either a sphingoid or a ceramide), neutral and acidic glycosphingolipids. Acid glycosphingolipids may include sialoglycosphingolipids (gangliosides), urono-glycosphingolipids, sulfoglycosphingolipids, phosphoglycosphingo-lipids, phosphonoglycosphingo lipids, containing 2-aminoethyl hydroxyphosphoryl groups, glycophosphatidylinositol lipids, which contain saccharides glycosidically linked to the inositol moiety of phosphatidylinositols or inositol residues.

This review on glycoimmunomics will be restricted to one class of glycoantigens, called sialylated glycosylceramides or gangliosides.

# 3  Gangliosides

## 3.1  Family of Glycoantigens

The term "ganglioside" is coined by Klenk (1942) for sialylated glycosphingolipids. Uniquely, they are abundant in neural tissues and unusually overexpressed by malignant extraneural tissues. On the surface of normal cells they are sparsely distributed, whereas on the tumor cell surface they exist in clusters. Structure, nomenclature, and tissue distribution of the gangliosides have been reviewed (Wiegandt 1985). They are amphiphilic molecules, with atomic mass units (AMU) ranging from 1300 to 2500. They have a hydrophilic head group of two or more sugars (glucose and/or galactose, neuraminic acid [sialic acid] with or without $N$-acetyl galactosamine or $N$-acetyl glucosamine), and a hydrophobic tail group of ceramide (sphingosine and a long-chain fatty acid). Sugar chain elongation of gangliosides involves a series of gene-specific glycosyltransferases (Basu, De, Das, Kyle, Chon, Schaeper and Basu 1987). The ganglioside head group of oligosaccharide chain extends up to 2.5 nm from the membrane/water interface. Its orientation on the cell surface depends on the nature of the tail groups, which are capable of rapid lateral and rotational diffusion. The chain length of fatty acids, the number of double bonds, and hydroxylation of fatty acids may differ for a particular ganglioside between normal and neoplastically transformed cells (Ladisch , Sweeley, Becker and Gage 1989). Gangliosides may not be evenly distributed on the cell surface, but rather exist in clusters (micro-domain of glycophingolipid) (Hakomori 1998, 2000, 2003, 2007, Hakomori, Handa, Iwabuchi, Yamamura and Prinetti 1998).

In the tumor microenvironment, in the body fluids or in the aqueous media, the gangliosides aggregate to form irregular micelles. On the other hand, in ethanol, the micelle formation of gangliosides requires a higher critical micellar concentration (CMC) and remains as independent micelles at lower concentration. This enables to simulate a monolayer or cluster on the polystyr-ene plates. This principle is employed to coat the microtiter plates for measuring the antiganglioside antibodies in an enzyme-linked immunosorbent assay (ELISA). The required or optimal ethanol suspension is evaporated to dryness on plates (Ravindranath, Ravindranath, Morton and Graves 1994b; Ravindra-nath, Muthugounder, Saravanan, Presser and Morton 2005b). During drying in vacuo, CMC increases causing the attachment of the hydrophobic tail group to the polystyrene and expose uniformly the hydrophilic sugar residues for interaction with antibodies. The objective is to simulate in vitro, the tumor cell surface expression of gangliosides.

## 3.2 Origin and Distribution During Tumor Formation

Structure, nomenclature, and tissue distribution of the gangliosides have been reviewed (Ando 1983; Wiegandt 1985). Table 1 summarizes the profiles of ganglioside antigens commonly associated with human cancer cells. The parent substance of the gangliosides is ceramides, the lipid molecules composed of sphingosine and a fatty acid. The ceramides function as cellular signals in regulating the differentiation, proliferation, programmed cell death and apop-tosis of cells. Kolesnick (2002) has summarized the factors promoting increased biosynthesis of ceramides. First, agonists and stress enhance ceramide levels; second, exposure to natural ceramide, exogenous spingomyleinase, or agents that interfere with enzymes of ceramide metabolism which mimics the effects of stress also induce ceramides. Third, genetic abnormalities such as in

**Table 1** Gangliosides commonly found in normal human tissues and cancer tissues

| Ganglioside | Structural formula | Atomic Mass Unit |
|---|---|---|
| $GM_1$ | Galβ1,3GalNAcβ1,4(NeuAcα2,3)Galβ1,1Cer | 1547 |
| $GM_2$ | GalNAcβ1,4(NeuAcα2,3)Galβ1,1Cer | 1385 |
| $GM_3$ | NeuAcα2,3Galβ1,1Cer | 1236 |
| $GD_3$ | NeuAcα2,8NeuAcα2,3Galβ1,1Cer | 1545 |
| $GD_2$ | GalNAcβ1,4(NeuAcα2,8NeuAcα2,3)Galβ1,1Cer | 1694 |
| $GD_{1a}$ | NeuAcα2,3Galβ1,3GalNAcβ1,4(NeuAcα2,3)Galβ1,1Cer | 1838 |
| $GD_{1b}$ | Galβ1,3GalNAcβ1,4(NeuAcα2,8NeuAcα2,3)Galβ1,1Cer | 1838 |
| $GT_{1b}$ | NeuAcα2,3Galβ1,3GalNAcβ1,4(NeuAcα2,8NeuAcα2,3)Galβ1,1Cer | 2144 |

Niemann-Pick Disease cells, *Asmase*$^{-/-}$ mice, and *Glucosylceramide synthase*$^{-/-}$ mice also manifest stress responses. Similarly, the biosynthesis of ceramides may be enhanced during initial phases of neoplastic transformation of the progenitor cells. The ceramides, thus induced, are known to promote apoptosis (Obeid, Linardic, Karolak and Hannun 1993; Mathias, Peña and Kolesnick 1998), and therefore it is speculated that increase in the level of ceramides by cancer cells may be a homeostatic mechanism to eliminate the formation of abnormal cells. To escape from apoptosis, tumor cells glycosylate ceramides to more complex sphingolipids. They can be neutral glycosphingolipids and acidic sialylglycosphingolipids or gangliosides. Gangliosides may be stored in the cytoplasm or in the outer layer of the bilayered lipid membrane.

Tumor proliferation is often associated with tumor cell death or necrosis. Often the central core of a surgically resected tumor tissue consists of such necrotized cells, indicating that tumor growth and proliferation unaccompanied by supply of required oxygen and nutrients may result in tumor necrosis. Probably, the neovascularization of tumor tissues may be an effort to prevent tumor necrosis to further promote tumor cell proliferation. Necrotic tumor releases or sheds tumor molecules, tumor antigens, and glycoantigens into the tumor microenvironment. Consequently, as tumor cells proliferate, the density of gangliosides, shed or released, may increase in the vicinity of tumor tissues. The released or shed antigens may enter into body fluid and circulation. The gangliosides can occur as micelles or as aggregates in circulation and in tumor microenvironment (Kong, Li and Ladisch 1998). They may bind to immunoglobulins (Hakansson, Fredman and Svennerholm 1985), serum proteins (ganglioproteins) (Tonegawa and Hakomori 1977), and lipoproteins (Valentino and Ladisch 1992) in circulation or in body fluids. Tumor-associated gangliosides may also be released as membranous exosome-like vesicles (Dolo, Li, Dillinger, Flati, Manela, Taylor, Pavan and Ladisch 2000). Such exosomes may result from pinching off of the cytoplasm (a process known as "clasmatosis"), a process not restricted to tumor cells, but also common to normal, embryonic cells, virally infected cells, physico-chemical and biotic stress exposed and starved cells (Kalashnikova 1985). The exosomes may contain free ribosomes or glycogen, or even mitochondria. In cancer cells, "clasmatosis" results in large membrane-enclosed fragments (>3 microns in diameter) in both liver and muscles ((Morris, MacDonald, Koop, Schmidt, Chambers and Groom 1993). These vesicles enter into blood, lymphatic channel, ascetic and amniotic fluid (Keller, Sanderson, Stoeck and Altevogt 2006). These membranous vesicles do carry clusters of gangliosides on the outer bilayered lipid membrane. Earlier experiments in animal models document that such membrane-bound gangliosides are more immunogenic than the micellar forms of gangliosides (Ravindranath, Morton and Irie 1994a).

All these observations point out that neoplastic transformation of normal cells may accompany overexpression of the gangliosides. The shed glycoantigens may perform paradoxically two different kinds of functions. (1) It may suppress cell-mediated immune functions; (2) they activate silent B cells to elicit IgM production against glycoepitopes of the gangliosides so that they may clear the shed antigens from circulation.

## 3.3 Antigenic Determinants

The antigenic determinants of gangliosides, like those of other carbohydrate antigens, are chains of sugars, such as glucose, galactose, $N$-acetylneuraminic acid, $N$-acetylgalactosamine, $N$-acetylglucosamine, and fucose. The specific epitope varies with the nature of these sugars and their glycosidic linkages ($\alpha$ and$\beta$; e.g., Gal$\alpha$1,3Gal, Gal$\alpha$1,4Gal, Gal$\beta$1,3Gal, Gal$\beta$1,4Gal, Gal$\beta$1,3GlcNAc, Gal$\beta$1,4GlcNAc, NeuAc$\alpha$2,3Gal, NeuAc$\alpha$2,6Gal, and NeuAc$\alpha$2,8NeuAc) (Table 1). Other glycoantigens may share ganglioside antigenic epitopes, and consequently an antibody generated in a patient against gangliosides or glycoproteins may cross-react with one another. The following observations support the hypothesis.

- Two murine monoclonal IgM antibodies, 2A3D2 and 2D11E2, developed with a ganglioside mixture prepared from human hepatocellular carcinoma cells as the immunogen, detected the ganglioside antigens, particularly ganglioside GM2, in carcinoma cells but also bound to an O-linked sialoglycoprotein related to a rare blood group antigen called Cad (Hiraiwa, Tsuyuoka, Li , Tanaka, Seno, Okubo, Fukuda, Imura and Kannagi 1990).
- Similarly, a murine antibody (KM696) specific for a ganglioside GM2 (GalNAc$\beta$1,4 [NeuAc$\alpha$2,3]Gal$\beta$1,4Glc$\beta$1, 1ceramide) may recognize a Cad-like blood group antigen (Nobu Hanai, personal communication).
- Antibodies to the carbohydrate determinant (sialyl Lewis[a]) of a ganglioside (121SLE [IgM], Neomarkers, Union City, CA; 1116-NS-19-9 [IgG1], Signet Laboratories, Dedham, MA; M8073022 [IgG1], Fitzgerald, Concord, MA; ZY-C09 [IgG1], Zymed, San Francisco, CA; MED-CLA 143, Accurate Chemical and Scientific Corporation, Westbury, NY; NS19-9, International CIS, Cedex, France) are directed against the sugar residues of mucin-type glycoproteins belonging to the MUC family, such as MUC-1.

Therefore, specific immune recognition of ganglioside epitopes requires the absence of cross-reactivity with other glycoantigens. Sugar-mimicking peptides may also affect the specificity of immune recognition (Krishnan, Lomash, Raj, Kaur and Salunke 2007). The sialylated sugar residues are the epitopes of the gangliosides forming contacts with the antibody. The complementary combining site (CCS) of the antibody is the paratope. The upper limit of epitope size is determined by the variable region, CCS of the paratope. Measuring the displacing activity of different oligosaccharides shows that the maximum physical capacity of the CCS is six sugar residues for antiglyco-antibodies and four amino acids for antipeptide antibodies. X-ray crystallographic studies of the oligosaccharide–antibody complexes have confirmed these estimates. The contact areas involve 255 A of the sugar residues and 304 A of the paratope. Fifteen amino acids of the antibody establish 90 van der Waal forces and nine hydrogen bonds (Rich, Fleisher, Schwartz, Shearer and Strober 1996). Antigen–antibody interactions may also involve salt bridges. The structures of some antibody–antigen complexes suggest rigid binding of both. The hydrophilic head group of gangliosides contains the antigenic determinant of CCS. **Table 1** summarizes the diversities in the glycoepitopes of gangliosides as illustrated from the structure of some of the most common gangliosides in normal and malignant human tissues. Ganglioside GM3 constitutes more than 90% of the total gangliosides in normal human cells such as melanocytes (Carubia, Yu, Macala, Kirkwood and Varga 1984), the level of other gangliosides ranges from 1% to 10% of total gangliosides. Upon neoplastic transformation of melanocytes, there is a remarkable change in the profile of gangliosides in that the ratio of GM3: GD3 reverses from 90:10 to 10:90, and the ganglioside GD3 tends to dominate (Ravindranath, Tsuchida, Morton and Irie 1991).

# 4 Glycoimmunomics of Gangliosides

Immunology of gangliosides was periodically reviewed (Ravindranath and Morton 1997b; Ravindranath and Morton, 2000; Ravindranath, Gonzales, Nishimoto, Tam, Soh and Morton 2000; Ravindranath, Muthugounder, Hannah and Morton, 2007a; Ravindranath, Yesowitch, Sumobay and Morton 2007c), since we first observed antibodies to GD3 and O-acetyl GD3 in melanoma patients receiving immunotherapy (Ravindranath, Morton and Irie 1989) Contrary to earlier notion that gangliosides are not immunogenic and require adjuvants to elicit immune response, we have observed the presence of IgM antibodies to different gangliosides in normal and "healthy" volunteers and these antibodies declined with age (see details in Ravindranath et al. 2000). Subsequent observations lead to the realization of cell-mediated immune response to gangliosides and the potential of antiganglioside IgM as early biomarkers of human cancer (see details in Ravindranath et al. 2007a). The properties of glycoantigens will be discussed from the perspective of gangliosides.

## 4.1 Gangliosides Are T-Independent Antigens

Gangliosides, like any other glycoantigens, are T-independent antigens that stimulate primary, but not secondary, B-cell responses and do not require T-help. There are two classes of T-independent antigens (Freimer, McIntosh, Adams, Alving and Drachman 1993). *T-independent-1 glycoantigens* at high concentrations stimulate polyclonal proliferation of B cells and at low doses stimulate the differentiation of B cells that secrete heteroglycan-specific antibodies. Response to these glycans occurs in cultured cells rigorously depleted of T cells, but can be modified by other cell types such as monocytes or natural killer cells. Secondary exposure to these antigens neither results in accelerated kinetics nor results in isotype switching. *T-independent-2 glycoantigens* stimulate antigen-specific responses, as observed in athymic nude mice, and do not stimulate polyclonal proliferation at high doses. In vitro responses may require interferon-$\gamma$, which may induce these T-I-2 antigen-stimulated B cells to switch isotype production to IgG3 in animal models. The observation that glycoantigens can trigger B cells to produce IgM antibody in T-cell-deficient mice indicates that glycoantigens are T-cell-independent.

## 4.2 Gangliosides Elicit IgM Response in Humans

IgM antibodies directed against glycoantigens may be hexameric or polymeric (Brewer, Randall, Parkhouse and Corley 1994), in contrast to conventional pentameric IgM, and may be without a J chain (Davis, Roux and Shulman 1988; Randall, King and Corley 1990). Polymeric IgM antibody without a J chain seems to fix complement 20-fold more efficiently than conventional pentameric IgM (Randall, Parkhouse, and Corley1992). Persistent IgM antibodies in patients with motor-neuron diseases are directed against carbohydrate residues of glycolipids (Ravindranath, Ravindranath and Graves 1997c). The B cells sharing pan-T cell antigen CD5 may augment the production of antiglycolipid antibodies. Immunogenicity of tumor-associated carbohydrate antigens is confirmed by immortalizing B cells with Epstein-Barr virus and developing human monoclonal antibodies. Monospecificity of human as well as murine monoclonal antibodies against specific carbohydrate epitopes is confirmed using sensitive enzyme-linked immunosorbent assay (ELISA) (Ravindranath et al. 1994b; 2005b).

Examining sera of healthy volunteers between the ages of 18 and 90, it was noted that antiganglioside IgM antibodies occur naturally at low levels in healthy individuals (Ravindranath, Ravindranath, Morton and Graves 1994b; Ravindranath et al., 2000) (Fig. 1). Most interestingly, the antibody levels decline after the age of 50 years, a finding that is of tremendous significance for most of the cancer patients and particularly prostate cancer patients since the- incidence of cancer increases above the age of 50. The levels of specific antiganglioside IgM antibodies also tend to increase with tumor burden and

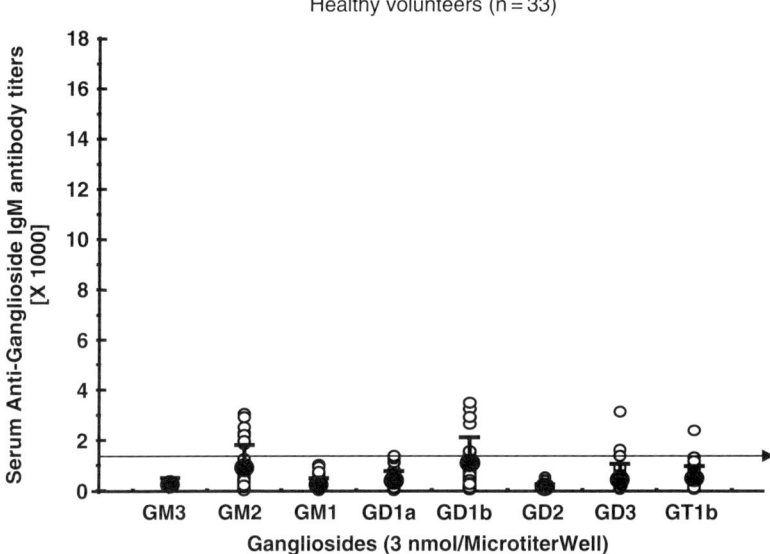

**Fig. 1** Titers of naturally occurring IgM antibodies in normal and healthy volunteers of varying ages against common gangliosides, which are also found in human cancers. (Reprinted from Ravindranath et al. (2002) Cryobiology 45, 10–21, with permission from Elsevier)

disease stage in patients with sarcoma and colorectal carcinoma (Perez, Ravindranath, Soh, Gonzales, Ye and Morton 2002; Ravindranath, Wood, Soh, Gonzales, Muthugounder, Perez, Morton and Bilchik 2002) (Fig. 2).

These observations lead to the hypothesis that tumor gangliosides released or shed from tumor cells may induce serum antiganglioside IgM antibodies, to clear them from the microenvironment. A number of earlier studies have reported that gangliosides are not capable of inducing antibody response and induction of anti-ganglioside antibodies may require exogenous adjuvants. While need for adjuvants to induce immune response against gangliosides have been extensively reviewed (Ravindranath et al. 1994; Ravindranath and Morton 1997b), it is far from clear as to how gangliosides released or shed from tumor can induce anti-ganglioside IgM without exogenous adjuvants. There are several possibilities.

- Reports document that the glycoantigens presented in the context of an intact membrane elicited a better immune response than soluble antigens (Gillard, Thomas, Nell and Marcus 1989; Ravindranath et al. 1994b; Ravindranath, Bauer, Amiri, Miri, Kelley, Jones and Morton 1997a).
- Gangliosides from tumor cells are released as vesicles and they may elicit immune response better than micelles.
- It is hypothesized that tumor necrosis in proliferating population of tumor cells may act as natural endogenous adjuvant.

This possibility was tested by determining the serum levels of total gangliosides and antiganglioside antibodies before and after inducing necrosis by

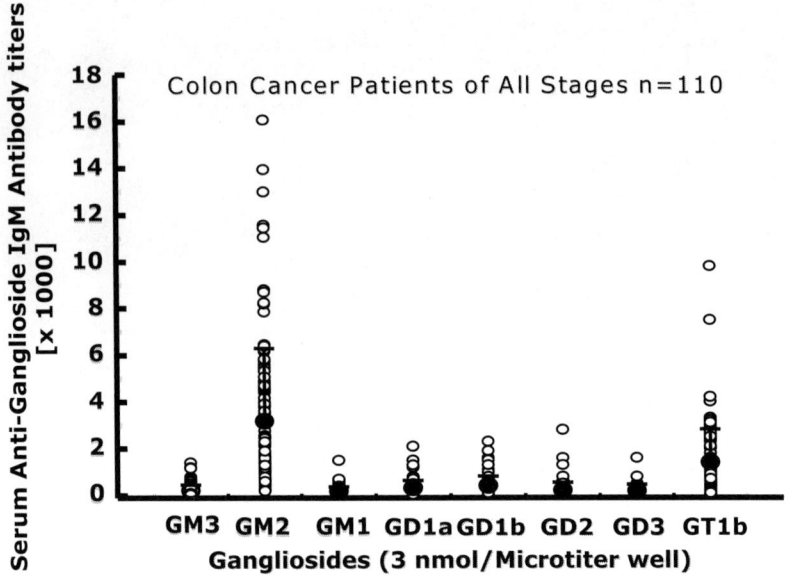

**Fig. 2** Titers of naturally occurring IgM antibodies in 110 patients with all stages of colorectal carcinoma. Compare with the profiles seen in normal and healthy volunteers and note high titers of IgM antibodies against GM2 and GT1b, the gangliosides commonly found in colon tumor tissues. (Reprinted from Ravindranath et al. (2002) Cryobiology 45, 10–21, with permission from Elsevier)

cryoablation in hepatic metastases from colorectal cancers (Ravindranath et al. 2002). Indeed, the serum total ganglioside level increased significantly after cryoablation; concomitant with this increase, without any exogenous adjuvants, there was a significant increase in the titer of antiganglioside IgM antibodies. Control patients, whose hepatic metastases were resected or destroyed by radiofrequency ablation due to coagulative necrosis that aggregates and denatures the glycolipids, did not show any change in serum ganglioside or antiganglioside antibody titers. The antibody titers associated with cryoablation-induced necrosis indicate that an antiganglioside immune response does not require an exogenous adjuvant. It appears that tumor necrosis may act as endogenous adjuvant to elicit antibodies against ganglioside shed or released from tumor. This could be a mechanism to eliminate gangliosides that are very well known to be immunosuppressive.

## 4.3 Gangliosides Suppress Immune Functions

Gangliosides have long been known to suppress immune functions of T and B cells (Ando, Hoon, Suzuki, Saxton, Golub and Irie 1987; Hoon, Ando, Viand, Touched, Oakum, Morton and Ire 1989; Grayson and Ladisch 1992;

Prokazova and Bergelson 1994; Prokazova, Dyatlovitskaya and Bergelson 1988; Kanda 1999; Kanda and Tamarisk 1999; Kanda and Watanabe 2000; Ravindranath, Gonzales, Soh, Nishimoto, Tam, Bilchik, Morton and O'Day 2001; Thornton, Kudo, Rayman, Horton, Molto, Cathcart, Ng, Paszkiewicz-Kozik, Bukowski, Derweesh, Tannenbaum and Finke 2004; Biswas, Richmond, Rayman, Biswas, Thornton, Sa, Das, Zhang, Chahlavi, Tannenbaum, Novick, Bukowski and Finke 2006). T cells from cancer patients are often functionally impaired, which imposes a barrier to effective immunotherapy. Neuroblastoma-derived gangliosides significantly inhibited the generation of dendritic cells in culture. The percentage of CD83+ cells decreased from $51.8 \pm 6.1\%$ in the control group to $12.9 \pm 2.7\%$ in cultures treated with the major neuroblastoma-associated ganglioside GD2 ($P < 0.05$) (Shurin, Shurin, Bykovskaia, Shogan, Lotze and Barksdale 2001). Gangliosides isolated from glioblastoma multiforme lines as well as purified fractions containing GM2 and GD1a are directly apoptogenic for T cells (Chahlavi, Rayman, Richmond, Biswas, Zhang, Vogelbaum, Tannenbaum, Barnett and Finke 2005). Most pronounced are the alterations characterizing tumor-infiltrating T cells, which in renal cell carcinomas includes defective NF-$\kappa$B activation and a heightened sensitivity to apoptosis (Thornton et al. 2004; Biswas et al. 2006). Co-culture of renal tumor cell line (SK-RC-45) or RCC tissue-derived gangliosides with Jurkat T cells and peripheral blood T lymphocytes induced apoptosis of T cells. Jurkat T cells overexpress a protein RelA, which protect the T cells against cell death. However, upon co-incubation of SK-RC-45 with the T cells caused a decrease in their RelA (p65) and p50 protein levels, which coincided with the onset of apoptosis. The disappearance of RelA/p50 protein was mediated by a caspase-dependent pathway, because pretreatment of T lymphocytes with a pan caspase inhibitor before co-culture with SK-RC-45 blocked RelA and p50 degradation. SK-RC-45-gangliosides also mediated this degradative pathway, as blocking ganglioside synthesis in SK-RC-45 cells with the glucosylceramide synthase inhibitor, PPPP, protected T cells from tumor cell-induced RelA degradation and apoptosis. Ganglioside GM2 purified from RCC promoted T cell dysfunction (Biswas et al. 2006). Importantly, anti-GM2 antibody (DMF10.167.4) blocked 50–60% of the T-cell apoptosis. Furthermore, the RCC-tissue-derived gangliosides also suppressed IFN-$\gamma$, IL-4 production by CD4+ T cells at concentrations (1 ng/mL-100 pg/mL) well below those that induce any detectable T cell death (4–20 µg/mL). Anti-GM2 antibody also suppressed the production of IFN-$\gamma$, suggesting that RCC-derived GM2 can downregulate cytokine production by CD4+ T cells. Ganglioside-mediated immunosuppression of T cells impedes the effectiveness of passive and active specific immunotherapies. The ganglioside-mediated immunosuppression may increase with increase in the level of shed ganglioside during tumor progression. Interestingly, the shedding of gangliosides from solid cancers is correlated with an endogenous (natural) antiganglioside IgM response.

Probably, the role of natural antibodies to gangliosides may be to eliminate the immunosuppressive gangliosides from circulation. The natural induction of antibodies to gangliosides as observed after cryoablation of colon cancer metastasized to liver (Ravindranath et al. 2002), increase with stages of tumor progression in sarcoma (Perez et al. 2002), and after vaccine administration in stage III melanoma (Ravindranath, Hsueh, Verma, Ye and Morton 2003) suggest that the primary role of naturally occurring antiganglioside antibodies may be to eliminate gangliosides released or shed from tumors to prevent immunosuppression and to restore immunocompetence.

## 4.4 Cell Recognition of Gangliosides

Specialized T cells known as natural killer T (NKT) cells may recognize glycoepitopes of glycoantigens, including neutral and acidic glycolipids such as gangliosides. Recognition of carbohydrate epitopes is precise, and lipid-reactive T cells alter systemic immune responses in infectious and autoimmune diseases (Wilson, Singh and Van Kaer 2002). These are innate lymphocytes that share receptor structures and functions with conventional T cells and natural killer cells. They may rapidly produce cytokines in response to T-cell receptor engagement, suggesting that activated NKT cells can modulate immune responses. Recent pre-clinical studies have revealed significant efficacy of NKT-cell ligands such as the glycolipid α-galactosylceramide for the treatment of metastatic cancers. These findings suggest that appropriate stimulation of NKT cells could be exploited for the prevention or treatment of human cancer (Young and Moody 2006).

Most importantly, the molecular recognition of gangliosides may involve insertion of the hydrophobic lipid portion of antigens into a hydrophobic groove to form CD1–lipid complexes, which then contact T cell receptors (TCRs) on NKT cells. These findings provide a previously unrecognized mechanism by which the cellular immune system can recognize alterations in many types of carbohydrate structures (Brutkiewicz 2006). The MHC class I-like CD1 family of antigen-presenting molecules is responsible for the selection of NKT cells. For example, CD1a is expressed on antigen presenting cells (APC), such as Langerhans cells (LCs) and dendritic cells (DCs), where it mediates T cell recognition of glycolipid and lipopeptide antigens that contain either one or two alkyl chains. Structural, biochemical, and biophysical studies support the view that CD1 proteins bind the hydrophobic alkyl portions of these antigens directly and position the polar or hydrophilic head groups of bound lipids and glycolipids for highly specific interactions with T cell antigen receptors. The ligands presented by CD1d to NKT or other CD1d-restricted T cells may include glycolipids from a marine sponge, bacterial glycolipids, normal endogenous glycolipids, tumor-derived phospholipids, neutral glycolipids, and gangliosides (Angata and Varki 2000). However, it is still far from clear whether the NKT cells are directly involved in cytotoxic killing of tumor cells, as observed with peptide antigens.

The sugar-binding lectins are present on the surface of NK cells. From the perspective of gangliosides, the sialylglycosylceramides, sialic acid binding lectin-like receptors on NK cells may play a major role. The sialic acid binding lectin-like receptors on immune cells and particularly on NK cells are called Siglec (Sialic acid-binding Ig superfamily lectins). Siglecs are transmembrane polypeptides, with a transmembrane domain and a cytoplasmic tail. The first set of Ig-like domain is the most important in carbohydrate recognition, and the second Ig-like domain may also contribute to the binding. NK-cell surface-Siglecs have the potential to recognize specific epitopes of tumor cell surface gangliosides, bind and mediate cytotoxicity, provided shed micellar counterparts of tumor gangliosides in tumor microenvironment do not engage Siglecs and block T cell–tumor cell interaction.

Eight members of the family have been described so far in humans, and each shows highly cell type–specific expression: Siglec-1/sialoadhesin (expressed on macrophages); Siglec-2/CD22 (on B lymphocytes); Siglec-3/CD33 (on myeloid precursors and monocytes); Siglec-4a/myelin-associated glycoprotein (on oligodendroglia and Schwann cells); Siglec-5 (on neutrophils and monocytes); Siglec-6/OBBP-1 (on B lymphocytes and placental trophoblasts); Siglec-7/AIRM1 (on natural killer cells and monocytes); and Siglec-8 (on eosinophils), Siglec-9 (monocytes and granulocytes), and Siglec-10 (mast cells, T and B cells) (Rapoport, Mikhalyov, Zhang, Crocker and Bovin 2003).

Several members of Siglec family may recognize gangliosides and the specificity of their recognition deserves careful scrutiny. The following Siglecs interact with gangliosides:

- Siglec-5 bound preferentially to GQ1b, but weakly to GT1b, whereas Siglec-10 interacted only with GT1b.
- Siglec-7 (as well as Siglec-9) displayed binding to gangliosides GD3, GQ1b, and GT1b bearing a disialoside motif, though Siglec-7 was more potent;
- Recombinant Siglec-7-Fc protein showed unusual binding preference for $\alpha2,8$-linked disialic acids of ganglioside GD3. It bound specifically to GD3 synthase-transfected P815 target cells expressing high levels of GD3, suggesting that GD3 expression on target cells can modulate NK cell cytotoxicity via Siglec-7-dependent and -independent mechanisms (Nicoll, Avril, Lock, Furukawa, Bovin and Crocker 2003). However, it remains to be seen whether CD1 family presents gangliosides to Siglec receptors on NK cells.
- Siglec-8 demonstrated low affinity to the gangliosides tested compared with other Siglecs.
- Siglec-9 displayed binding to gangliosides GD3, GQ1b, and GT1b bearing a disialoside motif, but also interacts with monosialoganglioside, GM3 and therefore lacks specificity.
- Siglec-9 also recognizes Neu5Ac$\pm$2-3Gal$^2$1-4[Fuc$\pm$1-3]GlcNAc structure (i.e., recognition is not disturbed by the fucose residue on the GlcNAc) and distinguishes between Neu5Ac$\pm$2-3Gal$^2$1-4GlcNAc (typically found in $N$-linked glycans) and Neu5Ac$\pm$2-3Gal$^2$1-3GalNAc (typically found in $O$-linked glycans and glycosphingolipids) (Angata and Varki 2000).

The functional relevance of this selectivity of Siglecs is still far from clear, although it appears that engagement of the Siglecs by polyvalent ligands may be able to elicit negative intracellular signals, thus silencing the cells that are inappropriately activated (Angata and Varki 2000). Siglecs may function "in a manner similar to that of killer cell Ig-like receptors expressed on natural killer cells" to send negative intracellular signals if engaged by MHC class I expressed on target cells (Angata and Varki 2000). The inhibitory effect of engaging Siglecs is well evident from the works of Crocker (Rapoport et al. 2003; Nicoll et al. 2003) and Paulson (Ikehara, Ikehara and Paulson 2004) team of investigators. Recently, it was demonstrated that NK cells were less able to kill target cells bearing the ganglioside GD3, a high-affinity Siglec-7 ligand, suggesting inhibition of NK activity by Siglec-7 (Nicoll et al. 2003). Both Siglec-7 (p70/AIRM) and Siglec-9 are "CD33"-related Siglecs expressed on natural killer (NK) cells and subsets of peripheral T cells. Like other inhibitory NK cell receptors, they contain immunoglobulin receptor family of tyrosine-based inhibitory motifs in their cytoplasmic domains. Sialic acid binding of Siglecs-7 and -9 are abrogated by mutation at the conserved Arg in the ligand-binding site (Siglec-7 (Arg$^{124}$ for Siglec-7) and (Arg$^{120}$ for Siglec-9) resulted in reduced inhibitory function in the NFAT/luciferase transcription assay, suggesting that ligand binding is required for optimal inhibition of TCR signaling (Ikehara et al. 2004). All these observations favor the view that binding of glycoconjugates with Siglecs may lower the activation and activities of immune cells.

## 4.5  Emerging Concepts of Clinical Glycoimmunomics of Human Cancer

Most importantly, one should be aware that the gangliosides such as GD3 or GM2 or GD2 or GD1a or GT1b, shed into tumor microenvironment either as micelles or as membrane vesicles, have the potential to bind to Siglecs and interfere or block T cell–tumor cell interaction. Observations made in patients with colorectal carcinomas (Ravindranath et al. 2002), pancreatic adenocarcinomas (Chu, Ravindranath, Gonzales, Nishimoto, Tam, Soh, Bilchik, Katopodis and Morton 2000), epithelial ovarian carcinoma (Santin, Ravindranath, Bellone, Muthugounder, Palmieri, O'Brien, Roman, Cannon and Pecorelli 2004; Ravindranath, Muthugounder, Presser, Selvan, Santin, Bellone, Saravanan an Morton 2007b), and melanoma (Ravindranath et al. 2003) clearly document that the gangliosides are shed with the progression of disease. The shed tumor-gangliosides are not only most prevalent in the tumor microenvironment but also enter into peritoneal cavity and systemic circulation. Although in vitro observations on (1) the binding of MHC-class I-like CD1 family of receptors to lipid moiety of gangliosides, (2) the interaction of Siglec molecules to sugar residues of glycoligands, and (3) the release of cytokines with potential anti-tumor activities document the potential for cell-mediated immune attack on tumor cells; the cell-mediated immune attack on tumor cells may be kept in abeyance by the glycoantigens shed from progressively necrotic tumor cells. There is a compelling need to eliminate the shed gangliosides and other immunosuppressive glycoantigens from tumor microenvironment. Indeed, the patients' first line of defense against shed gangliosides appears to be the production of anti-ganglioside IgM autoantibodies. It is this first line of defense which becomes a potential glycoimmunomic biomarker for early detection of prostate cancer. Possibly, a better understanding of the specific nature of innate anti-gangliosides IgM is required to develop strategies to eliminate the specific shed gangliosides in a particular type of cancer. This approach may pave way to generate anti-ganglioside IgM antibodies in patients by ganglioside-specific passive or active immunotherapy.

## 5 Glycoimmunomic Biomarkers

### 5.1 Glycoimmunomic Biomarker: Origin of the Hypothesis

As mentioned earlier, a serendipitous discovery was made while examining the amount of gangliosides in sera of patients before and after cryosurgical ablation of hepatic metastases from colorectal cancer (Ravindranath et al. 2002). The membrane-disruptive necrosis of tumor cells by the cryoablative (freeze-thaw) procedure released the gangliosides into the circulation. The mean sTG level before cryosurgical ablation ranged from 12.1 to 31.7 mg/dL. Immediately after cryosurgical ablation, sTG level ranged from 23.2 to 37.8 mg/dL (two-tailed $p < 0.03$). There was no statistically significant increase in mean sTG levels of patients treated by radiofrequency ablation (23.9 mg/dL before vs 24.9 mg/dL after) or surgical resection (17.0 mg/dL before vs 24.5 mg/dL after). It is interesting to note that an increase in sTG level was followed by an increase in titers of anti-ganglioside antibodies as evidenced by the immune response to the major gangliosides of colon carcinoma (GM2, GD1b, and GT1b), but not to the gangliosides of normal liver tissue (GM1 and GM3). There was no change in anti-GM1 IgM or anti-GM3 IgM titers after any type of treatment; however, titers of anti-GM2, anti-GD1b, and anti-GT1b IgM antibodies increased in most patients treated by cryosurgical ablation. This increase was not significant in patients treated with radiofrequency ablation and absent in patients treated with surgical resection. The sTG level remained high for approximately 18–54 days (half-life of approximately 25 days), but declined steadily. This unique pattern not only supported an endogenous immune response to elevated circulating gangliosides but also a homeostatic role for anti-ganglioside IgM antibodies in regulating sTG levels.

The following inferences emerge from this study. (1) Tumor necrosis may cause shedding of gangliosides from tumor during tumorigenesis and tumor progression; (2) the shed gangliosides may elicit specific anti-ganglioside

IgM antibodies; (3) the purpose of such induction of antibodies may be toclear the tumor-gangliosides from the circulation; (4) the adaptive value of such antibody-mediated clearance may be to downregulate ganglioside-mediated immunosuppression and restore immune competence in patients; (5) since shedding of gangliosides increase with tumor burden, there is a need either to lower tumor burden by resection of tumor or to boost the anti-ganglioside IgM antibodies with passive or active specific immunotherapy .

It is important to know at what stage of tumorigenesis and shedding of gangliosides, the antibody induction commences and how immune system is stimulated to produce the specific IgM antibodies. Furthermore, if antibody induction commences very early during tumorigenesis due to shedding of gangliosides, then the antibodies elicited against one or more specific tumor ganglioside can serve as early biomarker. During early stages of tumorigenesis, it would be easier to measure and monitor the antibody titers than measuring the shed gangliosides, which are often diluted and undetectable in the sera. The above tenet was extended to examine the precise nature of the early glycoimmunomic biomarker in prostate cancer.

## 5.2 Glycoimmunomic Early Biomarker of Prostate Cancer

The hypothesis is that during early-stage, clinically organ-confined prostate cancer (CaP) an endogenous IgM response may be elicited against gangliosides, released, or shed from tumor cells. After a careful analyses of the profile of tumor-associated gangliosides in CaP cell lines including clinically organ-confined CaP (HH-870), androgen-receptor positive (LNCaP FGC and LNCaP FGC-10) and negative (PC-3 & DU 145) cell lines, the titers of antiganglioside IgM antibodies in the sera were measured in a small cohort of patients ($n = 35$) with untreated early CaP (organ-confined; T1 and T2), patients with advanced CaP (stage T2 or T3) (Ravindranath et al. 2005). As controls, sera of patients with benign prostatic hyperplasia (BPH) and healthy volunteers were compared. Analyses of Variance (two-tailed $p$ value) showed that the titer values of anti-GM1 IgM, anti-GM2 IgM, anti-GM3 IgM, anti-GD2 IgM, anti-GD1b IgM, and GT1b IgM did not vary among the four groups. However, there were significant differences in anti-GD1a and anti-GD3 IgM antibodies at $p < 0.05$ level after the application of Fisher's least significant difference method. The log titers of anti-GD3 IgM were significantly lower in BPH and CaP groups than in healthy men. Although log titers of anti-GD1a IgM were similar between non-cancer groups, they were significantly higher in patients with early-stage CaP than in healthy men ($p = 0.02$), patients with BPH ($p < 0.008$), or patients with late-stage CaP ($p = 0.002$) (Table 2). The difference in the anti-GD1a IgM response between early-stage and late-stage CaP suggests that GD1a may be the most prevalent ganglioside

**Table 2** Anti-GD1a IgM as a potential early biomarker of clinically organ-confined cancer. The pairwise comparison by Fisher's least significant difference method shows that the anti-GD1a-IgM log titers significantly (*p* value) distinguished confined CaP (T1 /T2) from healthy controls and unconfined CaP (T3/T4). The observation made earlier (2005) is confirmed by increasing the sample size (*N*) (unpublished observations)

| | Anti-GD1a IgM | | | |
| | (IJC, 2005) | | (unpublished)* | |
| Pair-wise comparison | *N* | *p* | *N* | *p* |
|---|---|---|---|---|
| Healthy vs CaPT1/T2 | 11 vs 36 | 0.02 | 46 vs 66 | 0.005 |
| Healthy vs CaPT2/T3 | 11 vs 27 | 0.942 | 46 vs 33 | 0.47 |
| CaPT1/T2 vs CaPT2/T3 | 36 vs 27 | 0.002 | 66 vs 33 | <0.02 |

*includes *N* of earlier study (2005)

of early CaP and that its expression decreases during tumor progression. Recently, the validity of the findings was reexamined in a larger cohort of patients (data unpublished). A double-blinded analyses of antiganglioside-IgM titers were measured using ELISA on the sera obtained from (1) normal healthy volunteers ($N = 46$); (2) BPH-patients ($N = 58$), (3) CaP-patients with confined disease (stage T1/T2) ($N = 66$) and (4) CaP-patients with organ-unconfined disease (stage T3/T4) ($N = 33$). The pairwise comparison by Fisher's least significant difference method (Table 2) supported our earlier observations and also led to the following findings. Anti-GT1b-IgM log titers significantly distinguish CaP from healthy and BPH, whereas the anti-GD1a-IgM log titers significantly distinguish confined CaP (T1 /T2) from healthy controls and unconfined CaP (T3/T4) (unpublished observations). However, 22–29% of BPH patients showed a high titer of anti-GD1a-IgM, suggesting that these patients may have localized CaP. Although a much larger study is necessary to confirm these preliminary observations, it appears that an endogenous IgM response to CaP occurs during tumorigenesis, when the cancer is still contained within the prostate. The augmentation of anti-GD1a IgM antibody in patients with confined CaP is intriguing because the levels of anti-ganglioside IgM antibodies in healthy individuals tend to decline with age and remain low after the age of 50 (Ravindranath et al. 2000). Our recent observations also show that natural log titers of anti-GD1a IgM were also significantly higher in ascitic fluid of epithelial ovarian cancer patients ($n = 14$) than in the plasma of healthy volunteers ($n = 14$) ($p = 0.001$). Similarly, the plasma levels of anti-GD1a IgM were also higher ($p = 0.064$) in cancer patients ($n - 23$) than in the healthy volunteers. Clearly, GD1a is the dominant immunogenic ganglioside of ovarian epithelial cancer. Apparently, in these gender-based cancers as well as in melanoma, the GD1a that is shed from cancer cells elicits a specific endogenous IgM response. It was shown that GD1a has the propensity to induce specifically the production of IL-6 (Kanda and Watanabe 2000) and IL-10 (Kanda 1999) in monocytes. Both these cytokines can further augment IgM production by B cells. Possibly, induction of IL-6 and IL-10 by

GD1a might have generated an endogenous anti-GD1a IgM response early during tumorigenesis.


## 5.3 Glycomic Therapeutic Response Biomarker for Human Melanoma

If an endogenous IgM response against gangliosides can lower the serum total ganglioside (sTG) levels, then sTG levels might represent a useful biomarker to evaluate the response to treatments. To validate the above contention, a series of investigations were undertaken.

The first investigation elucidated the effect of an allogeneic melanoma vaccine on the sTG levels of 17 patients undergoing immunotherapy for in-transit melanoma (Ravindranath et al. 2003). In all patients, the sTG level increased significantly (two-tail $p < 0.0001$) between 1 and 3 months after the initiation of vaccine therapy. The sTG ranged from 15.2 to 31.4 mg/dL with a mean of $23.7 \pm 5.5$ mg/dL. At week 24, sTG level declined significantly ($18.1 \pm 2.3$ vs $20.4 \pm 3.2$, two-tail $p < 0.05$) in seven patients whose lesions completely regressed (responders), but remained higher than the pre-vaccine level ($23.3 \pm 5.1$ vs $17.2 \pm 2.7$ mg/gL, $p$ [two-tail] $< 0.002$) in 10 non-responders. When the sign test for non-normal distributions was used to analyze these changes, the $p$-value for the decrease in the responder group was 0.016 and the $p$-value for increase in the non-responder group was 0.002.

The second investigation was carried out in a cohort of 75 patients with melanoma metastatic to regional lymph nodes (Ravindranath et al. 2003). These patients received post-operative adjuvant therapy with the same allogeneic melanoma cell vaccine after complete surgical resection of melanoma. Patients were divided into three groups according to their duration of overall survival: group A ($<13$ months), group B (32 to 59 months), and group C ($>60$ months). The sTG level increased within 3 months (between week 2 and 16) in all three groups, as observed in patients with in-transit melanoma. In all three groups, the mean prevaccine sTG level ranged between 17. 3 and 18.5 mg/dL. In group A, the level of sTG increased to $21.6 \pm 4.4$ mg/dL (two-tail $p < 0.0001$) between weeks 2 and 16 and remained significantly higher ($21.3 \pm 5.7$ mg/dL) than the pre-vaccine level ($p < 0.01$). In groups B and C, the mean sTG level increased to 21.4 and 21.6 mg/dL, respectively, but decreased to the pre-vaccine level ranging from 15.9 to 16.9 mg/dL (two-tail $p < 0.0001$). Based on the changes in sTG level observed in patients belonging to groups A, B, and C, two patterns of post-treatment fluctuations emerged. In all patients, the sTG level increased between weeks 2 and 16 after the start of vaccine therapy. This increase in sTG level could be due to a vaccine-induced inflammatory reaction and/or tumor necrosis, resulting in the release of tumor gangliosides into circulation. These patients were further categorized into two groups, group 1 and group 2. The sTG level on week 24 remained much higher than its pre-vaccine level in group 1 ($n = 17$) but not in group 2 ($n = 53$). The median overall survival was 12 months for group 1 and 47 months for group 2 ($p < 0.001$, based on Kaplan-Meier Log Rank test). These findings suggest that the sTG increases within 3 months after the start of therapy; in patients who respond to the vaccine, the sTG level returned to pre-immune status or lower at week 24. The results of this investigation pointed out that either endogenous or induced IgM antibodies might play a role in downregulating the level of circulating TG, which in turn might reverse immunosuppression, restore immunocompetence, and improve survival.

The third investigation was carried out in cohort of stage IV melanoma patients ($n = 34$) immunized with dendritic cells co-cultured with irradiated, IFNγ-treated autologous tumor cells admixed with GM-CSF (Selvan, Dillman, Fowler, Carbonell and Ravindranath 2008). The sTG levels and anti-ganglioside-IgM antibody titers were measured in sera of vaccine-recipients at 0, 4, and 24 weeks of treatment. Based on sTG-level, whether lower (L) or higher (H) than the mean $\pm$ 1 SD of normal and healthy volunteers on weeks 0, 4, and 24, patients were categorized into cohorts-I (LLL, $n = 16$), II (HHL/HLL, $n = 4$), III (LLH/LHH/LHL, $n = 7$), and IV (HHH/HLH, $n = 7$). The cohorts were regrouped as sTG- downregulators (sTG-DR; $n = 20$) and upregulators (sTG-UR; $n = 14$). This study established a clinical correlation between the two cohorts, in overall or progression-free survival. The cohort of sTG-UR had a median survival rate of 39 months, and their survival rate during 48 months was significantly lower than that of sTG-DR. Six out of 14 sTG-UR died within 39 months. Progression-free median survival of sTG-UR was only about 2 months, whereas 70% of the sTG-DR group survived for 24 months and 61% survived for 48 months. The results presented in Fig. 3, which shows the difference in the progression-free survival (PFS) between two groups (sTG-UR

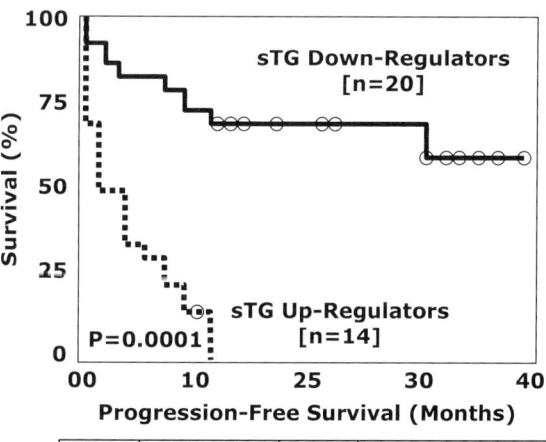

**Fig. 3** The Log Rank test showed significant ($p = 0.0001$) difference in the progression-free survival (PFS) between two groups (sTG-UR and sTG-DR) of vaccine recipients, indicating the survival relevance of post-vaccine sTG down- and upregulation in the patients with metastatic melanoma. (Reprinted from Selvan et al. (2008) Int. J. Cancer, with Permission from Wiley-Liss, Inc)

| Cohort | Progression/Total | PFS (month) (95% CI) | Survival Rate | | |
|---|---|---|---|---|---|
| | | | 24 mon | 36 mon | 48 mon |
| sTG-DR | 7/20 | n/a | 0.70 | 0.61 | 0.61 |
| sTG-UR | 13/14 | 2.0 | 0 | 0 | 0 |

and sTG-DR) of vaccine recipients. Indeed, the log rank test revealed that both OS [$p$ = 0.012] and PFS [$p$ = 0.0001] are highly significant, indicating the survival relevance of post-vaccine sTG down- and upregulation in the patients with metastatic melanoma.

This study documents that sTG levels could be a potential tool to monitor up- or downregulation within 24 weeks post-treatment, and sTG could be an ideal biomarker of therapeutic response for melanoma patients. Such a biomarker would be valuable to monitor the efficacy of a therapy for patients and would provide a scope to change the therapy after 24 weeks. Both endogenous and vaccine-induced anti-ganglioside-IgM antibodies appeared to regulate sTG levels. Non-responders had increased sTG with no or low IgM antibody-response. Any therapy that decreases sTG levels and acts as a sTG down-regulator would be the best therapeutic modality for melanoma. If sTG levels do come down 24 weeks after a therapy, intravenous anti-ganglioside human IgM antibodies can be contemplated as a passive immunotherapy. The sTG level is regulated within 24-weeks post-treatment and, therefore, may serve as an ideal biomarker for assessing therapeutic responses in patients. Clinical correlations of sTG indicate that sTG-downregulating therapy may be an effective treatment strategy for melanoma.

# 6 Conclusions

Every cell in our organs and tissues responds to physico-chemical and biologic stress. One of the responses to stress is to synthesize ceramides, which when accumulated beyond a threshold limit causes apoptosis. Differentiated cells when exposed to stress may undergo dedifferentiation. Such dedifferentiated cells may undergo neoplastic transformation. Ceramides may induce apoptosis, but the neoplastically transformed cells escape ceramide-mediated apoptosis by glycosylating the ceramides. The last step in the glycosylation process is addition of one or more sialic acids. The sialylglycosylceramides are gangliosides that are shunted to the outerlayer of bilayered lipid membrane. Over-accumulation of gangliosides may result in cluster formation on the tumor cell surface and eventually they are shed or released into the tumor microenvironment.

The tumorigenic events preceding immune response to the gangliosides are illustrated in Fig. 4, using ovarian epithelial carcinoma (OEC) as a model. This model is applicable to all the cancer types. The first event depicted in the figure is proliferation of OEC cells. Proliferation of tumor cells also accompanies infiltrating lymphocytes of various kinds. Tumor cell clustering during proliferation may result in lack or insufficiency of oxygen or hypoxia, which may consequently induce tumor cell necrosis. As observed in cryoablation studies, necrosis may release tumor-associated gangliosides (TAG) and other tumor-associated glycoantigens and molecules (TAM) into the stroma or tumor microenvironment and circulation. In OEC, we have observed the accumulation of tumor

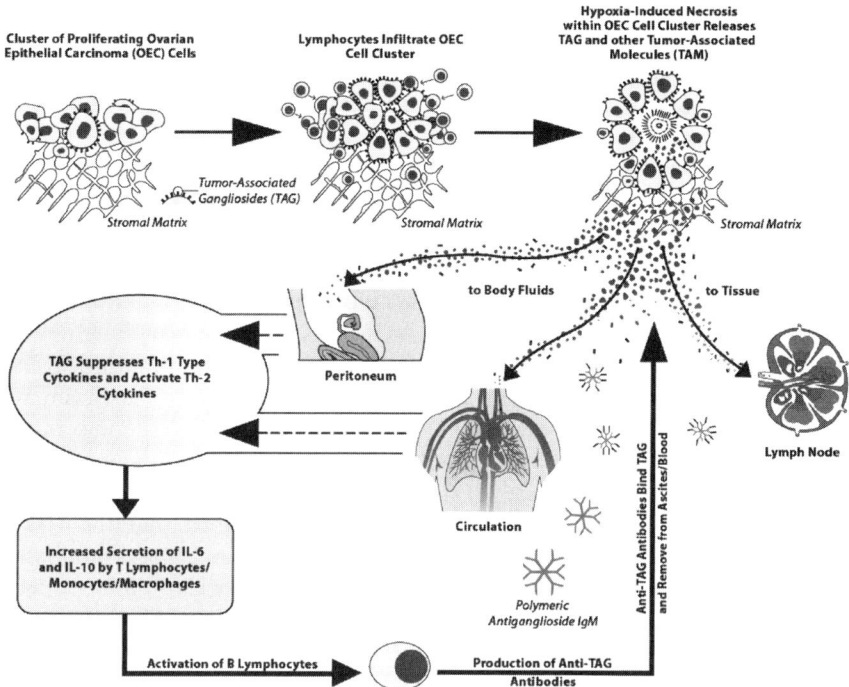

**Fig. 4** Events taking place during tumor progression, which are relevant to glycoimmunomics of human cancer. Using ovarian epithelial cancer as a model, the events leading to immune response to tumor-associated gangliosides (TAG) are narrated as follows: The 1st event is proliferation of tumor cells. The 2nd event is infiltration of a variety of lymphocytes into proliferating tumor cells. The 3rd event is release or shedding of tumor-associated molecules (TAM) and TAG consequent to necrosis resulting due to lack or insufficient oxygen or hypoxia. The 4th event is spreading of TAM and TAG into the stroma or tumor microenvironment, lymph nodes or ascetic fluid, and circulation. The 5th event is TAG-mediated suppression of a variety of Th-1 type cell-mediated immune functions, and at the same time activation of Th-2 cytokines. The 6th event refers to an increased secretion of IL-6 and IL-10 by T cells, monocytes, and macrophages in the sera of patients during tumor progression. The 7th event is that these cytokines activate B lymphocytes, to produce anti-ganglioside antibody response without T-cell help, a unique phenomenon in glycoimmunomics of human cancer. The 8th event is the production of IgM, which can be polymeric and without a J chain, due to involvement of CD5+ B cells. The 9th event is the elimination of immunosuppressive gangliosides by anti-TAG antibodies (*See* Color Insert)

gangliosides in ascetic fluid. In melanoma, tumor-associated gangliosides reach the sentinel lymph nodes even before tumor cells infiltrate the lymph nodes. The final destination of the gangliosides released or shed from tumor cells is portal circulation as depicted in the figure. While, the shed gangliosides suppress a variety of Th-1 type cell-mediated immune functions, they may activate Th-2 cytokines. Clearly, there is an increased secretion of IL-6 and IL-10 by T cells, monocytes, and macrophages in sera of patients during tumor progression. These cytokines have the potential to activate B lymphocytes, resulting in antiganglioside antibody response. These B cells are capable of producing the antibodies against glycoantigens without T-cell help. This is a unique phenomenon underlying glycoimmunomics of human cancer. Invariably, the antibodies produced against gangliosides in human are IgM. IgM can be polymeric and without a J chain. CD5+ B cells may be involved. Immune response to gangliosides does not require exogenous adjuvants because antiganglioside IgM antibodies occur in normal, healthy humans; while their level declines with age but steadily increases with the stage of tumor development. Cryoablation that induced tumor necrosis results not only in the release of gangliosides into the serum of patient but also in elevations in the titers of IgM antibodies to gangliosides.

The major ganglioside of gender-based cancers (Ravindranath, Muthugounder, Presser, Selvan, Portoukalian, Brosman and Morton 2004; Ravindranath et al. 2005; Ravindranath et al. 2007a, 2007b) and melanoma (Ravindranath, Muthugounder, Presser and Morton 2008) appears to be GD1a. Kanda (1999) has shown that the ganglioside GD1a induces interleukin-10 production by human T cells. Kanda and Watanabe (2000) have shown that the GD1a enhances immunoglobulin production by human peripheral blood mononuclear cells. These observations explain why the patients with organ-confined prostate cancer (stage T1 and T2), OEC patients (Ravindranath et al. 2007b), and patients with thin primaries (unpublished) have significantly elevated titers of IgM against GD1a. Interestingly, no such increase in anti-GD1a IgM is noticed in patients with advanced prostate cancer or in melanoma patients with thicker primaries or in normal and healthy volunteers. IgM antibodies against other cancer-associated gangliosides do not show such augmentation during early tumorigenesis. Therefore, anti-GD1a IgM could be the first glycoimmunomic signal for an early event of tumorigenesis. There is no doubt a deeper understanding of the glycoimmunomics of tumor-associated gangliosides may enable developing early biomarkers for human cancer, therapeutic response marker to assess the response to treatment, and also to effectively formulate glycoimmunotherapy of human cancer.

**Acknowledgments** The hypothesis of Glycomics of human cancer was conceived and developed at the Department of Surgical Oncology, University of California, Los Angeles, and at the Department of Glycoimmunotherapy, John Wayne Cancer Institute, under the guidance, tutelage, continuous support, and encouragement of Dr. Donald L. Morton to whom this chapter is dedicated with sincerity, love and respect.

# References

Ando, S. (1983) Gangliosides in the nervous system. Neurochem Int. 5, 507–537.
Ando I, Hoon DS, Suzuki Y, Saxton RE, Golub SH and Irie RF. (1987) Ganglioside GM2 on the K562 cell line is recognized as a target structure by human natural killer cells. *Int J Cancer*. (Erratum in: Int J Cancer) 40, 12–7.

Angata, T., and Varki, A. (2000) Cloning, characterization, and phylogenetic analysis of Siglec-9, a new member of the CD33-related group of Siglecs. Evidence for co-evolution with sialic acid synthesis pathways. J. Biol. Chem. 275, 22127–22135.

Basu, M., De, T., Das, K.K., Kyle, J.W., Chon, H.C., Schaeper, R.J and Basu S. (1987) Glycolipids. Methods Enzymol. 38, 575–607.

Biswas, K., Richmond, A., Rayman, P., Biswas, S., Thornton, M., Sa, G., Das, T., Zhang, R., Chahlavi, A., Tannenbaum, C.S., Novick, A., Bukowski, R. and Finke, J.H. (2006) GM2 expression in renal cell carcinoma: potential role in tumor-induced T-cell dysfunction. Cancer Res. 66, 6816–6825.

Blackhall, F.H., Merry, C.L., Davies, E.J. and Jayson, G.C. (2001) Heparan sulfate proteoglycans and cancer. Br J Cancer. 85, 1094–1098.

Bodey, B., Kaiser, H.E., and Siegel, S.E.(2004) *Cancer Growth and Progression: Immunological Aspects Of Neoplasia (The Role Of The Thymus)*. Springer Verlag, Netherland.

Brewer, J.W., Randall, T.D., Parkhouse, R.M. and Corley, R.B. (1994) IgM hexamers? Immunol. Today. 15, 165–168.

Brutkiewicz, R.R. (2006) CD1d Ligands: The Good, the Bad, and the Ugly. *J. Immunology*, 177, 769–775.

Carubia, J.M., Yu, R.K., Macala, L.J., Kirkwood, J.M., and Varga, J.M. (1984) Gangliosides of normal and neoplastic human melanocytes. Biochem. Biophys. Res. Commun. 120, 500–504.

Chahlavi, A. Rayman, P. Richmond, A.L., Biswas, K., Zhang, R., Vogelbaum, M., Tannenbaum, C., Barnett, G., and Finke, J.H (2005) Glioblastomas induce T-lymphocyte death by two distinct pathways involving gangliosides and CD70. Cancer Res. 65: 5428–38.

Chu, K.U., Ravindranath, M.H., Gonzales, A., Nishimoto, K., Tam, W.Y., Soh, D., Bilchik, A., Katopodis, N. and Morton, D.L. (2000) Gangliosides as targets for immunotherapy for pancreatic adenocarcinoma. Cancer 88:1828–1836.

Davies, E.J., Blackhall, F.H., Shanks, J.H., David, G., McGown, A.T., Swindell, R., Slade, R.J., Martin-Hirsch, P., Gallagher, J.T. and Jayson, G.C.(2004) Distribution and clinical significance of heparan sulfate proteoglycans in ovarian cancer. Clin. Cancer Res. 10, 5178–51186.

Davis, A.C., Roux, K.H. and Shulman, M.J. (1988) On the structure of polymeric IgM. Eur J Immunol. 18, 1001–1008.

Dolo, V. , Li, R., Dillinger, M., Flati, S., Manela , J., Taylor, B.J., Pavan, A. and Ladisch, S. (2000) Enrichment and localization of ganglioside G(D3) and caveolin-1 in shed tumor cell membrane vesicles. Biochim. Biophys. Acta, 1486, 265–274.

Freimer, M.L., McIntosh, K., Adams, R.A., Alving, C.R. and Drachman, D.B. (1993) Gangliosides elicit a T-cell independent antibody response. J Autoimmun. 6, 281–289.

Grayson, G. and Ladisch, S. (1992) Immunosuppression by human gangliosides. II. Carbohydrate structure and inhibition of human NK activity. Cell Immunol. 139, 18–29.

Gillard, B.K., Thomas, J.W., Nell, L.J. and Marcus, D.M. (1989) Antibodies against ganglioside GT3 in the sera of patients with type I diabetes mellitus. J Immunol. 142, 3826–3832.

Hakansson, L., Fredman, P. and Svennerholm, L. (1985). Gangliosides in serum immune complexes from tumor-bearing patients. J. Biochem. 98: 843–849.

Hakomori S.I.(1998) Cancer-associated glycosphingolipid antigens: their structure, organization, and function. Acta Anat (Basel). 161, 79–90.

Hakomori, S.I. (2000) Cell adhesion/recognition and signal transduction through glycosphingolipid microdomain. Glycoconj. J. 17, 143–51.

Hakomori, S.I. (2003) Structure, organization, and function of glycosphingolipids in membrane. Curr. Opin. Hematol. 10, 16–24.

Hakomori, S.I. (2007) Structure and function of lycosphingolipids and sphingolipids: Recollections and future trends. Biochim Biophys Acta. (in Press).

Hakomori, S.I., Handa, K., Iwabuchi, K., Yamamura, S. and Prinetti, A. (1998) New insights in glycosphingolipid function: "glycosignaling domain," a cell surface assembly of glycosphingolipids with signal transducer molecules, involved in cell adhesion coupled with signaling. Glycobiology. 8, 11–19.

Hanisch, F-G. and Müller, S. (2000) MUC1: the polymorphic appearance of a human mucin Glycobiology, 10: 439–449.

Hiraiwa, N., Tsuyuoka, K., Li, Y.T., Tanaka, M., Seno, T., Okubo, Y., Fukuda, Y., Imura, H. and Kannagi, R. (1990) Gangliosides and sialoglycoproteins carrying a rare blood group antigen determinant, Cad, associated with human cancers as detected by specific monoclonal antibodies. Cancer Res. 50, 5497–5503.

Hoon DS, Ando I, Sviland G, Tsuchida T, Okun E, Morton D.L. and Irie RF. (1989) Ganglioside GM2 expression on human melanoma cells correlates with sensitivity to lymphokine-activated killer cells. Int. J. Cancer. 43, 857–862.

Ikehara, Y., Ikehara, S.K and Paulson J.C. (2004) Negative Regulation of T Cell Receptor Signaling by Siglec-7 (p70/AIRM) and Siglec-9. J. Biol. Chem. 279, 43117–43125.

Kalashnikova, M.M. 1985. Appearance of clasmatosis in the normal and pathological liver. Bull. Eksp. Biol. Med. 100: 355–358.

Kanda, N. (1999) Gangliosides GD1a and GM3 induce interleukin-10 production by human T cells. Biochem. Biophys. Res. Commun. 256, 41–44.

Kanda, N., and Tamarisk. (1999) Ganglioside GD1b suppresses immunoglobulin production by human peripheral blood mononuclear cells. Exp Hematol. 27, 1487–1493.

Kanda, N. and Watanabe, S. (2000) Gangliosides GD1a enhances immunoglobulin production by human peripheral blood mononuclear cells. Exp. Hematol. 28, 672–679.

Keller S, Sanderson MP, Stoeck A. and Altevogt, P. (2006) Exosomes: from biogenesis and secretion to biological function. Immunol. Lett. 107, 102–108.

Kolesnick, R. (2002) The therapeutic potential of modulating the ceramide/ sphingomyelin pathway. J. Clin. Invest. 110, 3–8.

Kong, Y., Li, R. and Ladisch, S. (1998) Natural forms of shed tumor gangliosides. Biochim. Biophys. Acta, 1394, 43–56.

Klenk, E.Z. (1942) Gangliosides, a new group of sugar-containing brain lipoids. Z. physiol. Chem. 273, 76–86.

Krishnan, L., Lomash, S., Raj, B.P., Kaur, K.J. and Salunke, D.M. (2007) Paratope plasticity in diverse modes facilitates molecular mimicry in antibody response. J. Immunol. 178, 7923–7931.

Ladisch, S., Sweeley, C.C., Becker, H. and Gage, D. (1989) Aberrant fatty acyl alpha-hydroxylation in human neuroblastoma tumor gangliosides. J. Biol. Chem. 264, 12097–12105.

Mathias, S., Peña, L.A. and Kolesnick, R.N. (1998) Signal transduction of stress via ceramide. Biochem. J. 335, 465–480.

Morris, V.L., MacDonald, I.C., Koop, S., Schmidt, E.E., Chambers, A.F. and Groom, A.C. (1993). Early interactions of cancer cells with the microvasculature in mouse liver and muscle during hematogenous metastasis: videomicroscopic analysis. Clin Exp Metastasis. 11, 377–390.

Nakatsura, T., Komori, H., Kubo, T., Yoshitake, Y., Senju, S., Katagiri, T., Furukawa, Y., Ogawa, M., Nakamura,Y. and Nishimura Y. (2004) Mouse homologue of a novel human oncofetal antigen, glypican-3, evokes T-cell-mediated tumor rejection without autoimmune reactions in mice. Clin. Cancer Res. 10, 8630–8640.

Nicoll, G., Avril, T., Lock, K., Furukawa, K., Bovin, N. and Crocker, P.R. (2003) Ganglioside GD3 expression on target cells can modulate NK cell cytotoxicity via Siglec-7-dependent and -independent mechanisms. Eur. J. Immunol. 33, 1642–1648.

Obeid, L.M., Linardic, C.M., Karolak, L.A. and Hannun, Y.A.(1993) Programmed cell death induced by ceramide. Science. 259, 1769–1771.

Perez, C.A, Ravindranath, M.H., Soh, D., Gonzales, A., Ye, W. and Morton, D.L (2002) Serum anti-ganglioside IgM antibodies in soft tissue sarcoma:clinical prognostic implications. Cancer J. 8, 384–394.

Prokazova, N.V., Dyatlovitskaya, E.V. and Bergelson, L.D. (1988) Sialylated lactosylcera-mides. Possible inducers of non-specific immunosuppression and atherosclerotic lesions. Eur. J. Biochem. 172, 1–6.

Prokazova, N.V. and Bergelson L.D. (1994) Gangliosides and atherosclerosis. Lipids, 29, 1–5.

Randall, T.D., King, L.B. and Corley, R.B. (1990) The biological effects of IgM hexamer formation. Eur. J. Immunol. 20, 1971–1979.

Randall, T.D., Parkhouse, R.M., and Corley, R.B. (1992) J chain synthesis and secretion of hexameric IgM is differentially regulated by lipopoly-saccharide and interleukin 5. Proc. Natl. Acad. Sci. U S A. 89, 962–966.

Rapoport, E., Mikhalyov, I., Zhang, J., Crocker, P. and Bovin, N. (2003) Ganglioside binding pattern of CD33-related Siglecs. Bioorg. Med. Chem. Lett. 13, 675–678.

Ravindranath, M.H., Bauer, P.M., Amiri, A.A., Miri, S.M., Kelley, M.C., Jones, R.C., and Morton, D.L. (1997a) Cellular cancer vaccine induces delayed-type hypersensitivity reac-tion and augments antibody response to tumor-associated carbohydrate antigens (sialyl Le(a), sialyl Le(x), GD3 and GM2) better than soluble lysate cancer vaccine. Anticancer Drugs, 8, 217–224.

Ravindranath, M.H., Gonzales, A.M., Nishimoto, K., Tam, W.Y., Soh, D., and Morton, D.L. (2000) Immunology of gangliosides. Indian J. Exp. Biol. 38, 301–312.

Ravindranath, M.H., Gonzales, A., Soh, D., Nishimoto, K., Tam, W.Y., Bilchik, A., Morton, D.L. and O'Day, S. (2001) Interleukin-2 binds to ganglioside GD(1b). Biochem. Biophys. Res. Commun. 283, 369–373.

Ravindranath, M.H., Hsueh, E.C., Verma, M., Ye, W. and Morton, D.L. (2003) Serum total ganglioside level correlates with clinical course in melanoma patients after immunotherapy with therapeutic cancer vaccine. J. Immunother. 26, 277–285.

Ravindranath, M.H. and Morton, D.L. (1997b) Immunogenicity of membrane-bound gang-liosides in viable whole-cell vaccines. Cancer Invest. 15, 491–499.

Ravindranath, M.H., and Morton, D.L. (2000) Antigens: Carbohydrates. In *Encyclopedia of Life Sciences, MacMillan Co, United Kindom, London.*

Ravindranath, M.H., Morton, D.L. and Irie, R.F. (1989) An epitope common to gangliosides O-acetyl-GD3 and GD3 recognized by antibodies in melanoma patients after active specific immunotherapy. Cancer Res. 49, 3891–3897.

Ravindranath, M.H., Morton, D.L. and Irie, R.F. (1994a) Attachment of monophosphoryl lipid A (MPL) to cells and liposomes augments antibody response to membrane-bound gangliosides. J. Autoimmun. 7, 803–816.

Ravindranath, M.H., Muthugounder, S., Hannah, M.R. and Morton, D.L. (2007a) Signifi-cance of Endogenous augmentation of Antiganglioside IgM in Cancer patients: Potential Tool for Early Detection and Management of Cancer therapy. Ann. N.Y Acad. Sci. 1107, 212–222.

Ravindranath, M.H., Muthugounder, S., Presser, N. and Morton, DL. (2008) Ganglioside signatures of Primary and Nodal Metastatic Melanoma cell lines from the same patient. Melanoma Research (in Press).

Ravindranath, M.H., Muthugounder, S., Presser, N., Selvan, S.R., Portoukalian, J., Brosman, S., and Morton, D.L. (2004) Gangliosides of Organ-confined *versus* Metastatic Androgen Receptor-negative Prostate Cancer. Biochem. Biophys. Res. Commun. 324, 154–165.

Ravindranath, M.H., Muthugounder, S., Presser, N., Selvan, S.R., Santin, A.D., Bellone, S., Saravanan, T.S. and Morton, D.L. (2007b) Immunogenicity of gangliosides in human ovarian cancer. Biochem. Biophys. Res. Commun. 353, 251–258.

Ravindranath, M.H., Muthugounder, S., Presser, N., Ye, X., Brosman, S. and Morton, D.L. (2005a) Endogenous Immune Response to Gangliosides in Patients with Confined Pros-tate Cancer. Int. J. Cancer, 116, 368–377.

Ravindranath, M.H., Muthugounder, S., Saravanan, T.S., Presser, N. and Morton, D.L. (2005b) Human antiganglioside autoantibodies: validation of ELISA. Ann. N. Y. Acad. Sci. 1050, 229–242.

Ravindranath, R.M., Ravindranath, M.H. and Graves, M.C. (1997c) Augmentation of natural antiganglioside IgM antibodies in lower motor neuron disease (LMND) and role of CD5 + B cells. Cell Mol. Life Sci. 53, 750–758.

Ravindranath, M.H., Ravindranath, R.M., Morton, D.L. and Graves, M.C. (1994b) Factors affecting the fine specificity and sensitivity of serum antiganglioside antibodies in ELISA J. Immunol. Methods, 169, 257–272.

Ravindranath, M.H., Tsuchida, T., Morton, D.L. and Irie, R.F. (1991) Ganglioside GM3:GD3 ratio as an index for the management of melanoma. Cancer, 67, 3039–3035.

Ravindranath, M.H., Wood, T.F., Soh, D., Gonzales, A., Muthugounder, S., Perez, C., Morton, D.L., and Bilchik, A.J. (2002) Cryosurgical ablation of liver tumors in colon cancer patients increases the serum total ganglioside level and then selectively augments antiganglioside IgM. Cryobiology, 45, 10–21.

Ravindranath, M.H., Yesowitch, P., Sumobay, C. and Morton, D.L. (2007c) Glycoimmunomics of Human Cancer: Current concepts and future perspectives. Future Oncol. 3, 201–214.

Rich, R.R., Fleisher, T.A., Schwartz, B.D., Shearer, W.T. and Strober, W. (1996) *Clinical Immunology, Principles and Practice,* Vols I and II, CV Mosby, St. Louis.

Sanderson, R.D. (2001) Heparan sulfate proteoglycans in invasion and metastasis. Semin Cell Dev Biol. 12, 89–98.

Santin, A.D., Ravindranath, M.H., Bellone, S., Muthugounder, S., Palmieri, M., O'Brien, T.J., Roman, J., Cannon, M.J. and Pecorelli. S. (2004) Increased levels of gangliosides in the plasma and ascetic fluid of patients with advanced ovarian cancer. Int. J. Obstet. Gynaecol. 111, 613–618.

Selvan, R.S., Ihrcke, N.S. and Platt, J.L. (1996) Heparan sulfate in immune responses. Ann.N. Y.Acad. Sci. 797, 127–139.

Shurin, G.V., Shurin, M.R., Bykovskaia, S., Shogan, J., Lotze, M.T. and Barksdale, E.M Jr. (2001) Neuroblastoma-derived gangliosides inhibit dendritic cell generation and function. Cancer Res. 61, 363–369.

Selvan, S.R., Dillman, R.O., Fowler, A.W., Carbonell, D.J., and Ravindranath, M.H. (2008) Monitoring response to Treatment in Melanoma Patients: potential of a serum Glycomic marker. Int. J. Cancer, 122, 1374–1383.

Thornton, M.V., Kudo, D., Rayman, P., Horton, C., Molto, L., Cathcart, M.K., Ng, C., Paszkiewicz-Kozik, E., Bukowski, R., Derweesh, I., Tannenbaum, C.S. and Finke, J.H. (2004) Degradation of NF-kappa B in T cells by gangliosides expressed on renal cell carcinomas. J. Immunol. 172, 3480–3490.

Tonegawa, Y. and Hakomori, S.I. (1977) "Ganglioprotein and globoprotein": the glycoproteins reacting with anti-ganglioside and anti-globoside antibodies and the ganglioprotein change associated with transformation. Biochem. Biophys. Res. Commun. 76, 9–17.

Valentino, L.A. and Ladisch, S. (1992) Localization of shed human tumor gangliosides: association with lipoproteins. Cancer Res. 52, 810–814.

Wiegandt,H. Gangliosides. in Glycolipids (New Comprehensive Biochemistry, Vol. 10), pp. 99–260 (edited by H.Wiegandt, Elsevier, Amsterdam) (1985).

Wilson, M.T., Singh, A.K. and Van Kaer, L. (2002) Immunotherapy with ligands of natural killer T cells. Trends Mol Med. 8, 225–231.

Young, D.C., and Moody, D.B. (2006) T-cell recognition of glycolipids presented by CD1 proteins. Glycobiology, 16, 103R–112R.

# Translational Immunomics of Cancer Immunoprevention

Pier-Luigi Lollini

**Abstract** Cancer immunoprevention, a recent development of tumor immunology based on vaccines and other immunological maneuvers to actively reduce the risk of cancer development, will greatly benefit from immunomic approaches for its progress in basic research and for its translation to human applications. The main approaches discussed in this chapter are the use of genetically modified mouse models, microarray studies, and data mining approaches to define antitumor immune responses and to discover new oncoantigens, and agent-based mathematical models to simulate tumor–immune system interactions and to perfect vaccination protocols.

**Keywords** Translational research · Immunomics · Cancer · Immunoprevention

## 1 Cancer Immunoprevention

The immune system is very efficient in preventing disease, as clearly illustrated by the long-lasting immunity elicited by naturally occurring infections or by vaccines (Amanna, Carlson and Slifka 2007). However, until recently the sole application of immunology to cancer was immunotherapy, i.e. after tumor onset. The radical paradigm switch behind cancer immunoprevention of cancer is the attempt to activate the immune system *before* tumor onset, to prevent carcinogenesis.

The potential advantages of cancer immunoprevention are manifold, because this approach targets preneoplastic or early neoplastic lesions instead of established, malignant tumor masses. Preneoplastic lesions are not only smaller in size but more importantly have undergone fewer genetic changes and are less entrenched in the surrounding microenvironment than established tumors.

The perspectives are even more attractive in the case of tumors caused by viruses, because vaccines can induce protective immunity not against cancer

P.-L. Lollini
Section of Cancer Research, Department of Experimental Pathology, University
of Bologna, Viale Filopanti 22, I-40126 Bologna, Italy
e-mail: pierluigi.lollini@unibo.it

A. Falus (ed.), *Clinical Applications of Immunomics*,
DOI: 10.1007/978-0-387-79208-8_12, © Springer Science+Business Media, LLC 2009

cells, but against the causative agent itself, effectively preventing host exposure to the primary carcinogen. Moreover, the antigenic determinants on viral particles are more easily recognized by the immune system than the endogenous proteins expressed by tumor cells. For those reasons, the first successful clinical implementation of cancer immunoprevention are vaccines against oncogenic human viruses.

## 2 Immunoprevention of Viral Tumors

Hepatitis B virus (HBV) is not carcinogen per se, but in a minority of patients acute hepatitis B is not cured and evolves into chronic hepatitis, which in turn can foster the onset of hepatocellular carcinoma. The vaccine against HBV, currently in use worldwide, was the first human vaccine for which immunoprevention was clearly demonstrated (Chang et al. 2005). Tumor prevention afforded by the HBV vaccine might be regarded as a by-product of protection from hepatitis B. A more direct instance of cancer immunoprevention comes from recently approved vaccines against human papilloma viruses (HPV).

Unlike HBV, HPV directly cause various benign and malignant epithelial neoplasms, in particular cancer of the uterine cervix, also called cervical carcinoma, the second most common cancer in women worldwide, causing almost 300,000 deaths each year (Stewart and Coates 2005). After a highly successful series of clinical trials, two vaccines against HPV were recently found to be effective and approved in Western countries for use in the general population. In clinical trials, the HPV vaccines prevented almost completely both infection and tumor onset, raising the hope that similar levels of protection in the general population could lead to the disappearance of HPV-related cervical carcinoma, much as smallpox was eradicated worldwide by vaccination (Sawaya and Smith-McCune 2007).

Cancer preventive vaccines against viruses and other infectious agents can save million of human lives, because the current incidence of infection-related cancer worldwide is in excess of 1.5 million new cases per year, one third caused by HPV and one fourth by HBV. The success of HBV and HPV vaccines leads to the prediction that other vaccines currently under development could further the impact of cancer immunoprevention, for example, to hepatitis C virus and the remaining cancer-related infectious agents.

Infection-related tumors comprise between 15% and 20% of all human tumors – what can immunoprevention do for the remaining 80%?

## 3 Immunoprevention of Non-viral Tumors

The study of spontaneous tumor onset and progression was greatly enhanced in recent years, thanks to the advent of genetically-modified mouse models (GEM), in which induced genetic alterations increase or decrease the expression or function of specific genes (Green and Hudson 2005; Frese and Tuveson 2007). A major contribution of GEM to tumor immunology was the definitive demonstration that the immune system protects the host not only from infectious agents but also from the onset of all types of tumor. GEM lacking functional immunity invariably develops precocious tumors with a very high incidence, nearing 100% (Shankaran et al. 2001; Smyth, Dunn and Schreiber 2006).

However, the very existence of neoplastic disease in all vertebrates proves that the immune system is not fully efficient in tumor elimination; moreover, immunity is subject to aging, hence old immunocompetent humans and mice are prone to increasing tumor onset and development. This leads to the idea that a stimulation of antitumor immunity in young, healthy individuals could reduce tumor incidence later in life.

Early demonstrations of this concept came both from conventional mice exposed to chemical carcinogens and from GEM receiving interleukin 12 (IL-12), a cytokine that directly or indirectly stimulates T, B, and NK immunity (Noguchi et al. 1996; Boggio et al. 1998). Carcinogenesis was significantly delayed in all models, and in some cases a sizeable proportion of mice remained tumor-free for very long periods.

A number of studies of cancer immunoprevention was then done in various GEM models, using many different immune stimuli. Here we will focus our attention on HER-2/neu transgenic mice, which were the most popular model of immunoprevention (Lollini et al. 2006).

## 3.1 HER-2/neu Transgenic Mice

HER-2/neu is an oncogene amplified and hyperexpressed in about one fourth of human breast cancer, and in other human neoplasms. To obtain models of human breast cancer, the HER-2/neu gene was expressed in mice under the transcriptional control of murine mammary tumor virus (MMTV) long terminal repeats (LTR), or other promoters directing expression to the mammary gland, and eventually leading to the development of mammary carcinomas closely resembling the human counterpart (Di Carlo et al. 1999). This is a very flexible model of carcinogenesis in which various elements, such as the species of origin of the oncogene, the promoter, and the genetic background of the mouse, can be modularly interchanged to investigate different aspects of mammary carcinogenesis and tumor development. Here we will mainly refer to a transgenic mouse line named BALB-NeuT, harboring a mutant rat HER-2/neu gene driven by a MMTV LTR (Boggio et al. 1998). Mammary carcinogenesis is highly aggressive in BALB-NeuT mice; the first tumor invariably develops by six months of age, followed within a few weeks by the growth of multiple tumors in all ten mammary glands, and by metastatic spread to the lungs in a high percentage of cases. A further point of similitude with malignant human breast tumors is that overt tumor growth is preceded first by the development of preneoplastic lesions and then by the formation of carcinomas in situ. This process is eminently repeatable in all female mice, thus allowing the design of preventive protocols targeting the successive preneoplastic and neoplastic stages.

## 3.2 Mammary Carcinoma Immunoprevention in HER-2/neu Transgenic Mice

Various types of non-specific immune stimulants were tested in HER-2/neu transgenic mice, including IL-12, sphingolipids, and bacterial CpG DNA sequences (Lollini et al. 2006). Other researchers used various vaccines to enhance the specific immune response against the HER-2/neu oncoprotein p185 (Lollini et al. 2006). All the approaches mentioned above significantly delayed mammary carcinogenesis; however, in most cases mice eventually succumbed to progressive tumors, in particular whenever highly aggressive cancer models like the BALB-NeuT line were used.

The results indicated that cancer immunoprevention can indeed be obtained with different immunological approaches; however, it was also clear that more potent treatments were needed. To tackle this problem, our laboratory and that of Guido Forni and Federica Cavallo (University of Turin, Italy) designed new vaccines combining multiple effective immune stimuli. We formulated a cellular vaccine, called the Triplex vaccine, made of transgenic mammary carcinoma cells expressing the target oncoprotein p185 and two potent biological adjuvants, IL-12 and allogeneic major histocompatibility complex (MHC) class I antigens, which are known to stimulate multiple T cell clones. Forni and Cavallo used a plasmid DNA vaccine containing CpG sequences and encoding the extracellular and transmembrane domains of p185; electroporation in vivo enhanced DNA uptake and expression, and recombinant IL-12 was also used in some experiments as a further adjuvant. Both vaccines produced highly convergent results (Lollini et al. 2006), and the following discussion refers to both vaccines without distinction, when not otherwise specified.

For the first time, vaccinated mice were completely protected from mammary carcinoma onset (Nanni et al. 2001; De Giovanni et al. 2004; Quaglino et al. 2004). The proportion of tumor-free mice at one year of age, when experiments were usually concluded, was constantly above 85–90%. More importantly, the morphologic analysis of mammary glands showed that tumor progression was arrested by vaccination at the stage of atypical hyperplasia, that is the stage reached when vaccinations started. A striking finding was the presence of normal-looking mammary glands, devoid of HER-2/neu expression, in one-year-old mice. In summary, vaccinations effectively silenced the constitutive expression of an oncogene and blocked progression from preneoplastic lesions to overt malignancy.

# 4 Transcriptomics of HER-2/neu Mammary Carcinoma Development and Immunoprevention

The analysis of immune responses elicited by vaccination at the systemic level in the peripheral blood and spleen using standard immunoassays revealed an unexpected absence of cytotoxic T cell (CTL) responses and a strong induction of T cell cytokines, in particular γ-interferon (IFN-γ), and of anti-p185 antibodies (Nanni et al. 2001; De Giovanni et al. 2004). The absence of a significant CTL response, which in many immunotherapeutic contexts is the main effector of antitumor immunity, was probably a boon in our long-term prevention system, because chronic CTL responses are invariably toxic for the host, whereas our vaccinated mice did not show any sign of chronic toxicity.

To investigate local immunity within the mammary gland we initially used immunohistochemistry, which in other circumstances had greatly contributed to the interpretation of the immune responses induced by gene-transduced cell vaccines through the analysis of leukocyte infiltration (Musiani et al. 1997). However, in this case the vaccine did not induce a prominent infiltration of the mammary gland, despite the high potency of the preventive effect.

We then resorted to DNA microarrays to analyze changes in the transcriptome during mammary carcinogenesis in BALB-neuT mice and after prophylactic vaccinations. Here we will concentrate on immunological results, the reader interested in the progression of mammary carcinogenesis will find elsewhere a complete description of the results (Astolfi et al. 2005a). Microarray results provided a coherent picture in excellent agreement with more conventional technologies. In particular, we found in the mammary glands of vaccinated mice a strong activation of many IFN-responsive genes, attesting a functional local response to the circulating cytokine. It is interesting to note that transcriptomic analysis also revealed the unexpected activation of some lactation-related genes in the mammary glands of virgin vaccinated mice, possibly induced by the activation of Stat5 in mammary cells by IFN-γ, mimicking the physiologic effects of prolactin (Astolfi et al. 2005b). It is likely that we would not have uncovered this phenomenon by doing hypothesis-driven experiments. Microarray results also showed a strong activation of immunoglobulin genes in lymphnodes draining the mammary gland, thus confirming the copious antibody production previously found in the peripheral blood.

The major drawback of studying immunocompetent mice receiving powerful vaccines is that it is impossible to discriminate between protective immune responses and the various "bystander" immune responses equally elicited by vaccination, but not contributing to protection from tumor onset. One clear example for the Triplex vaccine are the immune responses directed against allogeneic MHC molecules expressed by vaccine cells, which attest the adjuvanticity of this vaccine component, but have no target in BALB-NeuT mice.

To precisely define protective immune responses, we crossed BALB-NeuT transgenic mice with various knockout mice selectively lacking immune responses. We found that the Triplex vaccine was completely ineffective if administered to mice lacking either IFN-γ or IgG antibodies. Interestingly, the Ig isotypes of IFN-γ knockout mice responding to the Triplex vaccine showed a selective deficit of IgG2a and other isotypes known to be IFN-dependent, but not of other isotypes like IgG1 (Nanni et al. 2001; Nanni et al. 2004). On the other side, IFN-γ was produced by antibody-deficient mice in response to vaccinations. In conclusion, the use of immunodeficient HER-2/neu transgenic GEM models allowed us to formally demonstrate the key role of IFN-γ and anti-p185 antibodies in the protection from mammary carcinoma afforded by the Triplex vaccine. Only some of the many activities of IFN-γ were mechanistically involved in cancer prevention; in particular, the induction of Ig isotype switch to cytotoxic immunoglobulins like IgG2a.

The global message is that the analysis of the complex immune response elicited by a potent multivalent vaccine can only be pursued through an immunomic approach combining classical immunoassays, transcriptomics, and GEM models, possibly with the further inclusion of proteomics and immunoproteomics.

# 5 Oncoantigens

The results of cancer immunoprevention in HER-2/neu transgenic mice represents a departure from classical paradigms of tumor immunology, based on CTL activity in a therapeutic context. This led us to reconsider the importance of the target antigen and its peculiarities *vis-à-vis* other tumor antigens.

Two features of HER-2 were clearly instrumental in determining the success of cancer immunoprevention in transgenic mice. Firstly, in this system HER-2 plays a fundamental role as the oncogene driving the carcinogenic process while at the same time being the target antigen of the immune responses elicited by vaccines (Lollini and Forni 2003). Most tumor antigens like CEA and PSA are not directly involved in neoplastic transformation and progression; hence tumor cells can easily downmodulate their expression without consequence for tumor growth, whereas downmodulation of HER-2/neu in transgenic mice resulted in the loss of in vivo tumorigenicity (Nanni et al. 2000). This implies that the selection of tumorigenic antigen-loss variants is highly unlikely, at least during the early phases of mammary carcinogenesis, when tumor cells are "addicted" to HER-2/neu. The second fundamental property of HER-2 is that its product p185 is an integral cell surface protein, accessible in its native form to antibodies. Many tumor antigens, like those of the MAGE family, are only present inside the cell and can be immunologically recognized only by T cells in the form of processed peptides brought to the cell surface by MHC molecules. However, downmodulation of MHC expression is almost ubiquitous in tumors (Garrido and Algarra 2001), thus preventing antigen recognition by cytotoxic T cells and killing of tumor cells. In this context, only surface molecules like HER-2 offer a persistent target to the humoral arm of the immune system.

We formalized the concepts outlined above in the definition of a new class of tumor antigens named *oncoantigens*, whose hallmarks are surface expression and a mechanistic role in tumor onset and neoplastic progression (Lollini et al. 2006).

# 6 Mining for New Oncoantigens

The current database of cancer antigens contains very few members fulfilling the definition of oncoantigens, among which we can cite MUC1 and the B and T cell idiotypes, in addition to HER-2 (Lollini et al. 2005b). This means that most known tumor antigens are unsuitable targets for cancer immunoprevention, being prone to tumor escape strategies that can affect immunotherapy as well.

We propose that an extensive de novo search should be conducted to identify novel oncoantigens. One straightforward way to attack this problem is by analogy, i.e., to investigate molecules conceptually similar to HER-2, such as other members of the HER/ErbB family, like EGFR/HER-1, or members of the receptor tyrosine kinase superfamily and their ligands, which includes promising candidates like the insulin-like growth factors and their receptors (Lollini et al. 2005b).

Federica Cavallo and colleagues have recently set forward an integrated immunomic strategy to search for novel oncoantigens (Cavallo, Calogero and Forni 2007). The first step is to use cancer-prone GEM models and transcriptomics to identify not only genes overexpressed in established tumors but also potentially more important genes expressed during the early stages of carcinogenesis, which might be more relevant targets for immunoprevention. In a translational perspective, comparative transcriptomics is then used to identify the subset of genes that have human equivalents. The choice of candidates to be tested in vivo can be further restricted through the analysis of coordinate gene expression in the course of carcinogenesis, and through a comparison of the cancer transcriptome with those of normal tissues, to include only potential oncoantigens expressed in neoplastic lesions during selected stages of tumor progression and not expressed by normal cells, to avoid the danger of auto-immune reactions. DNA or other types of vaccine are then used to immunize appropriate cancer-prone GEM and to verify if preventive immune responses can be elicited against each candidate. It should be noted that immune monitoring of anti-oncoantigen responses is very simple, because by definition it can be done by measuring circulating antibodies in serum. The final key step is the demonstration that the immune response against the candidate oncoantigen can indeed control tumor onset and progression.

# 7  Translational Development of Cancer Immunoprevention

The potency of cancer immunoprevention in preclinical mouse models indicates that this novel concept could be fruitfully applied to human conditions;, however, many hurdles have to be overcome to achieve this goal. We will devote the rest of this chapter to translational approaches implemented in silico, in vitro or in vivo.

Immunological maneuvers that were successful in preventing tumor onset in GEM models, most notably the Triplex vaccine, achieved maximal protection only when vaccinations were repeated chronically for the entire lifespan of the host. Shorter vaccination protocols elicited a highly significant long-lasting immune memory that protected the host from tumor onset for several months, but a progressive decline of immunity invariably paved the way to tumor growth. Only chronic administration of the Triplex vaccine completely prevented mammary carcinoma; however, it is doubtful that such an intensive vaccination protocol (more than 50 vaccinations over one year) would ever find application in humans.

The chronic protocol of Triplex vaccination might include a redundant number of vaccine treatments, but protocol optimization purely by means of in vivo experiments (each lasting at least one year) was clearly unfeasible. For this reason, we first constructed an agent-based mathematical model that faithfully reproduced in silico the results of cancer immunoprevention obtained with the Triplex vaccine in HER-2/neu transgenic mice (Pappalardo et al. 2005), then we implemented a novel genetic search strategy to minimize or reduce the number of vaccinations while keeping intact the level of cancer prevention (Lollini, Motta and Pappalardo 2006).

# 8  Mathematical Models of Cancer Immunoprevention

To build a solid and valid model that is both able to represent the biological scenario and to predict the result of an experiment conducted with a specific vaccination schedule, it is useful to follow a sequence of clearly defined steps. Firstly, one needs to understand the entities that, according to current biological knowledge, are relevant for the model. Then, one describes in detail the properties of the entities, the interaction rules, the internal states, and so on. Once the scenario is defined, one can model it using either a simplified mathematical model or a more detailed computational model. We chose the latter approach because it is very flexible and can be described in biological terms.

The model (Motta et al. 2005; Pappalardo et al. 2005) was implemented in a simulator called SimTriplex using a Lattice Boltzmann-like approach, including the entities (cells and molecules) of the adaptive and natural immune system, the cancer, and the vaccine (Fig. 1). The cytokines relevant in activating the immune response were also modeled. All of the various classes of immune

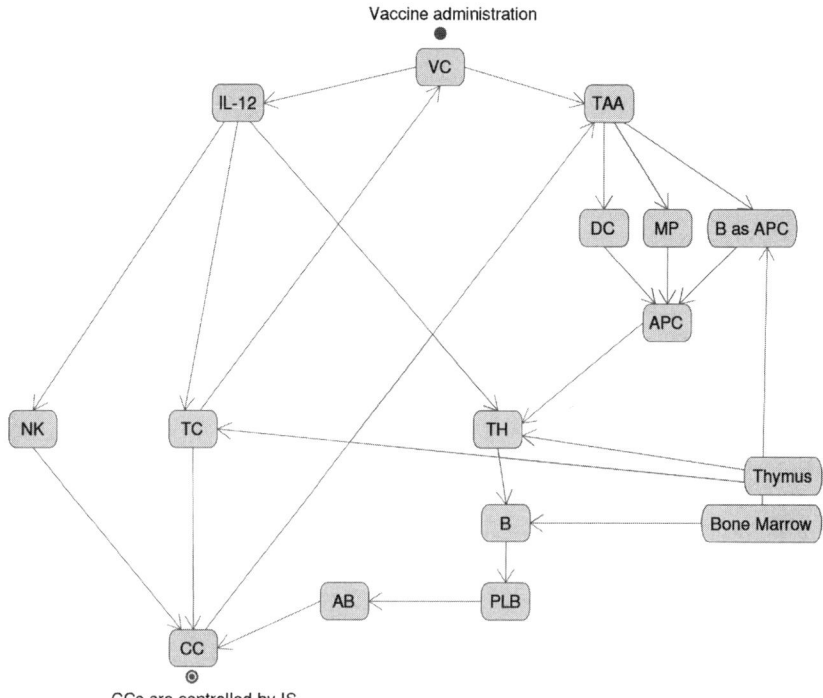

**Fig. 1** SimTriplex logic interactions chart. Vaccine Cells (VC) produce interleukin 12 (IL-12) and release Tumor Associated Antigens (TAA) when they die either spontaneously or as a consequence of cytotoxic T cell (TC) lysis upon recognition of allogeneic MHC on VC. TAA interact with antigen presenting cells (APC), i.e., macrophages (MP), dendritic cells (DC), and B cells, which internalize TAA and become able to stimulate helper T cells (TH). A TH↔B interaction induces differentiation of B cells into plasma cells (PLB), which produce antibodies (Ab) recognizing cancer cells (CC). IL-12 enhances the effectiveness of Natural Killer (NK) and TC against CC. The final state is that CC are controlled by immune responses activated by VC. All the interactions embed a probabilistic rule. which depends on biological affinity or other factors(Reproduced from Lollini, Motta and Pappalardo 2006, BMC Bioinformatics, BioMedCentral Ltd.)

functional activity, phagocytosis, immune activation, opsonization, infection, and cytotoxicity are described using probability functions and translated into computational rules.

An interaction between two entities is a complex stochastic event, which may end with a state change of one or both entities. Interactions can be specific or non-specific. Specific interactions need a recognition phase between the two entities. Recognition is based on the Hamming distance and affinity function, and is eventually enhanced by adjuvants. Both specific and non-specific interactions are stochastically determined using a probability function, which depends upon different parameters computed via random number generators. Changing the seed of the random number generator yields a different sequence of probabilistic events. This simulates biological differences between individuals who share the same event probabilities.

Once the model has been built, one needs to tune the free parameters against experimental data. This was done choosing randomly two sets of different *virtual* mice, S1 and S2, each one made of 100 different mice. A subset of 6 mice was then chosen from S1 and, with a trial-and-error method, free parameters were tuned to fit the experimental data. Finally, results were verified using all mice included in sets S1 and S2. Biological and immunological wisdom were used to drive the tuning phase.

At the end of the tuning phase model, results were in excellent agreement with in vivo experiments (Fig. 2).

SimTriplex simulator is able to simulate in $\sim$30 seconds an in vivo experiment lasting one year. The model can be used, as a virtual laboratory, to search for an optimal vaccination schedule. However, even if a virtual experiment is much faster than a real one (30 sec $\ll$ 1 year $\sim 3 \cdot 10^7$ secs), the number of possible schedules is still too high to be tested exhaustively. The evaluation of the computational time required for an exhaustive search is an interesting and instructive exercise. Taking into account the actual duration of one in vivo experiment ($\sim$400 days) and the fact that vaccinations are usually administered twice weekly, the number of days available for vaccine administration is $\sim$100 days. Hence the number of possible different schedules is $2^{100} \approx 10^{30}$. Considering that a virtual simulation takes 30 seconds, if we were to test all possible vaccination schedules, we would need $3 \cdot 10^{31}$ seconds $\approx 10^{24}$ years. This computational time exceeds any present or future computational capacity.

To approach this problem, one needs to use mathematical optimization methods.

The first step needed in order to solve the optimization problem is to translate the biological concept of effectiveness of a vaccination schedule in mathematical language. For this purpose, one considers the time interval $[0, T]$, where $T$ is the time length of the in vivo experiment. Then the given time interval is divided in $N - 1$ equally spaced subintervals of width $\Delta t$, i.e. $\{t_1 = 0, t_2, \ldots, t_i, \ldots, t_N = T\}$, where $t_k = t_{k-1} + \Delta t$ and $\Delta t = 1$ day, which corresponds to 3 time steps of the simulator. Vaccine can only be administrated at time $t_i$.

Consider a binary vector representing the sequence of vaccinations, $x = \{x_1, x_2, \ldots, x_i, \ldots, x_N\}$, where $x_i = 0/1$ means, respectively, administration/no-administration of the vaccine. The label $i$ represents the time $t_i$ of the vaccine administration and $t_N$ the end of the vaccination period. The number of vaccine administrations is given by

$$n = \sum_{i=1}^{N} x_i$$

The time of tumor formation, $\tau(x, \lambda_j)$, is a function of the vaccination schedule $x$ administered to the mouse $j \in S$ and of a parameter $\lambda_j$, which describes the biological diversity as described before. If $\tau < t_N$ the simulator

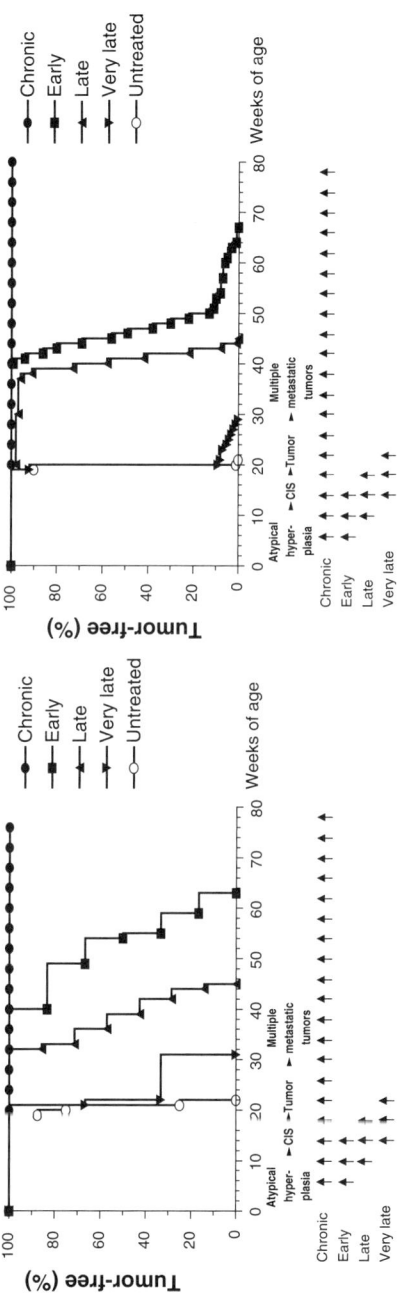

**Fig. 2** Tumor-free survival curves of real (on the *left*) and virtual (on the *right*) mice receiving the Triplex vaccine according to different protocols. Each *arrow* at the bottom of the graphs represents one cycle of vaccination. The sequence of neoplastic progression in untreated mice is outlined under the x axis; CIS, carcinoma in situ (Reproduced from Motta et al. 2005, Immunome Res., BioMedCentral Ltd.)

will return a time $\tau$, otherwise, as the simulation stops at $t = T = t_N$, the simulator will set $\tau = T_N$. A vaccine schedule is effective if $\tau = T_N$.

This definition can be applied either to a single mouse or to a set of mice.

Using the quantities defined above, we can formulate the optimization problem as the problem of finding an $\bar{x}$ such that:

$$\begin{cases} \tau(\bar{x}, \lambda_j) = \max\{\tau(x, \lambda_j)\} \\ (\bar{x}) = \min\{n(x)\} \end{cases}$$

where $n(x)$ represents the number of injections included in schedule $x$.

The function $\tau$ depends on the response of the stimulated immune system. Such a function cannot be expressed in an analytical form: it can only be computed through the simulator introduced previously. From the mathematical point of view, this is a multi-objective optimization problem. For this class of problems, a variety of methods and algorithms are available, i.e., branch and bound, general cutting edge, dynamic programming, heuristic algorithms, decomposition algorithms (Nemhauser and Wolsey 1998), and evolutionary algorithms (Dorne, Dorigo and Glover 1999). With respect to further biological requirements, genetic algorithms (GA) are the most flexible approach and thus were adopted for this search.

Generally speaking, the strategy used tried to reproduce the behaviour of entities elicited by the Chronic schedule using the minimum number of vaccine administrations. As the Chronic schedule is effective in vivo, this should guarantee that a schedule showing a similar kinetics of biological entities should also be effective.

The problem of finding a schedule suitable for all the virtual mice in the sets S1 and S2 was solved using the same strategy used in parameter fitting, i.e the problem was solved simultaneously for a randomly choosen sample of 8 mice and the result was then tested on the complete set. This strategy did work well and the solution found was effective for a high percentage of the mice in sets S1 and S2.

Finally, it is worth to point out that the optimal schedule search was also constrained by practical laboratory requirements, namely that vaccinations were allowed only during working days. This requirement, even if not biologically driven, is crucial to implement in vivo protocols found in silico. This constraint was easily added by an appropriate reduction of the search space, i.e. the cardinality of the vector $x$ previously defined.

The overall strategy search was able to find a schedule, which, according to the simulator, protects 85% of the mice of set S1 from tumor onset (Lollini, Motta and Pappalardo 2006). This vaccination is currently being tested in vivo.

Finally, it is worth to point out that a genetic algorithm with an attached simulator is a long and complex computational task. It requires a prohibitive amount of running time on a single-CPU machine: we needed high-performance computational resources to run the genetic search. Such a large computational resource may be available in research environments, but not for normal use in medical environments, like hospitals, therefore techniques to downsize the computational requirements will have to be investigated.

For this reason other optimal search techniques, like Simulated annealing, are under investigation. Preliminary results indicate that this method finds equivalent, but not equal, results using only few hours of computational time on multiprocessor workstations available today in the consumer market.


# 9 Toward Clinical Implementation, or Back to Immunotherapy

Modern studies of human cancer prevention, for example by means of drugs, (chemoprevention) entail treatment of thousands of individuals at risk for many years, followed by decades of monitoring (Cuzick et al. 2003; Bao, Prowell and Stearns 2006). It is quite obvious that a new biological modality of cancer prevention such as immunoprevention with vaccines will have to be first carefully tested in more limited and economical trials.

One reason because chemoprevention trials need to enrol thousands of individuals is that the risk of cancer in the general population is (luckily) comparatively low. One possible way to reduce the size of the study is to identify subgroups at high risk, for example owing to a genetic predisposition. It must be kept in mind that most hereditary neoplastic syndromes are primarily caused by the inactivation of tumor suppressor genes, which in many cases does not offer antigenic targets. However most hereditary tumors eventually arise as a consequence of combined somatic alterations of oncogenes and tumor suppressor genes. We showed that in a bigenic mouse model combining p53 heterozygous knockout and HER-2/neu mutation the Triplex vaccine prevented multiple tumor types (Croci et al. 2004). Most notably, anti-HER-2 immunity elicited by the vaccine prevented the loss of the functioning p53 allele, possibly by inhibiting the expansion of mammary cell populations that would subsequently be prone to p53 loss. The main difference between this bigenic model and humans carrying hereditary alterations of tumor suppressor genes is that in the former the target oncogen/antigen was predefined, whereas in the latter it is unknown a priori. Therefore, the definition of immunological targets in human hereditary cancer syndromes will again require data mining of microarray results, using strategies similar to that described above (*see* "Mining for new oncoantigens").

An alternative way to implement cancer immunoprevention in humans would be to follow the pathway traced by tamoxifen, a selective estrogen receptor modulator now approved for chemoprevention of breast cancer (Fabian and Kimler 2005). Tamoxifen was first used only for cancer therapy; however, the finding that, in addition to its antimetastatic effects, tamoxifen also prevented the onset of multiple primary breast tumors paved the way to large prevention studies.

To follow the same pathway, cancer immunoprevention should therefore be tested first in advanced cancer patients with conventional therapeutic study designs, starting with phase I trials (Lollini et al. 2005a). This type of clinical deployment will require a new round of ad hoc preclinical studies, because the preventive setup used so far is inappropriate for translation to human therapy. To this end we tested the Triplex vaccine in various therapeutic approaches against growing mammary tumors and lung metastases. As expected, we found that therapy of preexisting neoplastic lesions was less effective than prophylaxis, nonetheless the Triplex vaccine inhibited metastatic development by 90%; moreover a combination of Triplex with a monoclonal antibody against regulatory T cells resulted in a 99% inhibition of lung metastases (Nanni et al. 2007).

These results open up the possibility of early clinical testing of cancer immunopreventive strategies in small trials in metastatic cancer patients, clearly a more feasible approach to clinical implementation than large-scale preventive trials.

**Grant support**   EC contract FP6-2004-IST-4, No.028069 (ImmunoGrid); Italian Association for Cancer Research (Milan, Italy); Italian Ministry for Education, University and Research PRIN projects; Italian Ministry of Health; University of Bologna (Bologna, Italy); AlmaMedicina Foundation (Bologna, Italy); Italian Leukemia Association; "Pallotti" funds, Department of Experimental Pathology, University of Bologna.
A. Palladini is the recipient of a fellowship from the Italian Foundation for Cancer Research (Milan, Italy). S. Croci is the recipient of a research fellowship ("Assegno di Ricerca") from the University of Bologna. A. Antognoli is the recipient of a Ph.D. fellowship from the University of Bologna.

# References

Amanna, I. J., Carlson, N. E., and Slifka, M. K. 2007. Duration of humoral immunity to common viral and vaccine antigens. N. Engl. J. Med. 357:1903–1915.

Astolfi, A., Landuzzi, L., Nicoletti, G., De Giovanni, C., Croci, S., Palladini, A., Ferrini, S., Iezzi, M., Musiani, P., Cavallo, F., Forni, G., Nanni, P., and Lollini, P. L. 2005a. Gene expression analysis of immune-mediated arrest of tumorigenesis in a transgenic mouse model of HER-2/neu-positive basal-like mammary carcinoma. Am. J. Pathol. 166:1205–1216.

Astolfi, A., Rolla, S., Nanni, P., Quaglino, E., De Giovanni, C., Iezzi, M., Musiani, P., Forni, G., Lollini, P. L., Cavallo, F., and Calogero, R. A. 2005b. Immune prevention of mammary carcinogenesis in HER-2/neu transgenic mice: a microarray scenario. Cancer Immunol. Immunother. 54:599–610.

Bao, T., Prowell, T., and Stearns, V. 2006. Chemoprevention of breast cancer: tamoxifen, raloxifene, and beyond. Am. J. Ther. 13:337–348.

Boggio, K., Nicoletti, G., Di Carlo, E., Cavallo, F., Landuzzi, L., Melani, C., Giovarelli, M., Rossi, I., Nanni, P., De Giovanni, C., Bouchard, P., Wolf, S., Modesti, A., Musiani, P., Lollini, P. L., Colombo, M. P., and Forni, G. 1998. Interleukin 12-mediated prevention of spontaneous mammary adenocarcinomas in two lines of Her-2/neu transgenic mice. J. Exp. Med. 188:589–596.

Cavallo, F., Calogero, R. A., and Forni, G. 2007. Are oncoantigens suitable targets for anti-tumour therapy? Nat. Rev. Cancer 7:707–713.

Chang, M. H., Chen, T. H., Hsu, H. M., Wu, T. C., Kong, M. S., Liang, D. C., Ni, Y. H., Chen, C. J., and Chen, D. S. 2005. Prevention of hepatocellular carcinoma by universal vaccination against hepatitis B virus: the effect and problems. Clin. Cancer Res. 11:7953–7957.

Croci, S., Nicoletti, G., Landuzzi, L., De Giovanni, C., Astolfi, A., Marini, C., Di Carlo, E., Musiani, P., Forni, G., Nanni, P., and Lollini, P. L. 2004. Immunological prevention of a multigene cancer syndrome. Cancer Res. 64:8428–8434.

Cuzick, J., Powles, T., Veronesi, U., Forbes, J., Edwards, R., Ashley, S., and Boyle, P. 2003. Overview of the main outcomes in breast-cancer prevention trials. Lancet 361:296–300.

De Giovanni, C., Nicoletti, G., Landuzzi, L., Astolfi, A., Croci, S., Comes, A., Ferrini, S., Meazza, R., Iezzi, M., Di Carlo, E., Musiani, P., Cavallo, F., Nanni, P., and Lollini, P. L. 2004. Immunoprevention of HER-2/neu transgenic mammary carcinoma through an interleukin 12-engineered allogeneic cell vaccine. Cancer Res. 64:4001–4009.

Di Carlo, E., Diodoro, M. G., Boggio, K., Modesti, A., Modesti, M., Nanni, P., Forni, G., and Musiani, P. 1999. Analysis of mammary carcinoma onset and progression in HER-2/neu oncogene transgenic mice reveals a lobular origin. Lab. Invest. 79:1261–1269.

Dorne, C., Dorigo, M., and Glover, F. 1999. New ideas in optimization. McGraw-Hill.

Fabian, C. J., and Kimler, B. F. 2005. Selective estrogen-receptor modulators for primary prevention of breast cancer. J. Clin. Oncol. 23:1644–1655.

Frese, K. K., and Tuveson, D. A. 2007. Maximizing mouse cancer models. Nat. Rev. Cancer 7:645–658.

Garrido, F., and Algarra, I. 2001. MHC antigens and tumor escape from immune surveillance. Adv. Cancer Res. 83:117–158.

Green, J. E., and Hudson, T. 2005. The promise of genetically engineered mice for cancer prevention studies. Nat. Rev. Cancer 5:184–198.

Lollini, P. L., Cavallo, F., Nanni, P., and Forni, G. 2006. Vaccines for tumour prevention. Nat. Rev. Cancer 6:204–216.

Lollini, P. L., De Giovanni, C., Pannellini, T., Cavallo, F., Forni, G., and Nanni, P. 2005a. Cancer immunoprevention. Future Oncol. 1:57–66.

Lollini, P. L., and Forni, G. 2003. Cancer immunoprevention: tracking down persistent tumor antigens. Trends Immunol. 24:62–66.

Lollini, P. L., Motta, S., and Pappalardo, F. 2006. Discovery of cancer vaccination protocols with a genetic algorithm driving an agent based simulator. BMC Bioinformatics 7: 352.

Lollini, P. L., Nicoletti, G., Landuzzi, L., De Giovanni, C., and Nanni, P. 2005b. New target antigens for cancer immunoprevention. Curr. Cancer Drug Targets 5:221–228.

Motta, S., Castiglione, F., Lollini, P., and Pappalardo, F. 2005. Modelling vaccination schedules for a cancer immunoprevention vaccine. Immunome Res. 1:5.

Musiani, P., Modesti, A., Giovarelli, M., Cavallo, F., Colombo, M. P., Lollini, P. L., and Forni, G. 1997. Cytokines, tumour-cell death and immunogenicity: a question of choice. Immunol. Today 18:32–36.

Nanni, P., Landuzzi, L., Nicoletti, G., De Giovanni, C., Rossi, I., Croci, S., Astolfi, A., Iezzi, M., Di Carlo, E., Musiani, P., Forni, G., and Lollini, P. L. 2004. Immunoprevention of mammary carcinoma in HER-2/neu transgenic mice is IFN-gamma and B cell dependent. J. Immunol. 173:2288–2296.

Nanni, P., Nicoletti, G., De Giovanni, C., Landuzzi, L., Di Carlo, E., Cavallo, F., Pupa, S. M., Rossi, I., Colombo, M. P., Ricci, C., Astolfi, A., Musiani, P., Forni, G., and Lollini, P. L. 2001. Combined allogeneic tumor cell vaccination and systemic interleukin 12 prevents mammary carcinogenesis in HER-2/neu transgenic mice. J. Exp. Med. 194:1195–1205.

Nanni, P., Nicoletti, G., Palladini, A., Croci, S., Murgo, A., Antognoli, A., Landuzzi, L., Fabbi, M., Ferrini, S., Musiani, P., Iezzi, M., De Giovanni, C., and Lollini, P. L. 2007. Antimetastatic activity of a preventive cancer vaccine. Cancer Res. 67:11037–11044.

Nanni, P., Pupa, S. M., Nicoletti, G., De Giovanni, C., Landuzzi, L., Rossi, I., Astolfi, A., Ricci, C., De Vecchi, R., Invernizzi, A. M., Di Carlo, E., Musiani, P., Forni, G., Menard, S., and Lollini, P. L. 2000. p185(neu) protein is required for tumor and anchorage-independent growth, not for cell proliferation of transgenic mammary carcinoma. Int. J. Cancer 87:186–194.

Nemhauser, G. L., and Wolsey, L. A. 1998. Integer and combinatiorial optimization. New York, NY: Wiley.

Noguchi, Y., Jungbluth, A., Richards, E. C., and Old, L. J. 1996. Effect of interleukin 12 on tumor induction by 3-methylcholanthrene. Proc. Natl. Acad. Sci. U.S.A. 93:11798–11801.

Pappalardo, F., Lollini, P. L., Castiglione, F., and Motta, S. 2005. Modeling and simulation of cancer immunoprevention vaccine. Bioinformatics. 21:2891–2897.

Quaglino, E., Iezzi, M., Mastini, C., Amici, A., Pericle, F., Di Carlo, E., Pupa, S. M., De Giovanni, C., Spadaro, M., Curcio, C., Lollini, P., Musiani, P., Forni, G., and Cavallo, F. 2004. Electroporated DNA vaccine clears away multifocal mammary carcinomas in Her-2/neu transgenic mice. Cancer Res. 64 2858–2864.

Sawaya, G. F., and Smith-McCune, K. 2007. HPV vaccination–more answers, more questions. N. Engl. J. Med. 356:1991–1993.

Shankaran, V., Ikeda, H., Bruce, A. T., White, J. M., Swanson, P. E., Old, L. J., and Schreiber, R. D. 2001. IFN-gamma and lymphocytes prevent primary tumour development and shape tumour immunogenicity. Nature 410:1107–1111.

Smyth, M. J., Dunn, G. P., and Schreiber, R. D. 2006. Cancer immunosurveillance and immunoediting: the roles of immunity in suppressing tumor development and shaping tumor immunogenicity. Adv. Immunol. 90:1–50.

Stewart, B. W., and Coates, A. S. 2005. Cancer prevention: a global perspective. J. Clin. Oncol. 23:392–403.

# Index

Printed in the United States of America